DAVID HUGHES MCELREATH
DANIEL ADRIAN DOSS
BARBARA R. RUSSO
RAYMOND TESIERO
CARL J. JENSEN
JEFF VAN SLYKE
ROBERT NATIONS
LEISA S. MCELREATH
STEFAN E. SCHULENBERG
ASHLEY M. LINDSLEY
ASHLEY WELLMAN

DISASTERS THAT SHAPED EMERGENCY MANAGEMENT

CASE STUDIES FOR THE HOMELAND SECURITY/EMERGENCY MANAGEMENT PROFESSIONAL

Kendall Hunt
publishing company

Cover © Shutterstock.com

Kendall Hunt
publishing company

www.kendallhunt.com
Send all inquiries to:
4050 Westmark Drive
Dubuque, IA 52004-1840

ISBN: 978-1-5249-6202-9

Published in the United States of America

This book is dedicated to those we dearly love,
our families, our spouses, parents, children and
grandchildren.

CONTENTS

1 The Firestorm at Peshtigo or the Peshtigo Paradigm (1871)

Figure 1.1 The Area Impacted by the Peshtigo Fire
Source: National Weather Service

"A Firestorm knows no empathy, only hunger and never thirst."

—Denise Gess and William Luts

BRIEF OVERVIEW

It is hard to believe that anyone can truly understand the scope of the destruction and human tragedy that occurred when weather conditions, environmental mismanagement, and the encroachment of civilization came together creating a disaster almost beyond description in the region around Peshtigo, Wisconsin. The intensity of the fire cremated many of the victims' bodies and, in some places, turned sand into glass. Surprisingly, the 1871 Peshtigo Fire, or Peshtigo Paradigm, has been largely forgotten.

The firestorm that impacted the town of Peshtigo and the surrounding area remains the single worst wildfire in U.S. history, in both size and fatalities, is known as the Great Peshtigo Fire, which burned approximately 1.5 million acres (2,343 square miles), and purportedly killed between 1,200 and 2,500 people in northern Wisconsin and the Upper Peninsula of Michigan during the week of October 8–14, 1871 (Hipke, 1998).

Occurring the same night was the Great Chicago Fire, which has been described as a "fire tornado" that swept the area around Peshtigo, Wisconsin, and in its wake, destroyed 17 towns and communities, killed thousands, and left many more injured and homeless. The fire destroyed businesses and farms, which dramatically changed life in the area.

INTRODUCTION

In the late 1860s, and early 1870s, land in Wisconsin was available for those who were intrepid enough to face the challenges of this land. Covered with century-old timber, the rich land presented opportunities and attracted farmers and business men alike, the thousands of acres of timber proved appealing, and when harvested, the millions of feet of lumber would provide the materials critical in the construction of thousands of homes and businesses. Around the timber business, towns would emerge. Peshtigo was one of these towns, a small settlement about 6 miles northwest of Green Bay, built around timber industry with sawmills, wood-ware factory, hotel, taverns, churches, schools, and homes. The timber industry provided jobs for a large percentage of the thousands who had moved into the area. From the land cleared of timber, farms were established, homes constructed, crops planted, and families raised.

The town's location on the Peshtigo River made it an ideal area where harvested timber could be processed into lumber and transported down the river to Chicago. Investors recognized the value of extending rail lines into this part of Wisconsin which would further open the area and reduce the dependency on the river for the transportation of timber to the markets in Chicago (Pernin, 1971).

The Setting

In many ways, this part of Wisconsin was untamed and to many on the edge of the nation's frontier. Thousands of acres of virgin timber waited to be harvested and consumed by a growing nation. In the region, land was inexpensive and available. Many who moved to the area did so in the hope to gain employment in the timber industry and in doing so, accumulate for themselves enough money to purchase land, establish homes, and raise families. For the majority of those moving into the area, they understood life was difficult and at times, dangerous. These were strong individuals and for those who expected to prosper, they understood hard work would be a key element in their success.

During the summer and early fall of 1871, the Midwest was experiencing extreme drought conditions. The infrequent rain that fell did little to remove the danger of fire. Throughout the summer and fall, the creeks and bogs dried. In an area that so relied on the timber industry, the dry conditions proved hazardous. The area was dangerously dry, but the conditions benefited those constructing the rail line to link Peshtigo to Green Bay, and with it, to the rest of the nation. Few expressed concern.

Fire was undoubtedly a vital and commonly used tool by numerous industries. For example, fire was used by the timber industry to transform forests, farmers used fire to clear land, loggers cleared underbrush to allow for easier harvest, and those constructing the railroads used fire to quickly clear their intended path. Although the relative simplicity of fire became a transformative power, such action also contributed to reckless negligence by employees of many industries. Unfortunately, these methods were often recklessly employed across the regions, fires were often set and irresponsibly left unattended. The expectation being that such fires would, as in the past, burn themselves out. In addition, hunters would also leave campfires burning, which often allowed embers to drift that caused additional small fires.

In addition, the forest area was also being transformed, and fire was used as a vital tool. Eventually, however, the forest refurbishment process became reckless due to the irresponsible and haphazard practices by numerous people who would simply use fire in a traditional manner common during that period, which was simply flame-up and move on, and not realize the consequences of a vast and uncontrolled blaze. As such, few took notice of the fires burning throughout the Midwest in the weeks leading up to the disaster. All of this complacency coupled with the drought conditions led to the firestorm that erupted on October 8, 1871.

The Disaster Unfolds

Fire proved a great tool in the hands of man as well as a destructive instrument throughout history. In October 1871, fire would again demonstrate its destructive ability as it laid waste to much of Chicago and regions of Wisconsin near Green Bay. While the Great Chicago Fire burned itself into the collective memory of a nation, the firestorm that destroyed Peshtigo and the surrounding area became little more than a footnote in the history.

As stated in the introduction, weather conditions, environmental mismanagement, and the encroachment of civilization came together creating a disaster almost beyond description in the region around Peshtigo, Wisconsin. For those who escaped death, their lives were changed forever. Many carried the physical and psychological scars for the remainder of their lives.

The firestorm that proved so fatal was fueled by a fatal mix of unusually dry weather, shifting wind patterns, timber, underbrush, and reckless use of fire. A typical winter in this area of Wisconsin would easily experience 4–5 ft of snow followed by a wet spring, saturating the soil. Little snow fell during the winter of 1870–1871. The following spring was unusually dry. The last measurable rainfall for Peshtigo came on July 8, 3 months before the tragedy (Gordon, 2003).

The railroads and mills stored industrial chemicals in large quantities as well as paint and other flammable items. Peshtigo became a tinderbox just waiting for a source of ignition. Smaller fires had burned in and around the town for months leading up to the day of the fire, a common finding which failed to raise concern. Peshtigo did not have an organized fire department; rather, it relied on its residents with their buckets and hand tools to battle a fire. Residents were called to action by the sounding of the sawmill's steam whistle or by church bells. The firestorm that would destroy the area was well beyond their ability to stop with their limited resources and capabilities (Gess & Lutz, 2003).

October 8, 1871 began as another typical, hot, and smoky Sunday for the area around Peshtigo. Few knew two approaching weather fronts had collided to the west and were generating gale force winds that were quickly moving to the east. As the systems moved through Minnesota and into Wisconsin, the winds and dry conditions combined with the small fires, smoldering flames and embers in the vast forests.

In Peshtigo, residents felt changes in wind conditions prior to witnessing white ash raining down upon the town like a typical Wisconsin snowstorm. By 9 p.m., the windstorm provided the last element required to turn the forest into an unprecedented instrument of death and destruction. Once the fires began to sweep across the area, the magnitude of the event in combination with the lack of an organized fire response system made the fire(s) uncontrollable (Pernin, 1971).

Those living in outlying areas located in the path of the firestorm were the first to fall victim to this catastrophic event. Small farmers, travelers, and hunters were quickly overwhelmed as the fire moved rapidly. Countless heartbreaking stories told of attempts to escape the inferno, some successful, but in all too many cases, resulted in the loss of life.

The indiscriminate manner in which the fire enveloped the area thwarted attempts to flee the fire's deadly path of destruction as the fire's unrelenting pathway had transformed attempted escape routes into impassable death traps. Chance and luck often determined the difference between life and death. Survivors later reported that the intensity of the heat itself caused individuals in the path of the fire to burst into flames, experiencing horrific deaths. Others who managed to initially survive the fire but sustained injury lingered painfully and tragically until ultimately succumbing to their injuries.

The immediate death may have been a blessing in disguise as the injured survivors would have lingered in horrible pain (while likely receiving whatever remedies were available at that time in the area) until death occurred from electrolyte imbalances (fluid loss through the wounds) and/or sepsis from infection due to lack of sanitary conditions, unknown manner in which infection spread, and lack of body defense against infection due to open wounds.

The fire was indiscriminate in the destruction produced by its pathway. Natural barriers, including open fields or even rivers, failed to slow the firestorm's rapid rate of destruction. In the scramble to save their lives, many individuals sought protection in storm shelters built under or around homes or in wells. Ultimately, these attempts at survivorship were unsuccessful, often resulting in death from the intensity of the heat or suffocation.

Those who sought protection in the openness of farmland met an identical demise. Individuals seeking safety in streams or the river itself who were not killed by the intense heat or lack of oxygen ultimately perished of the hypothermia experienced as a result of entering the waterways in the middle of a Wisconsin winter (Pernin, 1971).

Impact

A fire tornado formed and demolished buildings, trees, and picked up burning debris—dropping it down on anyone or anything below generating more fire until the fuel was exhausted. The next morning survivors began to emerge from the river, wells, bogs, and others areas where they were fortunate enough to find relief. Peshtigo no longer existed. Death was everywhere. Human remains littered the recently thriving community. The area suffered from lack of resources to treat any survivors, thus resulting in increased fatalities.

RELIEF AND FIRST RESPONSE EFFORTS

As in all disasters, survivors were the first to take action to begin community recovery. The fire destroyed telegraph lines to Green Bay. Initial reports of the disaster by survivors seemed too great to believe. Additionally, Chicago had burned and the reports of damage and deaths in the city captured the attention of the nation.

As with so many relief operations, the Peshtigo postdisaster period proved challenging (McElreath, Doss, et al., 2014). Peshtigo was desperate for an immediate influx of every type of aid possible—including, but not limited to, food, clothing, medical assistance, and shelter. Temporary hospitals were established, proving what aid was possible. Despite these critical and time-sensitive needs, the reports coming from Chicago regarding the disaster experienced there slowed response to Peshtigo, therefore contributing to the large number of fatalities.

With the telegraph lines severed, 2 days passed before Governor Fairchild became aware of the incident (Knickelbine, 2012). When the news of the disaster arrived, the Governor of Wisconsin was absent from Madison. Fortunately, the Governor's wife, Francis Fairchild, took steps in her husband's absence to begin relief operations and the flow of supplies from Madison, calling for donations of food, clothing, and medical supplies. Donations flowed in and soon train loads of relief supplies were moving into the area. Over 9,000 people were soon receiving aid.

Supplies were sent from Green Bay to Peshtigo area by way of the Peshtigo River. Aid poured in from across the United States and abroad. Soldiers of the U.S. Army were soon deployed to the area to assist in the ongoing relief operations. Under the direction of the Secretary of War, the Army provided a wide range of operational and logistical support, especially related to the transportation and distribution of supplies and equipment to the area. Donations of tools, bedding, clothing, and other needed equipment came in from around the nation (Knickelbine, 2012).

Relief committees were established to assist in the distribution of aid, while survivors and volunteers underwent the difficult and tedious process of searching for survivors and, were possible, identifying and burying the dead who had not been cremated by the fire. Each individual who experienced the firestorm found his or her life changed. Some moved away while others began the painstaking process of rebuilding their lives.

AFTERMATH AND INVESTIGATION

Following the immediate response, questions began to arise, although an understanding of the firestorm would take time. Many blamed the Great Fire on crews constructing the railroad, their careless stacking, slashing, and burning of the timber and underbrush along the right of way, or the practice of those harvesting timber, but those were only part of the answer. It would take significant time before a thorough understanding of the complex nature which the fire and weather had to conform in order to produce this event could be understood. Unstable atmosphere resulting in tornados, dry conditions, fire, and vast amounts of fuel made the Peshtigo disaster the nation's deadliest fire.

LESSONS LEARNED AND POLICY IMPACT

By combining research methods currently available, analysis of more recent fires, and historical data from the Peshtigo disaster, we are able to identify lessons learned or reinforced. Intelligence related to the disaster, communications, and command and control are critical elements within any disaster response spectrum. The ability to gain an understanding of the event (intelligence) and initiate response steps are essential. The ability to communicate, directing the response effort, is also of the utmost importance to reduce confusion within the response effort in every way possible. Command and control and the ability to set and maintain an operational tempo related to the response are also key to response success (McElreath, Doss, et al., 2014).

The Peshtigo Fire contributed to important changes in fire safety. One important lesson drawn from this event was reassessing the long-held practice of leaving unattended fires burning whose embers could start a larger fire. Despite the practice of brush burning, we see the beginnings of the new practice of burn-offs, preset fires to burn up loose material that could start a forest fire.

Another important lesson is that large fires produce volatile gases that are both lethal and unpredictable. This was seen in the Yarnell Hill Fire that claimed the lives of the 19 Granite Mountain Hotshots as well as 2001's Thirty-mile Fire which claimed the lives of 14 entrapped firefighters. Inhaling superheated gases results in death, as evidenced by the fact that many Peshtigo victims met their demise by gases alone.

Safety zones and escape routes were another important lesson learned. During the Peshtigo Fire, residents were surrounded on multiple sides by the impinging fires. They did not understand that radiant heat could cross large areas and when seeking refuge in what they thought were safe, wide open areas, it became the complete opposite. Many were overcome merely by the radiant heat being given off by the fires and not the fire itself.

Another lesson learned is that fire behavior, especially involving large fires, can be capricious and unaccountable. This would explain why some houses burned to the ground while others next door survived. In Peshtigo, one group of residents may have been spared while others located only a short distance away burned to death. This typical fire behavior is seen time and time again when firestorms occur. Most notable is the 1991 firestorm in Oakland and Berkeley, California. The pattern of destruction often mirrors that of a tornado (McElreath, Jensen, et al., 2014).

A related lesson learned is that back-burns and under-burning do not guarantee safety or security. Before the Peshtigo Fire, numerous smaller fires had been burning and using up available fuels. Yet despite these efforts, plenty of fuel was still available for significant fires to occur. Partially consumed fuels are still available to fuel subsequent fires and this aided the fires in Peshtigo.

Evacuation routes are now implemented into emergency operations. Having designated routes allows for victims and response units to get in and out without confusion. Having designated routes allows for money to be delegated toward developing sound infrastructure to ensure these routes remain available throughout the emergency events (McElreath, Doss, et al., 2014).

Conservation considerations also stemmed from this incident. Because the land was cleared out by various farmers, the railroad company and other industries left no natural barrier for the fire to burn before hitting populated areas. During times of extremely dry weather, states and counties can impose burn bans which prohibit any type of burning while in effect. Burn permits were also developed in order to control fires (McElreath, Jensen, et al., 2014).

Finally, extreme fire behavior can develop abruptly (McElreath, Jensen, et al., 2014). While many in Peshtigo thought the smaller fires and a larger one in September were signs that the worst was over, in reality, the worst was yet to come. All it took was a change in the weather conditions and the arriving cold front ushered in the firestorm. In most wildfire, many individuals are often caught off guard or have been lulled into a sense that the worst is over.

IMPLICATIONS FOR MODERN EMERGENCY MANAGEMENT

The Peshtigo Fire served as a catalyst for the scientific study of fire behavior. Today, we continue to learn lessons from each major fire. Three factors must be considered in dealing with firestorms: wind, fuel, and topography. These individually can change the dynamics of how the fire behaves and the damage it can cause.

Emergency managers faced with wildfire threat are now able to incorporate weather experts into their emergency plans, allowing for more accurate weather predictions as fires grow and advance. This capability to anticipate weather conditions provide the responders with information leading to the development of improved response strategies. It is not uncommon for a member of the local National Weather Service to be on site in the command center of larger wildfires. A meteorologist is able to assist with important calculations that affect fire behavior, including wind speed and direction, the movement of fronts, and humidity levels, which all factor into fire behavior.

Today's emergency management officials need to determine the type and manner of warning systems to implement. Timely, credible warnings must be delivered to residents if action is to be taken ahead of largely unpredictable events such as wildfires. At the time of the fire, the town of Peshtigo relied on the sawmill's steam whistle or church bells. Research continues to demonstrate that the lack of warnings leads to increased death and destruction, as evidenced by the Gatlinburg Fire in 2016, and in other recent wildfires in California. In these instances, warnings were never given or were delayed, resulting in a lack of evacuations and what would be considered preventable deaths.

State and federal government disaster assistance can be requested during severe weather emergencies when local resources are exhausted or require additional support. This assistance, however, normally arrives only after executing a series of procedures, and that takes time. The extent and type of resources also depend on damage assessments. Communities must be self-reliant for at least the first few days, if not longer, until state and federal resources arrive.

Figure 1.2 Aftermath of the Gatlinburg Fire
Source: Joshua K Pruitt/Shutterstock.com

SUMMARY

Despite being often eclipsed by the Great Chicago Fire, the Peshtigo Fire remains a disaster many in Wisconsin cannot forget. The actual scope of this event and the lives lost cause many in modern emergency management to question how the practices so common at the time of reckless land, clearly through the use of unmonitored fire, could occur. As the wildland–urban interface continues to challenge many in this country who are prone to wildfires, the Peshtigo Fire remains a great example of the powerful forces behind wildfires and the manner in which conditions may quickly combine, leading to massive losses. We learned from this disaster how unpredictable wildland fires can be, yet we lose many wildland firefighters each season, along with countless properties and civilians, due to our lack of respect for Mother Nature.

REFERENCES

Gess, D., & Lutz, W. (2003). *Firestorm at Peshtigo: A town, its people, and the deadliest fire in American history*. New York: Macmillan. ISBN 978-0-8050-7293-8.

Gordon, J. (2003). Forgotten fury. *American Heritage, 55*(2). April/May 2003. Retrieved from http://www.americanheritage.com/content/forgotten-fury

Hipke, Deana C. (1998). *The Great Peshtigo Fire of 1871*. Retrieved February 13, 2018 from https://worldhistoryproject.org/1871/10/8/peshtigo-wisconsin-fire

Knickelbine, S. (2012). *The Great Peshtigo Fire: Stories and science from America's deadliest firestorm*. Madison, WI: Wisconsin Historical Society Press.

McElreath, D. H., Doss, D. A., Jensen, C. J., Wigginton, M. P., Nations, R., Van Slyke, J., & Nations, J. (2014). *Foundations of emergency management*. Debuque: Kendall Hunt.

McElreath, D. H., Jensen, C. J., Wigginton, M. P., Doss, D. A., Nations, R., & Van Slyke, J. (2014). *Introduction to homeland security*. Boca Raton: CRC Press.

Pernin, P. (1971, Summer). The Great Peshtigo Fire: An eyewitness account. *Wisconsin Magazine of History, 54*(4), 246–272.

Galveston Hurricane (1900)

2

Figure 2.1 Destruction Following the Galveston Hurricane
Source: NOAA

BRIEF OVERVIEW

The power and destructive ability of a Category 4 or Category 5 Hurricane is almost beyond imagination. With the advances in science and technology, the capability to anticipate and predict the strength and path has greatly increased the ability to provide warning of approach and with the warning, the time to take steps to enhance preparation. The capabilities of assessing the path and strength of hurricanes available today did not exist in 1900 when the city of Galveston, Texas was destroyed with extensive loss of life.

In 1900, Galveston, Texas was at ground zero for a devastating hurricane that left in its wake a destroyed urban infrastructure, devastated economic base, and thousands killed and injured. Yet, the outcome of destructiveness from the Galveston hurricane of 1900 would transform the city to become stronger and more resilient. Lessons were also learned on hurricane preparation and response which would prove to be a catalyst for improving communication, warning, and physical infrastructure.

The lessons learned from the Galveston incident are timeless, and have significance for modern emergency managers. Although times and technologies have changed societally throughout the last century, certain emergency management needs are unchanged, such as the necessity of effective and timely communication. Given these notions, this case study examines the Galveston event, and provides insight regarding lessons learned that serve to development and maintain principles and best practices for proactive emergency management.

INTRODUCTION

The 1900 Galveston, Texas Hurricane was the deadliest hurricane experienced by the United States (McElreath, Jensen, et al., 2014). While the true number of those killed and injured will never be known, it is safe to assume between 8,000 and 12,000 individuals (Longshore, 2008). South of Cuba, on August 27, 1900, the incident began as a tropical storm in the central Atlantic (Prothero, 2011). Days later, it worsened and was designated as a hurricane while it progressed in the Gulf of Mexico (McElreath et al., 2017). The storm gained intensity and momentum as it crossed the Gulf of Mexico. When it arrived at Galveston on September 8, it was a Category 4 hurricane with a storm surge of up to 15 ft (Prothero, 2011).

The Setting

In 1900, Galveston seemed to have everything going for it. A great location, it was a thriving center of commerce. In many ways, it was a modern city for the time. It is believed that Galveston is where the first European arrived about 1816. Within a decade, a port was established by the Congress of Mexico. In 1836, after independence from Mexico, Galveston was established and served as the capital of the Republic of Texas. By the mid-19th century, Galveston emerged as an international city with immigration and trade from around the United States and the world. The city became one of the nation's busiest ports, and was the world's leader of cotton exports. Galveston became one of Texas' largest cities and a prime commercial center. By the late 1800s, the city's population reached about 37,000, but danger awaited. The location and sea level of Galveston made it vulnerable to the major storms moving across the Gulf of Mexico. Low in elevation, and limited access to the mainland, a major storm from the Gulf could easily drive major amounts of water into and over the city.

The Disaster Unfolds

Although the National Weather Service maintained an office in Galveston, the residents of the town were unaware of the impending danger as the looming, imminent hurricane approached from the Gulf of Mexico. Thus, because of poor communication, Galveston was unaware of the severity of the incoming hurricane. Although various warnings were issued from a U.S. Weather Bureau outpost in Cuba to the Washington headquarters, the warning alerts were primarily unheeded and ignored because the U.S. and Cuban forecasts were incongruent (Keim & Muller, 2009). During 1900, ship-to-shore forms of communication were nonexistent (Wood, 2007). The local Galveston meteorologists and residents perceived little danger, or misread the major storm indicators until hours before the hurricane's landfall. Because of these shortcomings, any effective evacuation endeavor was impossible (Frantz, 2012).

Numerous efforts and attempts toward city evacuation failed miserably. Although many factors contributed to the great loss of life, a substantial factor resulting in many deaths was the submerging of bridges that connected Galveston with the mainland (Brennan, 2002). As a result, a large percentage of the populace was trapped within the municipality instead of evacuating in a timely manner to safety (Brennan, 2002). Additionally, the destruction of telegraph lines that prohibited communication, and storm surges that overcame and flooded the city, impeded the evacuation (Brennan, 2002).

Impact

The hurricane devastated the city upon landfall. During the storm, the water level was reported to have risen approximately 4 feet within 4 seconds (Rubin, 2012). At least 3,600 homes were destroyed on Galveston Island, and the added toll among commercial infrastructure generated a monetary loss of approximately $30 million (measured in period dollars; Smith, 2013).

On average, approximately 70 victims per day were discovered during the first month after the storm (Brennan, 2002). The last discovery of human remains was reported on February 10, 1901 (Ramos, 1998). An exact death toll was difficult for a variety of reasons. Many of the casualties were not counted throughout recovery operations, were washed out to sea, or were concealed by the rubble resulting from the destruction inflicted by the storm (Longshore, 2008). The initial attempts to eliminate the bodies of the deceased, including taking them into the Gulf of Mexico via barges and dumping them into the water, contributed to an increased amount of confusion when they washed ashore (Longshore, 2008). Until the following November, the bodies of many victims were cremated among funeral pyres, and the nauseating scents were smelled nearly 100 miles at sea by incoming relief personnel (Brennan, 2002).

Containing the Disaster

The scope of the disaster was beyond belief. As the wind subsided and the waters withdrew, survivors emerged from the wreckage to begin the steps of recovery. Major relief operation commenced immediately after the incident. The injured were treated were possible and the dead were collected. Search and recovery operations would continue for weeks. Efforts were made to establish temporary housing, food distribution, and medical care. Access to Galveston was soon opened, so relief support from outside the impacted area could flow into the area.

RELIEF AND FIRST RESPONSE EFFORTS

Sources of relief assistance involved both governmental and nongovernmental factions (Wall, 2011). Initially, operational goals involved forming committees, implementing sanitation activities, and instituting emergency medical facilities (Wall, 2011). Similar to modern times, the initial search and recovery efforts were performed by the survivors and factions from unaffected regions (Wall, 2011). Medical relief operations, including support from Clara Barton and the Red Cross, were conducted to provide healthcare, evaluate the conditions of survivors for immediate medical attention, distribute foodstuffs and clothing, distribute disinfectants, and prevent any incident of epidemic illness (Peak, 2013; Wall, 2011).

Private corporations, including Johnson & Johnson, provided relief support sending aid to Galveston physicians, druggists, and their families. The medical supplies donated by Johnson & Johnson were used by relief workers in Galveston to treat the injured and help survivors (Gurowitz, 2008).

Military units were deployed to establish order, assist in search, rescue, and recovery operations, establish temporary housing, support medical responses, and provide food for both survivors and relief workers (Wall, 2011). The services provided by the military significantly enhanced rescue efforts.

Relief materials consisted of various capital items, ranging from money to physical materials. A national appeal for assistance resulted in benevolence from New York, Maryland, Washington, D.C., Illinois, peer locales in Texas, and various other states, cities, and communities (Green, 2000). Financially, contributions were obtained from mere children (e.g., $1.50 from operating a lemonade stand), the goodwill of both cities and individuals ranging in thousands of dollars; profits from sporting events and concerts; and charitable organizations. Financial resources were supplemented by physical goods, including medicine, clothing, carbolic acid, charcoal, coffee beans, tea, rice, corn meal, beans, peas, lard, and baking powder (Green, 2000; Wall & Keeling, 2011).

AFTERMATH AND INVESTIGATION

The period of recovery was both dangerous and laborious. Galveston's leaders took several major steps to recover from the storm and to prevent a recurrence of the devastation. First, they developed a new form of municipal government, one with strong centralized control to handle the economic recovery of the city. Next, they built a massive seawall to turn back storm-generated waves. Perhaps the most amazing step they took was to raise the level of the entire city, by more than 16 feet in some areas, in order to keep flooding at a minimum, as major initiatives toward fortification were pursued as a means of the city's vulnerability to future storms. Eventually, over a period of approximately 12 years, substantial portions of the city were raised above sea level and barriers were established along the coastline to reduce the direct impact of tidal surges (Willett, 2013). The economic and physical recovery of the city would take decades (Wall and Keeling, 2011).

Safeguarding and bolstering Galveston against future storms necessitated a massive engineering effort involving "every home, every church, every school, every sidewalk, every fence, and every electric line, everything"

(Williams & Sheets, 2001, p. 68). This undertaking encompassed the raising of approximately 4,500 structures within the city (Williams and Sheets, 2001). The construction endeavor raised everything in the city at least 17 ft. above "mean low tide" (Williams and Sheets, 2001, p. 68). Completing this endeavor necessitated a period of 11 years, and involved the slurry of approximately 11 million cubic yards of sand underneath Galveston (Williams & Sheets, 2001).

Between 1902 and 1904, the erecting of an initial seawall occurred to guard Galveston against tidal surges. The construct had a base thickness of approximately 16 feet, was approximately 3 miles in length, and buttressed against the eastward area of the land mass (Williams & Sheets, 2001). After it was finished, a second wall was erected to protect the northern region of Galveston (Williams & Sheets, 2001).

The period of aftermath exhibited additional social dangers among the survivors and response personnel. Considerations of social class influenced the disaster response, and resulted in tense situations both societally and culturally (Wall & Keeling, 2011; Wellman, 2013; Wellman, 2017). Both looting and violence were reported during the aftermath of the storm (Green, 2000). In order to counter such dangers, police officers and soldiers were ordered to shoot looters or anyone else who attempted looting (Green, 2000).

Galveston's tragedy resulted in a stronger potential to mitigate future incidents. For instance, during 1909, while the city was being elevated, Galveston endured another hurricane (Williams and Sheets, 2001). Although its engineering safeguards were incomplete, the damages and destruction of the second hurricane were much less severe when compared with the devastation of the 1900 storm. This second hurricane resulted in the deaths of 275 individuals, "but none in the parts of Galveston behind the seawall, even though 6 feet of water washed into some parts of the city that hadn't been raised" (Williams and Sheets, 2001, p. 65).

LESSONS LEARNED AND POLICY IMPACT

Lessons gleaned from the Galveston storm are appropriate for any entity during modern times. Although the Galveston incident was tragic, its lessons influenced much of the emerging and maturing of preparedness and contingency planning. Initially, although an incident may be perceived as having little or no danger, it may develop into a substantial threat event quickly. Thus, calamities may occur with speed, surprise, and violence. Impeded communication, lack of planning, and overall unpreparedness combined to affect Galveston negatively. Given the Galveston experience, modern organizations and individuals must be mindful of preparedness, mitigation, response, and recovery—the emergency management cycle (EMC). Communicating effectively and efficiently throughout the EMC diminishes the dangerous effects of ignorance and misunderstanding. Any organization must have an efficient, effective plan and method of communication whereby concepts and information may be relayed and that contributes toward diminishing confusion (Doss, Glover, Goza, & Wigginton, 2015). Certainly, within the EMC context, communication methods must be well planned and have some levels of redundancy.

Nature may be extremely vicious, surprising, and unpredictable. Events may well be severe and extensive or mundane and limited. Galveston experienced the worst of the 1900 hurricane. Regardless, learned from the incident is the notion that preparedness measures can be taken toward safeguarding communities and mitigating the effects of cataclysms. Essentially, akin to the old saying, one may "expect the worst, and hope for the best" concerning the potential dangerousness of incidents.

Galveston's lessons are not unforgotten. Lessons gleaned from Galveston are profound reminders concerning the intensities of natural disasters and the relevancy of the EMC for modern organizations. During modern times, individuals, groups, and organizations may enact the EMCs phases toward safeguarding their assets and mitigating the effects of disasters. Although the Galveston incident was lethal and damaging, its lessons provided the basis for much of modern emergency management activities, ranging from preparedness planning to communication measures. In any case, the timeless lessons from Galveston provided a basis for other areas of the nation to better withstand future incidents. Contemporary emergency management planning now occurs among municipalities, states, regions, and nationally.

IMPLICATIONS FOR MODERN EMERGENCY MANAGEMENT

The EMC is a cyclical process for crafting strategies and policies for coping with calamities. The EMC consists of four phases: (a) mitigation/prevention, (b) preparedness, (c) response, and (d) recovery (McElreath, Doss, et al., 2014). Figure 2.2 delineates the EMC phases.

Mitigation and prevention involve diminishing or quashing the negative effects of a calamity. Preparedness involves threat identification, planning accordingly, and contingency planning. Response consists of executing a specific plan or course of action when incidents occur. Recovery consists of attempting to achieve much or all of the state of existence and lifestyle that occurred before the calamity happened. In other words, recovery attempts to achieve as much normalcy as possible. Throughout recovery, assessment and evaluation occur to determine how improvements could occur should a similar incident ever arise (McElreath, Doss, et al., 2014).

Regarding mitigation, Galveston was unprepared and relatively unwarned for the impending hurricane. Its populace existed in a state of ignorance regarding the incoming calamity and its potential lethality. Thus, its ability to mitigate the incident was lackluster. The city experienced the full magnitude of the storm without any means of diminishing the severity of its effects. The results of the storm and unpreparedness were unthinkable—many deaths, devastated infrastructure, and unrecovered economic capacity. During modern times, because of such ravages, mitigation practices exist for lessening the negative effects of incidents.

Regarding modern EMC policies, the Galveston incident provided lessons for modern preparedness and mitigation. The Galveston hurricane caused nations worldwide to develop mitigation and preparedness methods to reduce lethality (Longshore, 2008). Essentially, the Galveston storm spurred contingency planning in both the public and private sectors. Although envisioning all possible threats is impossible during emergency planning, conceptualizing and preparing for identifiable threats represents a sound investment from a strategic perspective (James & Gilliland, 2013). Lessons learned from Galveston's lethality provide insight regarding the necessity of modern preparedness and planning toward mitigating the dangers of calamities.

All incidents necessitate some response. The Galveston response marshaled and coordinated resources and capital locally, regionally, and nationally from individuals, groups, and organizations. An array of goods and services were necessary following the storm, ranging from clothing to medical services. Commensurately, during modern times, similar practices are witnessed during the aftermaths of disasters. After Hurricane Katrina, for instance, Walmart donated foodstuffs for 100,000 meals, approximately $20 million of cash resources, and "1,500 truckloads of free merchandise" (Penuel & Statler, 2011, p. 135).

Military units may be integrated as necessary components of response operations. Galveston necessitated military response after the hurricane as a means of restoring and maintaining societal order, facilitating services, and supplementing the overall relief endeavor (McElreath et al., 2017). During modern times, response

Figure 2.2 Emergency Management Cycle
Source: Arka38/Shutterstock.com

operations also incorporate military units. Both state and federal military resources may be used during the aftermath of an incident. Among states that possess their own state defense force (SDF), governors may deploy their respective SDFs in conjunction with their National Guard forces, thereby leveraging SDFs as force multipliers to better response effectiveness (McElreath et al., 2016). For instance, after Hurricane Katrina, SDFs were deployed during the aftermath of the storm to support response operations (Doss, Sumrall, & Jones, 2008).

After a disaster, some efforts toward achieving normalcy occur without any guarantee that full normalcy will ever be achieved. Simultaneously, when recovering from an incident, organizations must strengthen their resiliencies against future incidents. After the Galveston hurricane, the recovery period encompassed planning toward strengthening the city to withstand future incidents (Brennan, 2002). Despite its best attempts to recover from the hurricane and to resume some normalcy societally, Galveston never completely recovered either economically or with respect to the quality of life that existed before the storm (Brennan, 2002). Any consideration of the Galveston storm reveals dismal insight regarding the negative effects of disastrous incidents: no guarantees exist that recovery will occur completely and that any previous state of normalcy that existed before a calamity occurred will be regained. During modern times, this notion is applicable within the context of Hurricane Katrina regarding its population. For instance, Sastry and Gregory (2014) indicated that approximately 53% of the displaced adults had returned to New Orleans nearly a decade following the storm (Sastry & Gregory, 2014).

SUMMARY

Galveston was unprepared to experience the incoming storm, and was unaware of its magnitude and potential lethality. Despite the ability to communicate over great distances, communicating the dangerousness of the incoming storm was impeded, few assets existed that were capable of mitigation and response, and the recovery effort was massive. The experience was not moot; instead, Galveston heeded its lessons learned in 1909 by better mitigating the effects of the next hurricane. Regardless, Galveston never fully recovered its economic status.

During modern times, technologies and prediction methods have greatly improved. American society has experienced an array of events and situations that have influenced its abilities to prepare for disaster, mitigate the effects of an incident, respond effectively and efficiently to a calamity, and recover strongly from disasters. Although a century has passed since the Galveston hurricane and a decade has passed since Hurricane Katrina, organizations must be vigilantly aware of endangerments, and plan accordingly with respect to their respective threats.

Despite the best planning, preparedness, and mitigation efforts, nature can be severe and damaging. For instance, in 2017, the Texas coastline again experienced another devastating incident—Hurricane Harvey. At the time of this writing, initial estimates of damages are approximately $24 billion, thereby placing the incident between Hurricane Ike (2008—cost of $35 billion) and Hurricane Andrew (1992—cost of 48 billion; Crosthwaite, 2017). The economic impacts of the storm transcended the locality, and affected the region's gas prices because oil refineries became inoperative (Newsmax, 2017). This storm eliminated at least 10% of the U.S.'s ability to generate fuel (Newsmax, 2017). Flood waters in and around the affected area have submerged substantial areas in and around the affected region. In order to alleviate reservoir water quantities, the U.S. Army Corp of Engineers anticipates controlled releases of reservoirs near Houston, Texas. These actions are expected to cause additional flooding, and flood waters are anticipated to exist for 2 months . The devastation in Houston, the nation's fourth largest city, was unprecedented and represented the worst calamity in its recorded history.

Both hurricanes wrote tragedies in the history of not only Texas, but also American society. However, from the EMC perspective, differences existed between the events. The precursors for Hurricane Harvey consisted of improved tracking, forecasting, warning, and evacuating of residents. Modern technologies provided far better and reliable communication systems and messaging. Public and private sector entities collaborated earlier to facilitate preparedness, mitigation, and response pursuits. As recovery from Hurricane Harvey begins, time will reveal the level and amount of normalcy that may be achieved in due time.

Comparing the two hurricanes reveals the reality of emergency management: each incident is unique regardless of any perceived similarities. Each incident must also be treated as a local happening within the EMC context. When assessing potential incidents, organizations must uniquely identify their desired end state of existence with respect to the calamity, the methods and associated resources for achieving the desired end state, and the defining of courses of actions that, hopefully, will produce the identified end state. In other words, strategy must be crafted to address calamities.

After developing strategic initiatives, organizations may educate their personnel regarding potential disaster scenarios and corresponding policies. Training factions may enact goals (i.e., the end states), operators (i.e., necessary resources), methods (i.e., methods of achieving the end states), and selection rules (i.e., course of actions) among disaster training programs whereby personnel are prepared for possible incidents (Allen, McElreath, Henley, & Doss, 2014). Organizations may also consider the enacting of mock response drills in conjunction with other entities, such as law enforcement, logistics, medical, military, and similar entities. Performing such drills improves and strengthens the ability of an organization to withstand an actual disaster, and provides insight regarding potential weaknesses. It also provides valuable experience for improving the future phases and principles of preparedness, mitigation, response, and recovery.

Experience contributes toward improvement among emergency management environments. The Galveston experience was a catalyst for spurring improvements toward bolstering the ability to plan and withstand disasters. Although a tragedy, the Galveston incident was a starting point from which modern emergency planning and contingency formulation emerged and implemented. Through such improvements, it is hoped that the destructiveness and lethality of future incidents may be abated or avoided as a result of proactive preparedness.

REFERENCES

Allen, A., McElreath, D., Henley, R., & Doss, D. (2014, May). *The GOMS model and public safety management education.* Paper presented at the annual meeting of the Joint International Conferences 2014 Academic Business World International Conference & International Conference on Learning and Administration in Higher Education. Nashville, TN: Academic Business World.

Brennan, K. (2002). *The Galveston hurricane.* New York, NY: Chelsea House Publishing.

Crosthwaite, A. (2017). *'Historic Event': Harvey damages estimated at $24 billion.* Retrieved from http://www.newsmax.com/Finance/Economy/hurricane-harvey-damage-estimate/2017/08/27/id/810084

Doss, D.A., Glover, W., Goza, R., & Wigginton, M. (2015). The *foundations of communication in criminal justice systems.* Boca Raton, FL: CRC Press.

Doss, D., Sumrall, W., & Jones, D. (2008). Volunteer service: Quantitatively examining the organizational perceptions of commissioned officers versus non-commissioned personnel of a state defense force. *Proceedings of the Academy of Organizational Culture, Communications, and Conflict, 13*(1), 25–30.

Frantz, V. (2012). Deadliest hurricanes in U.S. history. *AccuWeather,* Retrieved from http://www.accuweather.com/en/features/trend/deadliest-hurricanes-in-us-his/64321

Green, N. (2000). *Story of the 1900 Galveston hurricane.* Gretna, LA: Pelican Publishing.

Gurowitz, M. (2008, August 6). *The origins of our disaster relief: The Great Galveston Hurricane of 1900.* Johnson & Johnson. Retrieved from https://www.kilmerhouse.com/2008/08/the-origins-of-our-disaster-relief-the-great-galveston-hurricane-of-1900

James, R., & Gilliland, B. (2013). *Crisis intervention strategies.* Belmont, CA: Brooks/Cole Publishing.

Keim, B., & Muller, R. (2009). *Hurricanes of the Gulf of Mexico.* Baton Rouge, LA: Louisiana State University Press.

Longshore, D. (2008). *Encyclopedia of hurricanes, typhoons, and cyclones.* New York, NY: Infobase Publishing.

McElreath, D., Doss, D., Jensen, C., Lackey, H., Jones, D., & Wigginton, M. (2017). Dangers from the sea: Considerations of the 1900 Galveston Hurricane. *International Journal of Maritime History, 29*(3), 529–543.

McElreath, D., Doss, D., Jensen, C., Lackey, H., Wigginton, M., & Jones, D. (2016). *State defense forces: Strategic resources for homeland security and emergency management.* Proceedings of the Southwest Academy of Management, 2016, 264–275.

McElreath, D., Doss, D., Jensen, C., Wigginton, M., Nations, R., & Van Slyke, J. (2014). *Foundations of emergency management.* Dubuque, IA: Kendall-Hunt Publishing.

McElreath, D., Jensen, C., Wigginton, M., Doss, D., Nations, R., & Van Slyke, J. (2014). *Introduction to homeland security* (2nd ed.). Boca Raton, FL: CRC Press.

Newsmax. (2017). *Gasoline surges, oil steady as Harvey shuts Texas refineries.* Retrieved from http://www.newsmax.com/Finance/StreetTalk/hurricane-harvey-gasoline-oil/2017/08/27/id/810083

Peak, D. (2013). Galveston relief effort was Barton's last disaster mission. *Galveston Newspapers.* Retrieved from http://www.1900storm.com/redcross

Penuel, K., & Statler, M. (2011). *Encyclopedia of disaster relief.* Thousand Oaks, CA: Sage Publishing.

Prothero, D. (2011). *Catastrophes: Earthquakes, tsunamis, tornadoes, and other earth-shattering disasters.* Baltimore, MD: John Hopkins University Press.

Ramos, M. (1998). Galveston's response to the Hurricane of 1900. Retrieved from http://www.texasalmanac.com/topics/history/galvestons-response-hurricane-1900

Rubin, C. 2012. *Emergency management: The American experience, 1900-2010* (2nd ed.). Boca Raton, FL: CRC Press.

Sastry, N., & Gregory, N. (2014). The location of displaced New Orleans residents in the year after Hurricane Katrina. *Demography, 51*(3), 753–775.

Smith, M. (2013). Post-storm rebuilding considered 'Galveston's finest hour.' *Galveston Newspapers.* Retrieved from http://www.1900storm.com/rebuilding

Wall, B., & Keeling, A. W. (2011). *Nurses on the front line: When disaster strikes, 1878-2010.* New York, NY: Springer Publishing.

Wellman, A. (2013). Grief in comparison: Use of social comparison among cold case homicide survivors. *Journal of Loss and Trauma: International Perspectives on Stress & Coping, 19*(5), 462–473.

Wellman, A. (2017). Understanding the surviving families of homicide. In B. van Eeden-Moorefield & S. Browning (Eds.), *Contemporary families at the nexus of research and practice* (pp. 165–183). Abingdon, UK: Routledge.

Willett, D. (2013). *Galveston chronicles: The queen city of the gulf.* Charleston, SC: The History Press.

Williams, J., & Sheets, B. (2001). *Hurricane watch: Forecasting the deadliest storms on the earth.* New York, NY: Vintage Books.

Wood, R. (2007). *The weather almanac: A reference guide to weather, climate, and related issues in the United States and its key cities.* Hoboken, NJ: John Wiley & Sons Publishing.

The San Francisco Earthquake and Fire (1906)

3

Figure 3.1 Decimated Building: The Smoking Aftermath of the 1906 San Francisco Earthquake and Fire
Source: Library of Congress

BRIEF OVERVIEW

The 1906 San Francisco Earthquake and Fire remains the deadliest disaster in the history of California. In many ways it was a defining moment in urban America. Prior to, and after the San Francisco Earthquake and Fire, American cities, both large and small, experienced the destruction of major fires. For example, New York City and Chicago experienced significant fires in the 1800s, while Atlanta was intentionally burned during the Civil War, and also suffered from a major urban inferno in 1917. The 1871 Peshtigo firestorm burned major areas in Wisconsin, which destroyed numerous communities, but the San Francisco Earthquake and Fire was different.

The San Francisco Earthquake and Fire, in many ways, redefined urban disasters and response in the nation. This event combined many different aspects of a disaster that included an earthquake, and subsequent fire with devastating consequences. The disaster was compounded by a confused and counterproductive response. For example, in addition to the aspects of the natural disaster, there were later reports of arson, which prompted confusion and panic and inhibited efforts to mitigate and respond to the disaster (McElreath, Jensen, et al., 2014).

INTRODUCTION

In the early 1900s, San Francisco was a west coast treasure. The earthquake struck the area on April 18, 1906, at 5:12 a.m., which had a magnitude of 7.8. The earthquake damaged many homes and buildings, particularly in Santa Rosa, California. However, the quake also damaged gas lines and toppled chimneys, which sparked many major fires throughout San Francisco. Water lines were also broken, and this hindered the emergency response. The initial death toll reported was around 700, but the subsequent fires increased that total to more than 3,000. While the quake itself caused a great deal of destruction to San Francisco, the fires led to more than 80% of the city being destroyed (Lawson, 1908).

The 1906 San Francisco Earthquake and Fire, as deadly and devastating as it was, gave birth to significant advances in both the study of earthquakes as well as fire suppression advances that might not otherwise have occurred had it not been for such devastation at the time. Seismology as a science was bolstered as were changes within the fire service. Auxiliary hydrant and pump systems became standard practice, fire boats were brought into service, and more stringent building codes evolved.

The Setting

In 1906, the population of San Francisco was about 410,000. It was the ninth largest city in the United States and largest on the west coast. The previous 50 years saw San Francisco become the center for trade, finance, and culture on the west coast. Following the disaster, much of the population relocated to Los Angeles and took with it industry and trade that once called San Francisco home.

San Francisco lies on the San Andreas Fault which runs approximately 750 miles through California. It is a tectonic boundary between the Pacific and North American Plates. It was initially discovered by Professor Andrew Lawson of UC Berkeley in 1895, when he identified its northern zone. Following the 1906 earthquake, Lawson determined that the fault extended all the way into Southern California. Today, the fault is divided into three segments—northern, central and southern. Each segment has different characteristics and as a result, different degrees of earthquake risk. It is believed that the southern segment is most at risk today—this segment includes Los Angeles. The fault passes just 35 miles to the northeast of Los Angeles, and a significant magnitude earthquake would cost countless lives and cause hundreds of billions of dollars in damage to the area (Collier, 1999).

San Francisco Fire Chief Dennis T. Sullivan assumed the position of chief in 1893, after serving as a fireman and working his way up through the ranks beginning in 1877. The *Guardians of the City Museum and Memorial* notes that he was "a fireman of the highest ability, and a firm disciplinarian, his leadership brought the Department to a very high standard of efficiency." It was Sullivan who warned of the need for more protective measures to be put in place after a string of smaller earthquakes and fires hit the San Francisco area (Guardians of the City, n.d.). Among his recommendations were to repair and maintain the city's cistern system that would help pump water should there be a supply issue to the hydrant system. He also advocated for the city to acquire and use fire boats, replace the smaller hydrant mains with larger ones, and housing explosives that firemen would be trained to use in order to create firebreaks if necessary (Szoenyi, 2017).

In a report he issued in July 1895, Sullivan noted that the city did not have enough water, adequate hydrants, or even enough water pressure to fight the fires that were burning in San Francisco at the time. In it, he said: "Our city is built almost entirely of wood," the chief declared, "and I think I can safely say that we have a smaller fire limit area than any other city of our size in the Union. We are far removed from any city with a fire department of sufficient size to give us the assistance we would require in case a calamity of this kind should visit us, and if our Department cannot quench it, there is nothing we can do but blow down the surrounding blocks or burn up" (San Francisco Municipal Reports, 1893).

However, his pleas went largely ignored. In 1903, Sullivan proposed the Auxiliary Water Supply System (AWSS). The AWSS was designed to house water from area reservoirs should the hydrant system fail. Again, Sullivan's requests for high-pressure water systems stemming from the bay were dismissed, and in 1905, during a speech Sullivan delivered the following warning: "This town is in an earthquake belt. One of these fine mornings, we will get a shake that will put this little water system out, and then we'll have a fire. What will we do then? Why, we'll have to fight her with dynamite." Those words unfortunately came true in 1906, when the earthquake struck (Szoenyi, 2017).

Ironically, Sullivan himself became a victim of the earthquake after rushing from his room to his wife who was sleeping in another room where they resided at the chief's quarters. Sullivan fell through the floor, dropping three floors below and received serious injuries. He died 4 days later on April 22 at Presidio General Hospital. When his remains were interred nearly a year later on April 15, 1907, an article in the *San Francisco Call* newspaper noted: "He died without knowing that the city had met the fate he had so frequently predicted and had so stubbornly and bravely battled against" (Szoenyi, 2017).

The Disaster Unfolds

There was very little warning ahead of the earthquake that struck San Francisco in the early morning hours of April 18, 1906. A strong foreshock hit the area about 20–25 sec ahead of the main shock, which struck at approximately 5:12 a.m., and lasted for about 42 sec. The 7.8 magnitude earthquake occurred in the early morning hours catching people by surprise as they slept. Initially it was believed that around 700 people perished due to the collapse of buildings and the initial quake. However, the fires that spread out across the city brought that number upward of 3,000 by most estimates.

As damaging as the earthquake and aftershocks were, it was the fires burning out of control that claimed the most lives and property, reducing the once great city to smoldering rubble. Nearly half of San Francisco's population of 410,000 was left homeless, and many ended up in refugee camps in the area lasting more than 2 years. Those who opted not to remain moved south to Los Angeles and surrounding areas while the city was being rebuilt (Aitken and Hilton, 1906).

Impact

What was most impressive about the disaster came as a result of the response and recovery efforts. While the *Posse Comitatus* Act prohibits the use of federal troops to enforce civilian laws, military personnel may still be called upon to support civilian officials who are diligently exercising their local authority under the *Immediate Response Authority,* when there is not enough time to seek immediate approval from higher authorities, and such action is imperative in an effort to "save lives, prevent human suffering, or mitigating great property damage" (Waugh, 2000).

Figure 3.2 San Francisco City Hall After the 1906 Earthquake
Source: Everett Historical/Shutterstock.com

After the death of Fire Chief John Sullivan, the deputy chief sent word to the Presidio, a U.S. Army fort at the time, and that he needed dynamite to create firebreaks within the city to slow down the spread of fires. General Frederick Funston who was stationed at the Presidio had already made the decision to use troops to take control of the situation. In doing so, he notified San Francisco's mayor Eugene Schmitz that army troops from nearby Angel Island would be mobilized to assist in the city, bringing with them dynamite (Greeley, 1906; Keeler, 1906).

Containing the Disaster

The U.S. military troops were used to patrol the streets in an effort to prevent looting and guarded buildings such as the U.S. Mint, post offices, and the county jail. They also assisted the firefighters in creating firebreaks with the dynamite they brought in, and also assisted the community by feeding, housing, and clothing many of the now homeless victims. More than 4,000 troops assisted in the efforts. By July 1, local authorities assumed relief efforts for the city and the army troops were withdrawn (Greeley, 1906). The U.S. Navy also assisted in the efforts to contain the disaster, providing ships and naval personnel to pump water from the bay into the city supplying water to support the efforts to fight the fires .

RELIEF AND FIRST RESPONSE EFFORTS

Disaster relief was also a new concept at this time, as no federal agency was in place to manage disasters. The federal government appropriated $1 million toward relief supplies that were shipped to the area, but this was not nearly enough. This included tents that would be used in the refugee camps and supplies to food kitchens to help feed the homeless. Relief efforts extended around the world and reached over $5 million at the time. This included London which raised thousands of dollars. Private businesses and businessmen also donated to the effort, including Standard Oil's $100,000 donation. Andrew Carnegie also gave $100,000 and the Dominion of Canada even made a special appropriation of $100,000 toward the cause (Greeley, 1906; Lawson, 1908).

AFTERMATH AND INVESTIGATION

Also hard hit by the disaster were the insurance companies. It is estimated that collectively they paid out between $235 and $265 million on policyholder claims for fire damage. Most policies of the period excluded shake damage caused by earthquakes. This flaw actually led to many people setting fires to their own homes just to be able to settle a claim. Almost 150 insurance agencies were involved in settling claims, causing many to ultimately go bankrupt in the process .

LESSONS LEARNED AND POLICY IMPACT

Numerous lessons were learned as a result of the earthquake and fires. First and foremost, Chief Sullivan's recommendations for an improved hydrant system, the AWSS, and addition of fire boats to fire department resources finally materialized. The department launched two fire boats in 1909, one named after the former fire chief. Then 3 years after the fire in 1913, the AWSS was completed (Szoenyi, 2017).

Three days after the earthquake, Governor George C. Pardee of California appointed an eight-member committee to analyze the event. Among its members was Andrew C. Lawson, chair of the University of California; Henry F. Reid of Johns Hopkins University; G.K. Gilbert of the I.S. Geological Survey; and five others. They delivered a statement of results and conclusions to the governor on May 31. It consisted of two volumes and an atlas edited by Lawson. This work highlighted the existence of what would become known as the San Andreas Fault, which gave official recognition to the elastic rebound theory. Ultimately this had an unusual significance in the development of the field of seismology.

At the time of the 1906 San Francisco earthquake, many California municipalities had building codes, but none considered seismic effects. The 1906 earthquake initiated discussions to improve earthquake engineering design and incorporating those improvements into regulatory codes, based on the modern scientific study of the San Andreas Fault system in California. Professional organizations, particularly the Seismological Society of America, which formed in 1906, and later, the Structural Engineers Association of California, were persistent advocates of code provisions for earthquake-resistant construction (Schulz & Wallace, n.d.). In 1925, Santa Barbara's building code required that structures be designed to withstand horizontal forces. This effort became the first explicit policy and legal consideration of the seismic safety of structures in California. Led by professors at Stanford, Palo Alto would also add seismic provisions to its building code in 1926. Today, it is typical to find that some of the most significant code changes have occurred after major earthquakes identified or emphasized structural deficiencies (Stanford, 2006).

IMPLICATIONS FOR MODERN EMERGENCY MANAGEMENT

Emergency managers include earthquakes as part of their comprehensive emergency management planning. While some of the country's most significant earthquakes have occurred in California and on the west coast, the east coast and areas in between are just as susceptible and must be prepared. The New Madrid Fault Line runs from New Madrid, Missouri toward the southwest, which poses a threat to parts of eight states, including Missouri, Mississippi, Illinois, Indiana, Kentucky, Tennessee, Oklahoma, and Arkansas. Emergency managers also serve as advocates for more stringent building codes, especially in seismic-prone areas.

Of the many types of disaster faced by emergency managers today, earthquakes are considered sudden-onset events because little warning is often received, as is the case with tornadoes. As such, those in seismic-prone areas often collaborate with the U.S. Geological Survey to draw upon the many resources offered, including real-time data information, maps, and software (McElreath, Doss, et al., 2014).

SUMMARY

The 1906 San Francisco Earthquake and Fire is a disaster that delivered many lessons and led to significant policy changes that are very evident today. But for this event, we may not have adopted the stricter building codes in California and other west coast states that we see today. We may not have seen cities such as San Francisco, Portland, Oregon and Seattle, Washington implement secondary hydrant systems. In this disaster, while there were significant losses, we fortunately made significant gains in areas such as engineering and building codes that have contributed to maintaining the safety and welfare of the public, and saved countless lives.

REFERENCES

Aitken, F., & Hilton, E. (1906). *A history of the earthquake and fire in San Francisco*. San Francisco: The Edward Hilton.

Guardians of the City. (n.d.). Bigger mains demanded: Chief Sullivan's scathing report on the city's lack of water. *Guardians of the City San Francisco Fire Department Museum*. Retrieved from http://guardiansofthecity.org/sffd/chiefs/dt_sullivan_1895_lack_of_water_report.html

Collier, M. (1999). *A land in motion*. Berkeley, California: UC Press.

Greeley, A. (1906). *Earthquake in California, April 18, 1906. Special report on the relief operations conducted by the military authorities*. Washington: Government Printing Office.

Keeler, C. (1906). *San Francisco through earthquake and fire*. San Francisco: Paul Elder & Company.

Lawson, A. (1908). *The California Earthquake of April 18, 1906: Report of the state earthquake investigation commission* (Vol. 2, p. 87). Washington, DC: Carnegie Institution of Washington Publication.

McElreath, D. H., Doss, D. A., Jensen, C. J., Wigginton, M. P., Nations, R., Van Slyke, J., & Nations, J. (2014). *Foundations of Emergency Management*. Debuque: Kendall Hunt.

McElreath, D. H., Jensen, C. J., Wigginton, M. P., Doss, D. A., Nations, R., & Van Slyke, J. (2014). *Introduction to Homeland Security*. Boca Raton: CRC Press.

San Francisco Municipal Reports. (1893). *San Francisco municipal reports, 1982-1893*. San Francisco, CA: City of San Francisco.

Schulz, S. & Wallace, R. (n.d.). *The San Andreas fault.* USGS. Retrieved from https://pubs.usgs.gov/gip/earthq3

Stanford University. (2006). *Evolution of codes.* Retrieved from: https://quake06.stanford.edu/centennial/tour/stop10.html

Szoenyi, V.A. (2017). SF's legendary fire chief Dennis T. Sullivan predicted the city's disaster. Retrieved from http://brokeassstuart.com/blog/2017/08/23/sfs-legendary-fire-chief-predicted-its-disaster-before-the-1906-earthquake-fire

Waugh Jr., William L. (2000). *Living with hazards dealing with disasters: An introduction to emergency management.* Armonk, NY: M.E. Sharpe.

4 Urban Tornadoes The Great Cyclone: 1896 St. Louis Tornado

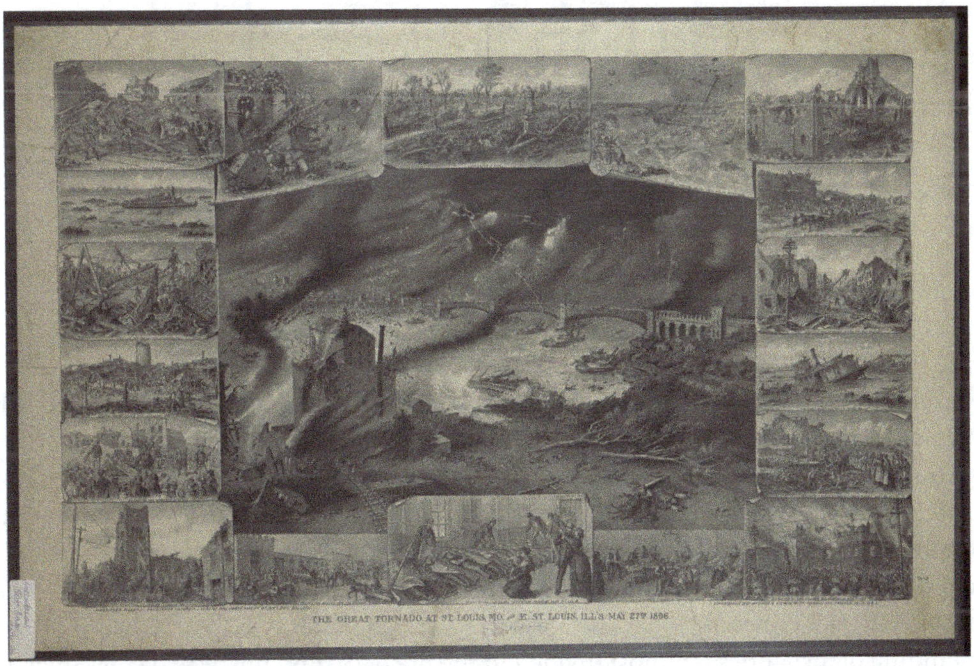

Figure 4.1 The 1896 St. Louis–East St. Louis Tornado
Source: Library of Congress

BRIEF OVERVIEW

Tornadoes are common in the United States and when they strike, the path of destruction is often wide and deadly. In 2017, over 1,500 tornadoes were reported across the country. The damage these storms inflict is often hard to imagine. Larger tornadoes easily create paths of destruction which might well exceed a mile in diameter and go great distances. Along their route, damage is inflicted by the effects of wind, rain, and displaced debris (McElreath, Doss, et al., 2014).

The frequency of tornadoes in the United States is closely tied with the progression of the warmer seasons when warm and cold air masses clash. Most of the early spring tornadoes in the United States tend to occur in the Southeast and South Central regions. Gulf States, such as Mississippi and Louisiana, are the frequent recipients of tornadoes from February to April. Late spring tornadoes generally spread farther north and west, often into Kansas, Nebraska, and the Tennessee Valley region (NOAA, n.d.-c, Historical Records and Trends). The central part of the United States experiences tornadoes in the early spring and summer. Weather conditions are

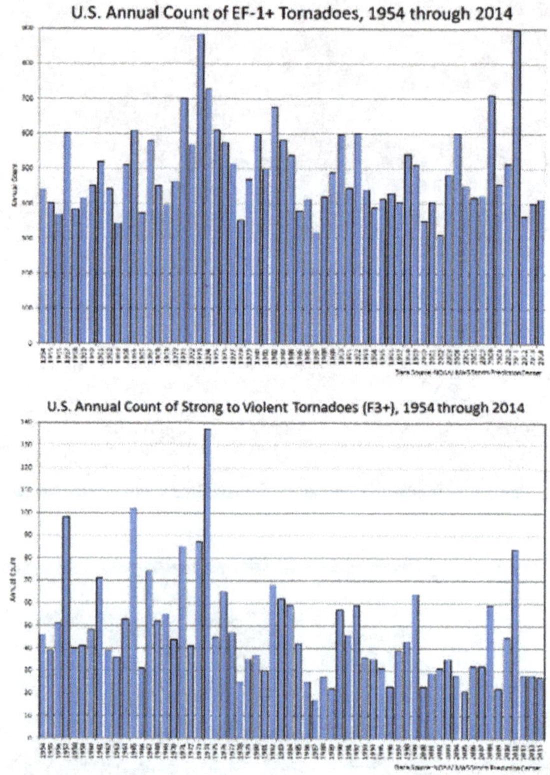

Figure 4.2 Tornadoes in the United States (1955–2015)
Source: NOAA

the source of origin of tornadoes; when the right atmospheric conditions exist, the wind can achieve a deadly intensity (McElreath, Jensen, et al., 2014).

Tornadoes know no boundaries. They might inflict little damage traveling in a rural area or their course might well impact a multi-jurisdictional or multistate area and in its travel, destroy the communities in its path. The damage to property is just one of the aspects of tornadoes that create response challenges for first responders, emergency management and homeland security officials. The injury and death linked to tornadoes is, of course, the most significant concern.

The understanding of tornadoes is almost divided in the pre-1950 and post-1950 eras. Since 1950, information pertaining to tornadoes has significantly improved. Advances in a wide range of technologies that can be applied to weather observation and prediction have allowed for increase accuracy in prediction, tracking, and warning.

The size, intensity, and scope of tornadoes vary significantly and nothing can be done to prevent them, but the lack of prevention does not preclude efforts to prepare for such events. Information and understanding are key to preparation.

Since 1950, about 2.5% of the tornadoes which have occurred in the United States have moved along a path of 25 miles or more with only about 0.1% of these traveling 100 miles or more. While many occur, few result in human death. Since 1950, about 2.5% of all tornadoes have resulted in deaths (Livingston, 2014).

INTRODUCTION

While the majority of tornadoes do not result in human deaths, the list of deadly tornadoes that have impacted our nation is significant. The deadliest tornado recorded in the United States was the 1925 Tri-State Tornado. Striking Missouri, Illinois, and Indiana, it was responsible for an estimated 695 deaths and injured approximately 2,027. This tornado traveled over 300 miles inflicting damage in Missouri, Illinois, and Indiana.

An 1840 tornado impacting the Mississippi River area near Natchez, Mississippi is considered to be the second deadliest tornado in the nation. One fascinating aspect of the Natchez Tornado is that the majority of the deaths occurred among the crews and passengers occupying the boats and barges which traveled the Mississippi River. The crews and passengers of these boats were shocked to find themselves at the mercy of the winds of the storm with no opportunity to escape (NOAA, n.d.-b, Deadliest Tornadoes). The 1840 tornado inflicted damage in both Mississippi and Louisiana resulting in the deaths of over 300 people (Staff, 2014). The third deadliest storm series in the nation was the 1896 Great St. Louis storm which we shall examine further.

As mentioned earlier, the scope and length of the path of a tornado may be short or extensive. A series of tornadoes in April 1965 which became known as the Palm Sunday Tornado Outbreak cut a path through and into six states—Iowa, Illinois, Wisconsin, Indiana, Michigan, and Ohio. Over 260 deaths and millions of dollars in damages were attributed to the storm (Berrington, 2015).

In other cases, the limited number of deaths proves remarkable. In 1966, a series of tornadoes impacted several cities, towns, and communities in Kansas. One of these, the Topeka Tornado, cut a wide path as it stayed on the ground for 26 min moving along a 22-mile course through the State Capital of Kansas, Topeka. The Topeka Tornado killed 17 people, injured over 500, destroying an estimated 800 homes, and damaging about 3,000 others. The damage was estimated to be over $200 million in 1966 dollars. It was surprising that the death toll was not significantly greater. Much of the credit for the limited deaths from the storm was due to the outstanding watch, warning, and storm spotter activation prior to the development of the storm and the rapid response by first responders and medical personnel (NOAA, n.d.-a, 1966 Topeka Tornado).

The second largest tornado outbreak on record for a single 24-hr period in the history of the United States occurred from April 3 to April 4, 1974. Within this storm cell were at least 148 tornadoes confirmed in 13 states, including Illinois, Indiana, Michigan, Ohio, Kentucky, Tennessee, Alabama, Mississippi, Georgia, North Carolina, Virginia, West Virginia, and New York; and the Canadian province of Ontario. Its extensive damage affected approximately 900 square miles along a total combined path length of 2,600 miles. This super outbreak of tornadoes remains the most severe convective weather episode of record in the continental United States.

In 2018, a tornado hit the campus of Jacksonville State University in Alabama. Fortunately, the storm and damage occurred during the academic spring break for the University and no death occurred. Our nation's history is full of stories of valor and resilience on the part of those impacted by tornadoes and the first responders who quickly react to these events.

The Setting

The study of tornadoes is a true science which has significantly advanced over the last 70 years. Prior to the 1950s, weather warnings depended on storm watchers who would warn of danger as quickly and efficiently as possible. According to Roger Edwards of the National Weather Service's Storm Prediction Center writes, "tornadoes were, for most, dark and mysterious menaces of unfathomable power, fast-striking monsters from the sky capable of sudden and unpredictable acts of death and devastation" (Giaimo, 2016).

For years, many believed it was impossible to predict the development and path of a tornado and even the forecasting of tornadoes was discouraged, but work in the prediction of weather was ongoing. Much of this early work was related to the military. Military commanders recognized the impact of weather on their operations. The foreknowledge of weather conditions might well prove advantageous.

Soon after the War Between the States, the United States Army Signal Service opened a weather forecasting office in 1870. These men assigned to the weather forecasting office took very early steps to understand the secrets of weather. The Signal Office's forecasting division made slow advancement and was moved to the Department of Agriculture, where it would later become the United States Weather Bureau (Giaimo, 2016).

Much early very important advancement in the area of science linked to tornadoes climatology in the United States was done by Colonel John Park Finley, an American meteorologist and Army Signal Service officer. It was Finley who is credited with being the first person to study tornadoes intensively and his work laid the foundation for the later forecasting of such weather events. Finley collected and analyzed climatological data,

Figure 4.3 Colonel John Park Finley
Source: : Library of Congress

established a nationwide weather observer network, later formed a private weather enterprise which was used by insurance companies to assess risk and even opened an early aviation weather school.

An accomplished scientist, he shared the results of his work in a range of publications related to weather to include his book *Tornadoes*, published in 1887. While some of Finley's safety guidelines have since been refuted as dangerous practices, his book remains a remarkable early work in tornado research (NOAA, n.d.-c, Historical Records and Trends).

While significant advancements were being made in other areas of weather and weather prediction, the science of tornado prediction moved slowly but a breakthrough was on the horizon. Soon after World War II, two air force meteorologists, Major Ernest J. Fawbush and Captain Robert E. Miller, were on duty at Tinker Air Force Base in Oklahoma where tornado struck inflicting damage to both Tinker Air Force Base and nearby Norman. The damage was extensive, destroying aircraft, equipment, buildings, and property. An investigative board of officers was convened which reviewed the vent and declared that "due to the nature of the storm, it was not forecastable given the present state of the art." Not satisfied, the area commander Major General Frederick S. Borum directed his weather forecasters to investigate the feasibility of forecasting tornado-producing thunderstorms (O'Connell, 2007).

Five days later, Fawbush and Miller noticed the same weather conditions that were present in the earlier tornado developing and briefed the General of their analysis. With the concurrence of General Borum, a forecast for heavy thunderstorms for later in the week was issued. As the anticipated weather situation further developed, the strong series of storms further displayed the characteristics of the earlier tornado. In discussion with General Borum, Captain Miller pointed out that even though the weather conditions were similar to those that fueled the earlier tornado, the odds against a tornado striking the same location within 5 days were extremely low. Also, no one had ever issued a tornado warning. As described by Major Fawbush and Captain Miller, General Borum said "Then, gentlemen, you are about to set a precedent" (Bernard, 2014).

As predicted, another tornado struck, but with the forecast and actions taken to secure the base, damage was limited. The work and forecast techniques of Major Fawbush and Captain Miller were recognized for their importance and their groundbreaking analysis. Their work proved to be an early model for prediction and was adopted by the meteorological community (Giaimo, 2016). Building upon this early work, scientists and meteorologists examined the characteristics of these storms and worked to further increase the body of knowledge

Figure 4.4 Major Ernest J. Fawbush and Captain Robert E. Miller
Source: USAF

FUJITA SCALE			DERIVED EF SCALE		OPERATIONAL EF SCALE	
F Number	Fastest 1/4-mile (mph)	3 Second Gust (mph)	EF Number	3 Second Gust (mph)	**EF Number**	**3 Second Gust (mph)**
0	40-72	45-78	0	65-85	**0**	**65-85**
1	73-112	79-117	1	86-109	**1**	**86-110**
2	113-157	118-161	2	110-137	**2**	**111-135**
3	158-207	162-209	3	138-167	**3**	**136-165**
4	208-260	210-261	4	168-199	**4**	**166-200**
5	261-318	262-317	5	200-234	**5**	**Over 200**

*** **IMPORTANT NOTE ABOUT ENHANCED F-SCALE WINDS:** *The Enhanced F-scale still is a set of wind estimates (not measurements) based on damage.* Its uses three-second gusts estimated at the point of damage based on a judgment of 8 levels of damage to the 28 indicators listed below. These estimates vary with height and exposure. **Important**: The 3-second gust is not the same wind as in standard surface observations. Standard measurements are taken by weather stations in open exposures, using a directly measured, "one-minute mile" speed.

Figure 4.5 The Enhanced Fujita Scale
Source: NOAA

related to tornadoes. With further knowledge and understanding of this weather event, it is not surprising that a scale to measure the strength of these storms was developed.

The strength of tornadoes today is measured according to the Enhanced Fujita Scale. Over the years, many studied the outbreak and conduct of tornadoes. The most significant researcher of this weather phenomenon was Dr. Tetsuya (Ted) Fujita. Dr. Fujita examined tornado features such as suction vortices. His investigation led, in 1971, to the development of the original Fujita Scale in collaboration with Allen Pearson. The Fujita Scale of measurement is organized on a sliding scale ranging from 0–5 with 5 being the most violent. Prior to the F-Scale, tornadoes were rated based on newspaper and eyewitness accounts, including questionable personal accounts, rather than storm data descriptions with tornadoes prior to 1971 which were rated retroactively.

In 1973, the Fujita Scale was modified to take into account a tornado's path length and width when classifying a tornado's intensity (Wheatley, 2017). The Enhanced Fujita Scale which is used today was developed between 2000 and 2004 and officially replaced the original Fujita Scale in 2007 in the United States and 2013 in Canada. This scale uses the same concept as the original Fujita Scale, but with revised wind speeds and damage indicators considered as factors in classifying these storms.

The Disaster Unfolds

The lack of scientific understanding of tornadoes placed communities in jeopardy. Prior to the 1950s, tornado warnings heavily depended on the observations of weather watchers who were volunteers or the private citizens who happened to see the storm system form and move. As time passed and technology evolved, scientists and especially meteorologists gained a better understanding of the weather trends. As they related to these storms and with the advancement in areas such as weather radar, their ability to make accurate storm predictions significantly increased (McElreath, Doss, et al., 2014).

The disaster we will examine occurred in 1896. In the 1890s, storm warnings, as discussed, depended on citizens who spread an alarm when the danger was recognized. Often church bells were sounded to alert of the danger. While the 1896 St. Louis–East St. Louis Tornado occurred well prior to the modern advancement in the science in storm analysis, the event serves well as a case study examining the impact and destructiveness of an urban tornado from which lessons learned may be extracted.

By the late 1890s, St. Louis, Missouri was a city experiencing what many have called its "Golden Age." It has survived many challenges. Its population had suffered during the cholera epidemics of the 1830s, 1840s, and 1850s. A major fire spread through the city in May 1849 after a fire which originated in a docked steamboat White Cloud spread from the riverfront into the city. Driven by strong northeast winds, the fire spread jumped from the White Cloud to other boats docked nearby. With the docked boats ablaze, the buildings occupying the riverfront near the docked barges soon caught fire. In less than an hr, 23 steamboats and the buildings along the shore were in flames. Spreading quickly, it would take 11 hs for responders and citizens to put out the fire, but the damage had been done. The fire destroyed 418 buildings over a 15-block area (O'Neil, 2014). The city quickly began rebuilding reflecting the resilience of the city and its people.

Neither the fire nor the epidemic did little to slow the growth of the city. St. Louis prospered in the mid and late 1800s. The city was on the move, economically, socially, and politically. In the decades from the 1860s into 1900, the city's population increased to over 575,000 making St. Louis the nation's fourth largest city. The city was growing, full of promise and benefiting from a robust economy. It was a major stop for many moving west during the nation's great expansion west and with it a center of trade and a major hub for railroads.

In 1896, St. Louis was also a cultural melting pot. The population was diverse with many neighborhoods in the city reflecting the cultural background and national identities of those who resided in the area. By the late 1890s, approximately half of the city's population was foreign-born or first-generation Americans (O'Neil, 2014). While the city grew and prospered, a weather-related disaster soon would become part of its history.

Disasters of the 1800s and early 1900s reveal the primitive nature of weather forecasting and with it, the development of effective disaster warning networks. From 1840 to the beginning of the 20th century, volunteers were observing and recording data from different areas across the country and submitting their findings to the Smithsonian Institute. "By 1860, 500 stations were furnishing daily telegraphic weather reports to the *Washington Evening Star*, and as the network grew, other existing systems were gradually absorbed, including several state weather services" (Ewen, 2017). Prior to the 1890s, there was little advancement in terms of forecasting weather for informational purposes to the public.

As discussed, economically, businesses and industries were flushing. The city remained a transportation hub linking railroads to roads and a major river port. From its founding, it was a passageway to the west. Now it was a major port on the Mississippi River as well as an important point railroad hub linking the north and south and east and west. St Louis was the site of the Eads Bridge, the first bridge in St. Louis to span the Mississippi River.

St. Louis embraced the new wonders that would become characteristic of a modern city. Wires crossed the city providing electricity and linking the increasingly popular telephone. By 1880, 600 telephones were in use in the city (St. Louis City Plan Commission, 1969). A public transportation system had been established and linked various parts of the city. This public transportation system initially included cable cars, which were replaced beginning in 1887 by electric trolley cars. The public system provided inexpensive transportation while relieving some urban congestion.

The urban infrastructure included brick streets, with new construction ongoing. The city would claim to be the location of one of the nation's first skyscrapers, the 10-story Wainwright Building. Sports and music flushed, ragtime and baseball became major pastimes in the city. The city had truly become an example of the modern America of the soon to arrive 20th century. The people of St Louis were rightfully proud of their city.

The St. Louis metropolitan area has a history of tornadoes. The 1896 tornado season has the distinction of being one of the deadliest in U.S. history. There were at least 40 tornadoes from April 11 to November 26 that resulted in deaths, including the one that earlier struck the St. Louis area.

For St. Louis, May 27, 1896 would prove to be significant. The St. Louis, Missouri metropolitan, as a small part of the Missouri/Kansas weather corridor area, has a painful history of destructive tornadoes. On that day a major storm cell developed west of the city. Intense and deadly, it gained power as it moved, fueled by the weather conditions. By today's standards, the tornado that struck the St. Louis area was powerful. The storm, which would become known as the "Great Cyclone," headed toward St. Louis, would cross the city and the Mississippi River and into Illinois where it would strike East St. Louis as it continued its eastward journey. It would be a deadly storm, killing at least 255 people, injuring hundreds more as it cut a wide path of death and destruction (O'Neil, 2014).

In less than an hour, the tornado that traveled across the city would inflict its damage. As it developed west of the city, it was later reported that a meteorologist by the name of Irl Hicks watched the black clouds and green skies. He had watched the barometer in his office throughout and realized the air pressure in St. Louis was dangerously low (Collins, 2012). The weather system approaching St. Louis began reflecting the dangerous sign of forming into a tornado. The tornado that would form, believed to have been an EF-4, moved along a three-mile wide path. The storm damaged or destroyed more than 8,000 buildings and homes.

The destruction was compounded as several fires began in the wreckage of homes and businesses along its path. Warnings were sounded by citizens as they rang fire alarm boxes, church bells, and telephones. In many ways it was fortunate that the storm front brought with it significant rain that helped in suppressing the fires within the city (NOAA Photo Library, 2009, the Great Cyclone at St. Louis, 2009).

Impact

In reviewing the records of the storm, it would be considered an EF-4 or possibly an EF-5 with estimated wind speeds between 165 and 200 miles per hr (NOAA, 2017). During the time the tornado was on the ground, it tracked a three-mile-wide path of death and destruction across the area.

In St. Louis, the tornado completely destroyed blocks of residential housing and businesses. Hundreds of miles of electric wires and thousands of telephone and telegraph poles were torn down by the fierce winds. The tornado uprooted trees and hurled them with other wreckage over a wide area.

Similar to the tornado that impacted Natchez, Mississippi, in 1840, the St. Louis Tornado did not spare the riverfront. As the tornado reached the city's riverfront, the boats and those onboard were especially vulnerable. Along the Mississippi River, steamboats and other vessels were destroyed, often broken into pieces and scattered along the shores or sunk.

Over 20 steamboats, tug boats, and ferries were ripped from their moorings and destroyed. The steamboat "Anchor Line" was hurled across the Mississippi, crashing into pieces on the eastern shore of the river. Although the official death toll on this day is 255, many believe the number is much higher. On the riverfront, scores of people lived in shanty boats. Since their bodies were washed downriver, perhaps as many as 150 deaths were unaccounted for (Collins, 2012). From the riverfront, the tornado continued east into Illinois and East St. Louis, Illinois, where the damage continued.

The tornado then crossed the Mississippi River into East St. Louis, Illinois retaining great power continuing to damage and destroy buildings along the river and into East St. Louis where damage was estimated to reach $2,000,000 in 1896 value and about 118 additional deaths (Project, 2015).

Figure 4.6 Destruction of the Eads Bridge, St. Louis
Source: NOAA

Figure 4.7 Remains of St. John Nepomuk Church District Destroyed by the 1896 Tornado
Source: US Department of the Interior/National Park Service

Containing the Disaster

Little could be done to actually contain the tornado. For many in its path, they were at the mercy of the weather system. Others who had information about the impending disaster took shelter in basements and cellars which were common in many of the homes and building in the area. As soon as the storm passed, recovery efforts commenced.

Volunteers from across the city and surrounding areas came together to help those injured. City employees, the military, and volunteers worked to find the injured, recover the dead, and clear the streets. As they searched through the wreckage and debris, others were setting up a makeshift hospital, a temporary morgue, and emergency housing. Faith-based groups opened their doors and offered housing and provided food. Countless acts of valor occurred as those responding worked to help the injured as quickly as possible. Work related to search and recovery would continue for weeks. Rebuilding would start immediately (NOAA, 2017).

RELIEF AND FIRST RESPONSE EFFORTS

Response and recovery would begin almost immediately and involve not only the entire community, but also the nation. Firefighters, police, and the militia organized with citizens to conduct search and rescue operations for those injured and killed. Hospitals quickly received those injured. Shelters were established for those who had lost their homes. Business and civic leaders on both sides of the river began damage assessment. As people searched for the missing, the city morgues worked to identify the dead. At the St. Louis Morgue, the crowd became so large and demanding that the police were called in to restore order (Collins, 2012). This was an expensive event in terms of life and property (NOAA, 2017).

AFTERMATH AND INVESTIGATION

This would not be the first or the last tornado to strike an urban area in the nation, but the 1896 St. Louis Tornado proved to be one of the deadliest and most destructive tornadoes in U.S. history. This large and violent tornado was the most notable of an outbreak which produced other large, long-track, violent, killer tornadoes. It caused over $10,000,000 in damage in 1896 dollars ($4.35 billion in today's dollars). A total of 255 people died and over 1,000 were injured. More than 5,000 people were homeless and lost all of their possessions. The storm would not deter the growth of St. Louis. By 1900, the city's population surpassed 575,000, and, in 1910, it surpassed 687,000. The new century would witness the city host a World Fair and the Olympic Games, both in 1904. The city would also again be the victim of nature as significant tornadoes would return to inflict damage in 1927, 1959, 1967, and 2011.

LESSONS LEARNED AND POLICY IMPACT

The lessons learned from the St. Louis Tornado are the same lessons we continue to learn with more modern tornadoes. More recent tornado-related events continue to indicate the power of this event and the destructive force it can unleash. On May 20, 2013, an EF-5 tornado struck the central Oklahoma communities of Newcastle, Oklahoma City, and Moore. The destructions required the full engagement of local, state, and federal emergency management, and captured the attention of the nation. The tornado was on the ground for about 39 min. During that time, local television news stations provided event coverage and warnings to supplement those broadcast through emergency management channels. An estimated 2,393 structures were impacted overall, 1,128 of which were destroyed, including a hospital. Though destruction was extensive, approximately 24 deaths and less than 400 injuries were identified by authorities (Riley & Krautmann, 2016).

IMPLICATIONS FOR MODERN EMERGENCY MANAGEMENT

Much has changed in the last century when it comes to the prediction of weather and the anticipation of the impact of weather-related events. In the 1960s, the assets available to the weather service were often based on World War II era technology. Today, state-of-the-art equipment is available to identify weather paths and trends. Additional public and agency warnings benefit from today's technology. In examining the impact of such events and the steps necessary to initiate the response and recovery phases, significant lessons may be drawn from these events.

- There is a need for communities to have an active and long-standing community preparedness plan with sheltering options which may well prove instrumental in reducing potential casualties and injuries.
- Outstanding education, cooperation, and understanding between governmental agencies, the media, and public are important.
- Mutual aid agreements with their regional partners must be established and tested to determine that the level of support anticipated will be available in time of need.
- Planning is based on the concept of the development of a "whole community" response capability in which risk and vulnerability assessments are identified.
- Response capabilities are built on preparedness structures and activities.
- Establish a multilevel warning system to alert those to the anticipated event.
- Effective logistics planning and support capabilities to include food, fuel, shelter, and medical support to meet the needs of both victims of the event as well as first responders providing response and recovery efforts.
- EMS, medical personnel, and mutual aid agencies have the capability to conduct field triage and medical treatment for survivors in the midst of major damage and significant casualties.
- Establish and exercise command and control to include, but not limited to, organizational and command structure, operational sustainability, and redundancy.
- Form community-based support teams as part of the "whole community" approach that creates functional response partnerships.
- Place procedures in place to conduct accurate and timely fiscal accountability.
- Work with partners not only in adjoining jurisdictions, but also on other levels in the governmental structure.
- Advancements in technology, to include supercomputers, have allowed quick and thorough analysis and prediction of weather data which, when combined with Doppler weather radars, provide the ability to look inside thunderstorms for areas of rotation, which are sometimes a precursor to the tornado.
- Doppler weather radar present state-of-the-art high detail color images of both reflectivity and velocity zoomed to city and county level. Additionally, 3D interrogation of storms can pinpoint critical areas for storm spotters to observe before tornado development.
- Direct broadcasts to the public by the NOAA Weather Radio network allow the average citizen to receive warning information anytime during the day or night, and at the same time as media outlets.
- The explosion of the Internet and use of mobile devices has allowed a multitude of weather data and alerts to be available to nearly everyone, especially through the use of social media.

SUMMARY

The tornado that struck St Louis, Missouri, East St Louis, Illinois, and the surrounding area on May 27, 1896, was considered as one of the deadliest to occur in the United States. The tornado, which ranked as an F4 tornado, was considered to be among the top three deadliest tornadoes in the nation. Nature is extremely dangerous despite any docile appearance. The resulting damages showed the ferocity of natural destructiveness. Overall, the incident itself was a testament to the uncontrollability of nature. The tornado killed a total of 255 people and injured at least 1,000 others. At least 5,000 people were homeless and lost all of their possessions after the storm.

During modern times, tornadoes are just as uncontrollable as they were generations ago. Since the St. Louis Tornado, only technologies and emergency management methods have changed whereby such incidents now have some chance of mitigation. Despite such improvements, modern tornadoes can be just as dangerous and destructive as their historical predecessors. Any calamity may occur with speed, surprise, and violence. For instance, over the period of approximately 39 min, the Oklahoma communities of Newcastle, Oklahoma City, and Moore experienced a tornado that gained national attentiveness. It necessitated the full engagement of local, state, and federal emergency management entities. News media and emergency management resources broadcast warnings across appropriate communication channels. The tornado affected approximately 2,393 structures overall, and destroyed 1,128. Although extensive damages occurred, a total of 24 deaths and less than 400 injuries were identified by authorities (Riley & Krautmann, 2016).

Modern emergency management methods, techniques, and resources are indispensable aspects of addressing both man-made and natural calamities. Comparing the 2013 Oklahoma incident with the 1896 tornado shows marked improvements toward incident and emergency management. Although both incidents resulted in the destruction of large quantities of physical structures, the death tolls were substantially different. The 1896 tornado killed a total of 255 people and injured at least 1,000 others, whereas the 2013 tornado resulted in the deaths of 24 people and fewer than 400 injuries. Although some may question the population density regarding the incident locations with respect to the differences in casualties, the benefits of improved forecasting and warning mechanisms were undeniable. Through the use of modern emergency management paradigms and resources, chances are higher for preserving human lives.

REFERENCES

Bernard, H. W. (2014, March 21). *A one in twenty-million chance–The true story of the first tornado forecast–part II.* H. W. Bernard. Retrieved from http://buzzbernard.com/2014/03/one-twenty-million-chance-true-story-first-tornado-forecast-part-ii

Berrington, A. (2015, April 11). *Palm Sunday 1965: Southern Great Lakes ravaged by one of the worst tornado outbreaks on record.* Tornado History. Retrieved from http://www.ustornadoes.com/2015/04/11/palm-sunday-1965-southern-great-lakes-ravaged-by-one-of-the-worst-tornado-outbreaks-on-record

Collins, C. (2012, November 12). *The Great Cyclone of 1896.* Distilled History. Retrieved from https://www.distilledhistory.com/1896cyclone

Ewen, E. L. (2017). *The 1896 East St. Louis tornadoes and how they changed the United States.* Indiana Disasters. Retrieved from http://indianadisasters.blogspot.com/2014/05/the-1896-east-st-louis-tornadoes-and.html

Giaimo, C. (2016, March 15). *Until 1950, U.S. weathermen were forbidden from talking about tornadoes.* Slate. Retrieved from http://www.slate.com/blogs/atlas_obscura/2016/03/15/until_1950_u_s_weathermen_were_forbidden_from_talking_about_tornados.html

Livingston, I. (2014, April 11). *Long-track tornadoes: Historical clues about intensity, where, and when they occur most.* Tornado Climatology. Retrieved from http://www.ustornadoes.com/2014/04/11/long-track-tornadoes-historical-clues-about-intensity-plus-where-and-when-they-occur-most

McElreath, D. H., Doss, D. A., Jensen, C. J., Wigginton, M. P., Nations, R., Van Slyke, J., & Nations, J. (2014). *Foundations of emergency management.* Debuque: Kendall Hunt.

McElreath, D. H., Jensen, C. J., Wigginton, M. P., Doss, D. A., Nations, R., & Van Slyke, J. (2014). *Introduction to homeland security.* Boca Raton: CRC Press.

NOAA. (n.d.-a). *1966 Topeka Tornado.* National Weather Service (NOAA). Retrieved from https://www.weather.gov/top/1966TopekaTornado

NOAA. (n.d.-b). *Deadliest tornadoes.* National Oceanic and Atmospheric Administration. Retrieved from https://www.ncdc.noaa.gov/climate-information/extreme-events/us-tornado-climatology/deadliest

NOAA. (n.d.-c). *Historical records and trends.* National Oceanic and Atmospheric Administration. Retrieved from https://www.ncdc.noaa.gov/climate-information/extreme-events/us-tornado-climatology/trends

NOAA. (2017, May 27). *On this day: The Great St. Louis Tornado of 1896.* National Centers for Environmental Information. Retrieved from https://www.ncei.noaa.gov/news/great-st-louis-tornado-1896

NOAA Photo Library. (2009, October 1). *The Great Cyclone at St. Louis.* NOAA Photo Library. Retrieved from http://www.photolib.noaa.gov/nws/tornado.html

O'Connell, A. (2007, June 22). *Tinker Airmen first to predict tornadoes.* Tinker Air Force Base. Retrieved from http://www.tinker.af.mil/News/Article-Display/Article/388450/tinker-airmen-first-to-predict-tornadoes

O'Neil, T. (2014, August 2). *Look Back 250: Massive tornado, landmark buildings help shape fast-growing 1890s*. St. Louis Post Dispatch. Retrieved from http://www.stltoday.com/news/local/govt-and-politics/look-back-massive-tornado-landmark-buildings-help-shape-fast-growing/article_af52c298-d148-5a6b-baff-a4b405fc7623.html

Project, T. T. (2015). *3: The St. Louis/East St. Louis Tornado of 1896*. Tornado Project Online. Retrieved from http://www.tornadoproject.com/alltorns/3.htm

Riley, R., & Krautmann, A. (2016). *Managing disaster: 20 May 2013 Central Oklahoma Tornado*. Oklahoma: Southern Climate Impacts Planning Program.

Staff, C. (2014, April 27). *The 10 deadliest U.S. tornadoes on record*. CNN. Retrieved from http://www.cnn.com/2013/01/30/us/deadliest-tornadoes/index.html

Wheatley, K. (2017, May 10). *An overview of the modern tornado record, 1950 through present (maps)*. Tornado Climatology. Retrieved from http://www.ustornadoes.com/2017/05/10/united-states-tornadoes-by-year-1950-through-present-maps

The Mississippi River Flood (1927)

5

Figure 5.1 Mississippi River
Source: U.S. Army Corps of Engineers

BRIEF OVERVIEW

The Mississippi River is one of the great rivers of the world. For the United States, it has an essential role in the commerce of the nation. The river is alive, powerful, and often unpredictable. It can bring and support life, as well as demonstrate a destructive side, raging over its banks and flooding thousands of acres, destroying homes, business, and crops. Periodic flooding occurs along the Mississippi River and in spite of man's efforts to control the river with levees and reservoirs, the power of the river has never been full contained.

In 1927, the Mississippi River escaped its banks and in doing so was responsible for the greatest flooding incident in American history. Although a tragedy, the 1927 flood proved to be the event that not only generated a massive federal response to the disaster, but also served as the catalyst for the enactment of federal legislation and infrastructure development that affects modern emergency management regarding flooding incidents. This chapter examines the 1927 incident from the cyclical emergency management phases advocated by the U.S. Department of Homeland Security, and provides some commentary regarding lessons learned with respect to modern homeland security incident management.

INTRODUCTION

Fires and floods have been the most common disasters in the United States, and they were the focus of the earliest emergency management efforts (Waugh, 2000). Specifically, throughout history, American society has experienced a variety of calamities arising from natural and human origins along the Mississippi River. Endangerments along the Mississippi River often provide a basis for regional danger and national attention, especially since the Mississippi River divides the U.S. mainland. The Mississippi River represents the "largest river basin" within the United States, and is the "third largest river basin" globally (Kusky, 2005, p. 152). More specifically, Johnson, Holmes, and Waite (2004, p. 1) indicate that the Mississippi River continuously "drains approximately 40% of the continental United States (approximately 1.25 million square miles)—all or part of 31 States, and two Canadian provinces, Ontario and Manitoba."

The Setting

Historically, the Mississippi River has flooded numerous times. A perusal of the National Oceanic and Atmospheric Administration (NOAA) data reveals that it has experienced 39 recorded incidents since de Soto's discovery (NOAA, 2015b). The first recorded event was observed by Hernando de Soto in 1543 (Hoyt & Langbein, 1955). The duration of the flood was at least 40 days, and the expanse of flooding most likely reached into the lower areas of the river (Hoyt & Langbein, 1955). Since then, a variety of floods and high-water events have been associated with the Mississippi River. For instance, during 1809, mass flooding deluged the lower Mississippi River (Hoyt & Langbein, 1955). Although several private levees were erected to control flooding, they failed multiple times during the period between 1858 and 1890 (NOAA, 2015b). It wasn't until 1893 that the levees did not succumb to the forces of nature, and held steadfast, thereby preventing yet another flood resulting from high waters (NOAA, 2015b).

The 20th century witnessed a variety of floods and high-water incidents along the Mississippi River. In 1908, for at least 100 days, the area of lower Mississippi exceeded the flood stage below the city of Arkansas City, Arkansas (Hoyt & Langbein, 1955). Since then, 22 other incidents occurred along the Mississippi River (NOAA, 2015b). Notably, during 1937, flooding of the Mississippi River created a lake within the river basin that was approximately the size of Lake Superior (Green, 2005). In the terms of dollars, when it occurred, the 1993 Mississippi River flood represented the costliest flood in U.S. history (Johnson, Holmes, & Waite, 2004).

The 2011 flooding affected several states along the Mississippi River. This flood resulted from a combination of spring rains and melting snow within the Ohio Valley (NOAA, 2015b). Much devastation occurred along the Mississippi River, and inland among its adjacent states. For instance, in the town of Tunica, located at Tunica Cutoff, Mississippi, population of 10,000 residents, at least 300 homes were destroyed because of flooding (CBS, 2011). Within the locale of Greenville, Mississippi, over 800 homes were destroyed by the flood's waters (CBS, 2011). The extensiveness of the flood's waters was unparalleled during modern times. Governor Haley Barbour indicated that a vast amount of water existed in places where there had been no water in his lifetime (Doss, Glover, Goza, & Wigginton, 2015).

The 2011 flood was tragic and costly. A total of 20 deaths occurred during the flood (NOAA, 2015a). The cataclysm resulted in approximately $208 billion in damages among 119 counties along the states bordering the lower Mississippi River (Sainz, 2013). Damage occurred in the states of "Arkansas, Illinois, Kentucky, Louisiana, Mississippi, Missouri, and Tennessee" (Sainz, 2013, p. 1). The flood affected at least 21,000 residences and businesses, and approximately 1.2 million land acres (Sainz, 2013). Similar to Governor Barbour's observations, many Mississippians had never witnessed such destruction during modern times. However, the 2011 flood, despite its severity, was not the most destructive flood along the Mississippi River. During 1927, a previous generation witnessed even greater devastation wrought by the flood's waters along the Mississippi River.

The Disaster Unfolds

The Mississippi River flood of 1927 remains the greatest flood during modern history along the lower Mississippi River (NOAA, 2015b). Excessive amounts of rain among tributaries contributed to swelling the Mississippi

River across lengthwise expanses of 80 miles throughout the floodplains (NOAA, 2015b). Among some areas, waves were measured with heights of 12 ft, and the water had a depth of 100 ft (Kirkpatrick, 2013). Although a natural calamity and wonder of nature, the flood also altered the course of American history by providing an opportunity for Herbert Hoover to advance toward the presidency (NOAA, 2015b).

Its severity and damage resulted in legislative action nationally. Specifically, the 1927 flood resulted in the passing of the 1928 Flood Control Act that mitigated and controlled the Mississippi River through the erecting of various levees, spillways, and flood control projects for alleviating cataclysmic flooding (NOAA, 2015b). A total of 246 individuals perished during the flood, and approximately 700,000 were displaced (Sherman-Morris, Wax, & Brown, 2012). Another estimate of the death toll indicates that 500 people died during the flood (Julien, 2002).

Although usually benign, the Mississippi River becomes dangerous when its water levels increase. The occurring of much rain during the fall of 1926, and the melting of large amounts of snow during the spring of 1927, generated a voluminous quantity of water within the Mississippi River (Kusky, 2005). Massive storms and strong rains occurred within the Southern United States on April 15, 1927, and were the catalysts for the Great Flood of 1927 (O'Daniel, 2013). The breaking of levees, stretching from Cairo, Illinois to Louisiana, contributed to the flooding (Weston, 2007). The severity of the devastation was enhanced by the cascading effects of bodies of water that had been "confined between levees" (Weston, 2007, p. 166).

In short, most of the levee system, from Illinois downward to the Gulf of Mexico, was destroyed. Along the river, breaches occurred among 145 locations (Kirkpatrick, 2013). Damages from the flood were excessive and penetrating throughout the Mid-South. For instance, in Mississippi, the entirety of the Yazoo Delta area was underneath varying levels of between 2 ft and 18 ft of water (Mathur & Da Cunha, 2001). In terms of modern dollars, the state of Mississippi experienced approximately between \$8 and \$9 billion worth of residential damages (O'Daniel, 2013). Commercially, in terms of modern dollars, the damages in Mississippi were between \$3 and \$4 billion (O'Daniel, 2013). Cumulatively, the 1927 flood remained the costliest flood in U.S. history until the 1993 flooding of the Mississippi River (Johnson, Holmes, & Waite, 2004).

Figure 5.2 Secretary of Commerce and Later President Herbert Hoover

Source: Library of Congress

Impact

Herbert Hoover, who held the position of Secretary of Commerce, and James Fieser, vice-chairman of the American Red Cross, oversaw the disaster relief operations (Jones, 2013). Their responsibilities included overseeing the evacuating of residents and the providing of shelter, food, and medical resources (Jones, 2013).

President Calvin Coolidge empowered Hoover with authority for commanding any federal agency or resource that was necessary (Clements, 2010). It was the first time that a civilian organization, the Red Cross, had a primary responsibility for rescue efforts instead of the military (Jones, 2013). Despite campaigns to generate relief monies for the flood refugees, practically no funding actually made its way directly to the survivors of the flood (Hopper, Baker, & Browning, 2008).

Containing the Disaster

The scope of the disaster was enormous, about 16.5 million acres flooded in seven states, an estimated 637,000 people displaced, over 160,000 homes flooded, 41,000 buildings destroyed, and surprisingly only an estimated 250 to 500 deaths. Emergency assistance and management occurred in many fashions. Agricultural assistance was provided in the form of farming implements, seed, feed, and livestock (Jones, 2013). The National Guard provided tent shelters for some refugees, and many others were sheltered in churches (Jones, 2013). The U.S. Coast Guard provided seaplanes that were used to ferry medical supplies and perform reconnaissance (Daniel, 1996). Despite the best efforts of emergency management, the emergency response was hampered and complicated because much infrastructure was destroyed. The ability to communicate was practically absent because the flooding destroyed telegraph and telephone lines (Sherman, 2011).

RELIEF AND FIRST RESPONSE EFFORTS

The Red Cross played a major role in the response to this disaster, establishing 154 Red Cross camps to provide housing, food, and medical care. Memphis, Tennessee, was a critical hub for emergency management during the aftermath of the flood. The city established areas for housing evacuees and distributing emergency materials (Sherman, 2011). Refugee (technically evacuee) camps were established within the city at the 168-acre fairgrounds (Dowdy, 2014).

Aviation resources from Memphis were necessary for ensuring primal modalities of distribution of goods and communication. For instance, medical personnel, medicines, and mail were ferried via aircraft to the affected regions (Sherman, 2011). Food was dropped from airplanes to rooftops among the flooded areas (Sherman, 2011). Messages were ferried by airplanes among affected regions, and aerial photographs were used to identify potential ruptures among levees (Sherman, 2011). The 1927 flood solidified the reputation of the city of Memphis as a primary distribution point for materials and resources despite the lack of a formal plan for response operations (O'Daniel, 2013).

AFTERMATH AND INVESTIGATION

In many fashions, the aftermath of the flood affected society within the Mid-South. Entire cities and towns were demolished by the flood's waters (Beaulieu, Everitt, & Dehm, 2010). Instances of disease and illness occurred among refugee camps (Hopper, Baker, & Browning, 2008). The refugee camps were makeshift, segregated, and unequipped to satisfy the needs of their respective residents (Hopper, Baker, & Browning, 2008). As the flood's waters receded, flood victims experienced homelessness, starvation, and unemployment (Hopper, Baker, & Browning, 2008). Despite the receding waters, much farmland was utterly destroyed, and deemed unsatisfactory for the planting of crops because of water impurity (Hopper, Baker, & Browning, 2008).

The National Flood Control Act of 1928 resulted from the incident. This legislation provided a basis for federal management of the Mississippi River, including redesigning of levees (Kirkpatrick, 2013). Not only did it affect the Mississippi River, but it also affected the entirety of controlling floods within the South (Davis, 2006). Flood control improvements along the river included deepening channels, crafting floodways, bolstering levees, and constructing systems of reservoirs (Davis, 2006). As a result of the flood, during its aftermath, construction began toward the erecting of artificial levee systems (Vega, Grymes, & Rohl, 2013). Later, in 1933, the Tennessee Valley Authority was created as a federal organization for managing water flows and levels via the erecting of spillways and dams upstream from the course of the main river (Davis, 2006). Essentially, given these constructs, the resulting legislation from the 1927 flood continues to affect modern flood policy (Black, 2006).

Before the 1927 flood, American policy emphasized the use of levees as the primary, systemic method of controlling and managing floods (Julien, 2002). The 1928 Flood Control Act created the Mississippi River and Tributaries (MRT) project which represented a new perspective and approach regarding flood incidents (Julien, 2002). Within the MRT, instead of employing levees as the primary focus of controlling and managing floods, the U.S. Army Corps of Engineers incorporated a paradigm that improved both levees and floodways, and that used a spillway for diverting water into Lake Pontchartrain near the city of New Orleans (Julien, 2002). The MRT project represented the first national approach toward navigation and comprehensively managing flooding (Beaulieu, Everitt, & Dehm, 2010).

Lessons Learned and Policy Impact

The 1927 flood affected American society both intangibly and tangibly. Intangibly, the flood resulted in the crafting of legislation and policies that enhanced public safety, and which continue to affect American society during modern times. Tangibly, the 1927 flood resulted in the enhancing of flood control and management via the erecting of levees, dams, reservoirs, and spillways along the Mississippi River and among its tertiary waterways. The combination of intangible and tangible resources provides a basis for controlling and diverting flood waters whereby modern society experiences some amount of protection against flooding. However, despite the best efforts to maximize public safety, no guarantee exists that flooding will always be averted. The 2011 flooding along the Mississippi River and its adjacent regions certainly shows the potential of floods to occur during modern times despite the presences of bulwarks, levees, spillways, dams, and reservoirs.

Implications for Modern Emergency Management

The U.S. Department of Homeland Security advocates an all-hazards approach to accommodating endangerments and managing incidents that may arise from either human or natural origins (McElreath, et. al., 2013). Within this approach exists the emergency management cycle (EMC). During the 1927 flood, the EMC did not exist, and strategic plans for addressing the calamity were absent. However, following the 1927 flood, change was necessary for improving methods of addressing emergency incidents. During the 1930s, a phase-based perspective of emergency management emerged for describing, understanding, and organizing emergency management practices (Baird, 2010). Decades later, during the 1970s, the National Governors' Association fashioned a four-phase construct that represents the foundational stages of the modern EMC (Baird, 2010).

The EMC is a resource through which organizations may craft strategies and policies for coping with calamities (McElreath, et. al., 2014a). The EMC consists of four phases: (a) mitigation/prevention, (b) preparedness, (c) response, and (d) recovery (McElreath, et. al., 2014a). Mitigation or prevention involves diminishing or quashing the negative effects of a calamity. Preparedness involves various aspects of identifying threats, planning accordingly, and examining contingency courses of actions. Response consists of enacting a specific plan or course of action when threats arise. Recovery consists of invoking courses of actions that attempt to achieve much or all of the state of existence and lifestyle that occurred before the calamity happened. As recovery progresses, methods of assessment and evaluation occur to determine how the event could be handled better if a like incident ever occurs again (McElreath, et. al., 2014a). The 1927 flood may be considered from the perspective of the EMC regarding its lessons for modern organizations.

Regarding mitigation and prevention, a formal plan was absent for such a mass flooding incident. As a result, the response was makeshift and used resources as they became available. Before 1927, the approach toward potential floods consisted of using levees as the primary method of controlling or averting incidents. However, the erecting of levees was unplanned systemically; their locations along the Mississippi River and its tertiary areas occurred with respect to the perceived needs of localities without any large-scale coordination and consideration of how best to maximize the efficacy of a cumulative levee system with respect to the cumulative attributes of the Mississippi River and its adjacent regions.

Modern contexts are derived from the historic flood. The 1928 Flood Control Act provided the basis for erecting numerous bulwarks that protect modern society against the dangers of flooding along the Mississippi River. Fortifying and erecting new levees, erecting reservoirs and spillways, and constructing diversionary routes represent the modern bases for prevention and mitigation. The improvements showed their value when the Mississippi River again flooded in 2011 because the severity of damage was much less than that in 1927 (Memphis, 2015).

Preparedness for the magnitude and scope of the 1927 flood was poor. Nationally, before 1927, a federal approach to systemically and methodically addressing large-scale flooding incidents was absent. During the aftermath of the 1927 flood, refugee camps were makeshift, communication was almost impossible, transportation was impeded because roadways and bridges were destroyed, and the necessary emergency goods and services were provided as they became available. Essentially, an emergency plan for addressing such a severe incident simply did not exist. Despite the lack of planning, O'Daniel (2013) indicates that the city of Memphis ingeniously fomented resources and activities that established its relevance as a distribution center.

Lessons learned from the 1927 flood provide a basis for modern emergency strategic planning activities as a preparedness mechanism. Preparation for calamities necessitates the crafting and maintaining of emergency plans to maximize the effectiveness of responses to flooding incidents (Sene, 2008). Planning also contributes toward effective recovery efforts that succeed incident responses. For instance, in Vermont, planning exists to enhance the resiliency and effectiveness of recovery endeavors (EPA, 2015). Such planning must incorporate facets of long-term, strategic vision and creativity for public benefit (Farazmand, 2014). Planning accommodates a variety of issues, including land policies, chains-of-command; positions and personnel responsibilities; necessary equipment; conditions for plan activation; communication; cooperation among emergency organizations; contingencies; evacuation routes, processes, and procedures; resource allocation and distribution; and so forth (EPA, 2015; Farazmand, 2014; McElreath, et. al., 2014a; McElreath, et. al., 2014b).

Certainly, both federally and locally, the response to the 1927 flood was affected by a lack of planning regarding mass-scale incidents. Resources were marshaled and leveraged as they became available. Innovation and ingenuity were necessary for managing and leveraging available resources to enhance the response effort. For instance, aircraft were fitted with pontoons to facilitate access to flooded regions (Sherman, 2011). By doing so, aviation resources were leveraged as tools for rescuing marooned refugees, and delivering medical personnel, delivering supplies, and medicines (Sherman, 2011).

During modern times, responses are executed according to a specific plan. Adherence to the emergency plan provides a basis for controlling, leading, organizing, staffing, and directing emergency responses (Farazmand, 2014). For instance, in Pearl River, Mississippi, near the Gulf Coast, the county maintains a plan for mitigating and protecting against the dangers of flooding (Bigelow & Chamberlain, 2015). When enacted, the plan contains methods for proactively attempting to minimize structural damages, processes, and procedures for enhancing public safety, provisions for emergency shelters, addressing the needs of various population demographics, performing roadway evacuations, and so forth (Bigelow & Chamberlain, 2015).

Recovering from the 1927 flood necessitated much change along the Mississippi River. Recovery attempts occurred to reinstate the normalcy that existed before the flood, but normalcy was unachieved completely among many of the affected regions. In a sense, a new normal emerged among the affected regions as towns, cities, families, and organizations established themselves (Jacobs, Wellman, Fuller, Anderson, & Juardo, 2016). Entire towns and cities were destroyed, personal possessions were forever lost, both agricultural and economic capacities were devastated, and many people either died or were injured (Campbell, 2008). Surviving individuals and organizations adapted to such conditions. Ignorance and apathy abated with the loss of innocence regarding the dangerousness of nature. Recovery necessitated federal legislation via the 1928 Flood Control Act and the erecting of numerous dams, spillways, reservoirs, and levee systems along the Mississippi River. These constructs continue to affect modern society both intangibly and tangibly.

During modern times, recovery is an essential aspect of emergency management. Recovery adheres to some type of plan(s) that may occur throughout long-term periods. Thus, achieving some semblance of normalcy does not always occur rapidly. For instance, after a flood occurred in Nashville, Tennessee, during 2010, recovery efforts are still ongoing (Nashville, 2015). After flooding of the Missouri River in 2011, recovery efforts are still underway 4 years later (KMA, 2015). Such efforts are not sporadic. Both federally and among the states, various assistance plans and programs exist to facilitate recoveries among affected areas. Federally, the National Flood Insurance Program is a resource for "reducing the impact of flooding on private and public structures" via the provision of "affordable insurance for property owners and by encouraging communities to adopt and enforce floodplain management regulations" (FEMA, 2015). As an example of state resources, in New York, assistance exists for residents and business owners who experience damages from flash floods (New York, 2015).

The experience of a calamity provides insight and knowledge that are necessary for improving emergency strategies and plans to accommodate any future instances of disaster. Throughout the recovery phase of the EMC, various assessments and evaluations occur regarding the efficacy of the emergency response, and their outcomes are formally detailed within after-action reports (Fagel, 2014). This information is incorporated within existing emergency plans to improve their robustness and strengthen any future emergency response that may be necessitated during an emergency. Often, the results of these assessments necessitate the altering of existing infrastructure or the crafting of new infrastructure to improve the resiliency of a community against disaster. For instance, after the 1927 flood, contemplating the event resulted in the improving of existing infrastructure and the erecting of new dams, spillways, reservoirs, levees, and bulwarks as well as national policies and strategies for flood management.

Incorporating change for improvement signifies the commencing of a new instantiation of the EMC. Essentially, communities and localities acknowledge the lessons learned from an incident, and enter a new phase of mitigation/prevention with a greater level of knowledge and preparedness. Although one cannot predict with full certainty future events, the preceding experience provides a basis for better enduring a new instance of calamity. During modern times, protective infrastructures and emergency management strategies and plans are reflective of this notion because they represent an aggregating of lessons learned from former disaster experiences.

SUMMARY

The dangerousness of nature must never be ignored or discounted casually. Disasters may occur at any time at any location with little or no warning. Natural calamities may inflict severe and long-lasting damages that destroy infrastructure and hamper the economic performance of society. Catastrophes arising from human origins may be just as deadly as those generated by nature. In any case, the EMC exists as a resource whereby modern emergency management crafts strategies for addressing calamities arising from human and natural origins.

The EMC did not exist before or during the 1927 Mississippi River flood. However, during modern times, it is an essential, practical resource whereby incident management strategies are crafted and honed through time. Use of the EMC enhances public safety, contributes toward sparing human life throughout a calamity, and helps to mitigate the damaging effects of cataclysms. Before and during the 1927 flood, the ability to survey potential danger areas and integrate multiple pieces of information was limited, especially by the technologies of the period. However, during modern times, such activities occur quickly through the use of satellite imagery and radiometric resources. For instance, levees along the Mississippi River can be screened for weaknesses by using such technologies (Aanstoos, et al., 2011).

Neither nature nor humans have changed since 1927. Modern tragedies have the potential of being just as dangerous, deadly, and damaging as their predecessors. The 2011 flooding of the Mississippi River inundated and destroyed entire towns in a fashion similar to the events of the 1927 flood. Modern humans possess the same physical and mental attributes that existed among preceding generations. People died just as easily in the 2011 flood as they did during 1927. However, when comparing and contrasting the floods, stark differences are observable:

(a) a total of 246 people died in the 1927 flood, whereas a total of about 20 perished during the 2011 incident (NOAA, 2015a); (b) the EMC did not exist in 1927, whereas it was advocated by the U.S. Department of Homeland Security in 2011; and (c) strategic emergency planning regarding flood management was absent in 1927, whereas it was present and employed during 2011. Although the basic characteristics of both nature and humans have remained static since 1927, effective resources for managing emergency incidents exist during modern times.

Central among these modern resources are the EMC and various technologies that enhance societal robustness and resiliency against endangerments. Through using the EMC to craft effective strategies and by using modern technology to maintain and improve preventive infrastructures, society better prepares for, mitigates the effects of, responds to, and recovers from disasters than it did in 1927. Despite the best efforts, use of the EMC, strategic planning, or modern technology is not a panacea for averting or mitigating completely the dangers of any incident. Nature is often unpredictable, and its events may become uncontrollable. Regardless, when examining the 1927 flood versus the 2011 flood, the spared lives and reduced damages both exhibited during the 2011 event are testaments to the efficacies of the EMC, strategic planning, and modern technologies as essential resources for homeland security management.

During modern times, the EMC permeates disasters whether locally, regionally, or nationally. Although each locality is unique, it may incorporate the basic EMC concept within its emergency planning endeavors whereby policy is influenced and crafted. By doing so, localities may maximize their ability to respond to and recover from calamities, including floods. Flooding is often devastating. But, through the experience of such devastation, lessons are learned that better the ability of both individuals and organizations to withstand future disasters of a similar scope and nature. Along with such improvement may come knowledge and understanding of what not to do—which may be just as important as what to do when disaster strikes. Thus, when crafting policies and plans, learned from the experiences of tragedies, administrators and leaders have better insight regarding how best to leverage their resources strategically.

These notions are tautologies when considered from the perspective of river floods, especially along the mighty Mississippi. Modern flooding insurance owes much of its existence to the events of the 1927 Mississippi River flood. Contemporary infrastructures for managing waterways owe much of their design and resilience to the tragedy of 1927. The necessities of timely warnings and evacuations also have antecedents among the 1927 events. Essentially, when considering emergency management from the context of flooding events, much of the modern policies and practices owe their existence to the Mississippi River flood of 1927.

REFERENCES

Aanstoos, J., Hasan, K., O'Hara, C., Prasad, S., Dabbiru, L., Mahrooghy, M., Gokaraju, B., Lee, M., & Nobrega, R. (2011). Earthen levee monitoring with synthetic aperture radar. *2011 IEEE Applied Imagery Pattern Recognition Workshop*, October 11–13, 2011. Washington, DC.

Baird, M. (2010). The "phases" of emergency management. *Vanderbilt center for transportation research.* Retrieved from http://www.vanderbilt.edu/vector/research/emmgtphases.pdf

Beaulieu, R., Everitt, J., & Dehm, D. (2010). *National smart water grid: Integrated solutions for sustainable fresh water.* Las Vegas, NV: Global Beau Publications.

Bigelow, E., & Chamberlain, S. (2015). *Pearl River County, Mississippi hazard mitigation and flood protection plan.* Retrieved from http://www.pearlrivercounty.net/planning/files/PRC%20HAZ%20MIT%20PLAN.pdf

Black, B. (2006). *Nature and the environment in twentieth-century American life.* Westport, CT: Greenwood Press.

Campbell, B. (2008). *Disasters, accidents, and crises in American history.* New York, NY: Infobase Publishing.

CBS. (2011). Miss. town could be wiped off map by flood. *CBS News.* Retrieved from http://www.cbsnews.com/news/miss-town-could-be-wiped-off-map-by-flood

Clements, K. (2010). *The life of Herbert Hoover: Imperfect visionary, 1918-1928.* New York, NY: Palgrave-Macmillan.

Daniel, P. (1996). *Deep'n as it come: The 1927 Mississippi river flood.* New York, NY: Oxford University Press.

Davis, D. (2006). *Southern United States: An environmental history.* Santa Barbara, CA: ABC-CLIO Publishing.

Doss, D., Glover, W., Goza, R., & Wigginton, M. (2015). *The foundations of communication in criminal justice systems.* Boca Raton, FL: CRC Press.

Dowdy, G. (2014). *On this day in Memphis history*. Charleston, SC: The History Press.

EPA. (2015). *Planning for flood recovery and long-term resilience in Vermont*. U.S. Environmental Protection Agency.

Fagel, M. (2014). *Crisis management and emergency planning: Preparing for today's challenges*. Boca Raton, FL: CRC Press.

Farazmand, A. (2014). *Crisis and emergency management: Theory and practice* (2nd ed.). Boca Raton, FL: CRC Press.

FEMA. (2015). The national flood insurance program. *Federal Emergency Management Agency*. Retrieved from http://www.fema.gov/national-flood-insurance-program

Green, J. (2005). *Disasters: 1993 Mississippi river floods*. Milwaukee, WI: Gareth Stevens Publishing.

Hopper, S., Baker, T., & Browning, J. (2008). *An Arkansas history for young people* (4th ed.). Fayetteville, AR: The University of Arkansas Press.

Hoyt, William G., & Langbein, W. B. (1955). *Floods*. Princeton, NJ: Princeton University Press.

Jacobs, K., Wellman, A., Fuller, A. M., Anderson, C. P., & Jurado, S. M. (2016). Examining family dynamics of cold case homicide survivors. *Journal of Family Studies, 22*(3), 256–271.

Johnson, G., Holmes, R., & Waite, L. (2004). The great flood of 1993 on the upper Mississippi River – 10 years later. *U.S. Geological Survey*. Retrieved from https://www.leopold.iastate.edu/sites/default/files/events/USGS.pdf

Jones, M. (2013). *The American red cross from Clara Barton to the new deal*. Baltimore, MD: Johns Hopkins University Press.

Julien, P. (2002). *River mechanics*. New York, NY: Cambridge University Press.

Kirkpatrick, M (2013). *It happened in Mississippi*. Guilford, CT: Morris Book Publishing.

KMA. (2015). Four years later, flood recovery progresses. *KMA Radio*. Retrieved from http://www.kmaland.com/news/four-years-later-flood-recovery-progresses/article_436f8eec-1049-11e5-a5e5-b7b5ce70a7f2.html

Kusky, T. (2005). *Encyclopedia of earth science*. New York, NY: Facts on File.

Mathur, A., & Da Cunha, D. (2001). *Mississippi floods: Designing a shifting Landscape*. New Haven, CT: Yale University.

McElreath, D., Doss, D., Jensen, C., Wigginton, M., Kennedy, R., Winter, K., Mongue, R, Bounds, J., & Estis-Sumerel, J. (2013). *Introduction to law enforcement*. Boca Raton, FL: CRC Press.

McElreath, D., Doss, D., Jensen, C., Wigginton, M., Nations, R., & Van Slyke, J. (2014a). *Foundations of emergency management*. Dubuque, IA: Kendall-Hunt Publishing.

McElreath, D., Jensen, C., Wigginton, M., Doss, D., Nations, R., and Van Slyke, J. (2014b). *Introduction to homeland security* (2nd ed.). Boca Raton, FL: CRC Press.

Memphis. (2015). *Memphis district, U.S. army corps of engineers: A brief history of the Memphis district*. U.S. Army Corps of Engineers. Retrieved from http://www.mvm.usace.army.mil/About/History.aspx

Nashville. (2015). Nashville continues recovery 5 years after historic flood. *WKRN News*. Retrieved from http://wkrn.com/2015/05/01/nashville-continues-recovery-5-years-after-historic-flood

New York. (2015). NYS disaster recovery assistance programs. *State of New York*. Retrieved from http://www.nyshcr.org/programs/nys-flood

NOAA. (2015a). Flood history of Mississippi. *National Oceanic and Atmospheric Administration*. Retrieved from http://www.srh.noaa.gov/media/jan/Hydro/Flood_History_MS.pdf

NOAA. (2015b). High flows and flood history on the lower Mississippi river. *National Oceanic and Atmospheric Administration*. Retrieved from http://www.srh.noaa.gov/lix/?n=ms_flood_history

O'Daniel, P. (2013). *When the levee breaks: Memphis and the Mississippi Valley flood of 1927*. Charleston, SC: The History Press.

Sainz, A. (2013). Mississippi river flood of 2011 caused $2.8B in economic damage: Army corps. *Insurance Journal*. Retrieved from http://www.insurancejournal.com/news/national/2013/02/27/282875.htm

Sene, K. (2008). *Flood warning, forecasting and emergency response*. New York, NY: Springer Publishing.

Sherman, J. (2011). *Walking on air: The aerial adventures of Phoebe Omlie*. Oxford, MS: The University of Mississippi Press.

Sherman-Morris, K., Wax, C., & Brown, M. (2012). *Mississippi weather & climate*. Oxford, MS: University Press of Mississippi.

Vega, A., Grymes, J., & Rohl, R. (2013). *Louisiana weather and climate*. Sudbury, MA: Jones and Bartlett Publishing.

Waugh, W. L., (2000). *Living with hazards dealing with disasters: An introduction to emergency management*. Armonk, NY: M.E. Sharpe.

Weston, D. (2007). *Infection prevention and control: Theory and practice for healthcare professionals*. Hoboken, NJ: John Wiley & Sons Publishing.

The Great Dust Bowl (1930s)

6

Figure 6.1 Dust Cloud Sweeping Fields and Houses
Source: USDA

"It was the worst man-made ecological disaster in American history when the irresistible promise of easy money and the heedless actions of thousands of individuals, encouraged by their government, resulted in a collective tragedy that nearly swept away the breadbasket of the nation."

—Ken Burns—The Dust Bowl

BRIEF OVERVIEW

It seemed almost beyond belief that the rich and fertile land of our Nation's Midwest had been blown away by the wind due to an extensive period of no rainfall. Consequently, the dreams of a generation who had invested their livelihood in the farming occupation suffered exceedingly within the Midwest region of our Nation. For example, an enormous amount of topsoil was swept away, leaving the fields barren, as winds turned countless tons of soil into storm clouds of dust, which carried the soil hundreds of miles eastward (Ganzel,

2003). Subsequently, the deadly dust storms killed people and livestock, destroyed crops, and devastated the expectations of many people. The catastrophic events were unlike anything experienced before in the history of our Nation.

INTRODUCTION

During the latter part of the 19th century, much of our Nation's economy consisted primarily of the agricultural business that was based on family farms, due to the exceeding amount of either free or inexpensive land. As a result, the evolution of our Nation's economy seemed unlimited as millions of people were linked either directly to the land and cultivation process, or involved with the harvest, distribution, and sale of the produce within our Nation. The land and the climate resulted in great harvests, but by the early 1930s, poor agricultural practices, combined with a sustained drought, caused what became known as the Great Dust Bowl.

Great Plains

The Great Plains of our Nation consisted mostly of grasslands, which had become very appealing to farmers and ranchers who were part of our Nation's migration of the mid and late 1800s. Initially, during the formative years of the agricultural business, farmers deeply plowed and planted crops that included corn and wheat, and were extremely successful because of the adequate and necessary rainfall.

Unavoidably, weather patterns shifted over the Atlantic and Pacific oceans during the 1930s, which became the basis for a disaster to unfold. The Pacific Ocean cooled and the Atlantic warmed. Such temperature differentials contributed to a change in the direction of the jet stream, which significantly altered the rain patterns over the Midwestern part of our Nation.

The air current that usually carried moisture from the Gulf of Mexico toward the Great Plains had altered significantly. With such a change, the rain that was so important to the central part of our Nation never reached the Great Plains (Amadeo, 2017). By the early 1930s, the Great Plains were experiencing drought conditions, and few realized that the absence of rain would impact the region for years and as such stimulate great change in agricultural practices, economics, migrations, and the role of the federal government in disaster relief (McElreath, Doss, et al., 2014).

The Setting

During our Nation's first century, it appeared that the majority of people believed that America possessed unlimited lands and resources, and that land could also be obtained cheaply, or even free. Such attitude prompted millions of people to migrate westward, and settling in lands across our Nation, establishing new homes, farms, businesses, communities, and cities.

In an effort to encourage the migration, the national government enacted a series of laws that allowed access to land, and offered the opportunity to purchase property inexpensively, or obtain it for free. The Homestead Act of 1862 encouraged Western migration by providing settlers 160 acres of public land. In exchange, homesteaders paid a small filing fee, and were required to reside on the land continually for 5 years. As a result of the Homestead Act of 1862, and subsequent land-related legislation, an estimated 270 million acres of public land, or nearly 10% of the total area of the United States, was provided to an estimated 1.6 million homesteaders.

Soon, millions of acres went into production on farms and ranches due to the onset of World War I in 1914, as the demand for wheat, both domestically and internationally, increased significantly. As a result, farmers profited and pushed to place more land into production. Upon the conclusion of World War I in 1918, the demand for food continued, but soon the impact of the Great Depression would change the global economic balance and the revenues from farming.

The Disaster Unfolds

Rising wheat prices in the 1910s and 1920s, and increased demand for wheat from Europe during World War I, encouraged farmers to plow up more acres of native grassland in their efforts to produce more wheat, corn, and other row crops. From an economical perspective, it was very advantageous to be a farmer during this period of time.

Figure 6.2 Devastated Farm Lands
Source: USDA, "Sourthern Plains Climate Hub Helps Land Managers Build Resilience to Climate Variability" (2018)

Then, the Great Depression led to an economic downfall, which resulted in high levels of unemployment and fiscal uncertainty, to the extent that the global industrial and agricultural areas were significantly affected, which resulted in the plummeting of wheat prices and farm revenue. In the American Midwest, farmers continued the practice of land over usage by placing additional land into production, deeply plowing the land, and transforming more grassland of the Great Plains in an attempt to increase production.

During the years of adequate rainfall and good weather, the farms of the Nation were extremely productive, but for the Midwest, changes in rain patterns beginning in the early 1930s, would create the right conditions for a national disaster. Dry conditions and unregulated farming practices made the topsoil that was essential for crop production vulnerable to the wind. As a result, the lack of rain and the abhorrent farming practices, coupled with the wind crossing eastern Colorado, eastern New Mexico, Kansas, Oklahoma, and Texas, would create a series of "dust storms," and earn the region the unwelcomed title of the *Dust Bowl*.

Initially, farmers believed that such dry conditions were temporary and would soon pass, but the land continued to suffer the effects of the drought. In 1932 and 1933, there were over 50 dust storms. Then, by 1934, the drought covered 75% of the country, and affected 27 states. Consequently, an estimated 35 million acres of formerly cultivated land had been rendered useless for farming, while another 125 million acres, an area roughly three-quarters the size of Texas, was rapidly losing its topsoil (Wilhite, 2009). Eventually, by 1937, the effects of the drought resulted in more than one out of five farmers having to receive federal emergency relief (Amadeo, 2017).

Throughout the droughts, many of the farmers who could continue to farm attempted to plow and plant their fields, but such efforts failed. Due to the extreme lack of rain, many farmland areas lacked the necessary topsoil ground cover. As the extreme winds would sweep across the fields, dust debris would completely fill and darken the sky for days. The thick layers of dust completely covered homes and business, and also created healthcare issues caused from the tremendous amounts of dust that entered the lungs and inflicted damage, which resulted in numerous deaths and lifelong health issues.

Containing the Disaster

The containment of the disaster would not be simple. The scope of the disaster was not only great from a geographical sense, but also from an economic, social, political, and personal level. Changes would be required

Figure 6.3 Dust storm, Elkhart, Kansas, May 1937
Source: Everett Historical/Shutterstock.com

in each of these areas, and leadership and money required to even attempt to mitigate the disaster. In each of these areas, economic, social, political, and personal, changes would be far-reaching (Nelson, n.d.). From an economic perspective, farming methods had directly contributed to the disaster, which prompted the need to better understand the relationship dynamics between the land, weather, and farm procedures.

Changes in the way land was farmed, along with the incorporation of improved farming techniques, would be necessary steps in the prevention of a future disaster of this type. These changes involved a combination of education, training, and financial assistance for the farmer. The federal government took steps to provide financial assistance to farmers, in an effort to avoid widespread loss of farming due to bank foreclosure. The migration westward that was stimulated by the Dust Bowl changed the American West. In the short run, many of those who left their farms seeking a better life in the west encountered a struggle of another type. However, the lack of employment, shortage of housing opportunities, and in many areas resistance to the migration propelled the personal tragedy. Eventually, for many struggling farmers, the entry of the United States into World War II became the economic juggernaut that would greatly expand business, industry, and agriculture, as well as provide some degree of long-term stability on behalf of those who had moved to the west coast.

RELIEF AND FIRST RESPONSE EFFORTS

During the initial phase of the dust storms and drought, the role of the federal government in relief operations was limited. President Hoover, who had played such an important role in the relief operations during the Mississippi River Flood of 1927, for some unknown reason did not take action to activate national assets on behalf of those affected by the drought and dust storms.

As President Franklin Roosevelt assumed office following President Hoover, the federal government began taking steps to provide relief to those impacted by the disaster. The Farm Credit Act of 1933 established local banking and credit associations to support the agricultural sector. Then, in May 1933, the Emergency Farm Mortgage Act allotted $200 million for refinancing mortgages to help farmers facing foreclosure (Ganzel, 2003).

Additionally, in an effort to stabilize prices in the livestock market, in October 1933, the federal government purchased and slaughtered over 6 million hogs. Unfortunately, there was no plan in place to process and distribute the meat to the benefit of the public. With this waste, outcry led to the creation of the Federal Surplus Relief Corporation, which was tasked with the diversion of federally acquired agricultural commodities to relief organizations (Amadeo, 2017). Soon, agricultural products to include apples, beans, canned beef, flour, and pork products were processed and distributed through local relief channels (Ganzel, 2003).

In June 1934, the Frazier–Lemke Farm Bankruptcy Act was enacted with the intent of providing further protection to the farmer by restricting the ability of banks to dispossess farmers in times of distress. Originally effective until 1938, the Bankruptcy Act was renewed four times until 1947, when it eventually expired. Also in 1934, President Roosevelt signed the Taylor Grazing Act, which allowed the government to remove up to 140 million acres of federally owned land from the public domain and establish grazing districts (Ganzel, 2003).

In January 1935, the federal government created a Drought Relief Service to coordinate relief activities. The Drought Relief Service bought cattle in counties that were designated emergency areas, for $14 to $20 a head. In addition, those cattle that were unfit for human consumption, which amounted to more than 50% at the beginning of the program, were destroyed. The remaining cattle were given to the Federal Surplus Relief Corporation to be used in food distribution to families nationwide. Although it was difficult for farmers to sell much of their herds, most realized that the decline of beef prices, combined with the impact of the drought on the grazing lands, made the sale of their cattle one of the only options available to them to avoid bankruptcy and the loss of their land.

In April 1935, President Franklin Roosevelt approved the Emergency Relief Appropriation Act that provided $525 million for drought relief, and authorized the creation of the Works Progress Administration (WPA), which eventually resulted in the employment of over 8.5 million Americans. Also in 1935, Congress declared soil erosion "a national menace," and established the Soil Conservation Service in the Department of Agriculture (McElreath, Jensen, et al., 2014). Under the Soil Conservation Service, farming techniques such as strip cropping, terracing, crop rotation, contour plowing, and cover crops were advocated. In an effort to provide an incentive to change their farming techniques, farmers were paid to practice soil-conserving farming techniques (Ganzel, 2003).

In another attempt to prevent a future wind-related disaster, President Roosevelt directed the Civilian Conservation Corps (CCC) to plant a huge belt of more than 200 million trees from Canada to Abilene, Texas, which sought to break the wind, keep water in the soil, and hold the soil in place. By 1938, extensive work re-plowing the land into furrows, planting trees in shelterbelts, and other conservation methods, resulted in an estimated 65% reduction in the amount of soil blowing.

AFTERMATH AND INVESTIGATION

In the fall of 1938, the rain returned, bringing an end to the drought. During the next few years, with the coming of World War II, the Nation's economy escaped the Depression. The Great Plains of our Nation also recovered, and once again, began producing food for the Nation.

The aftermath of the Dust Bowl rendered millions of acres of farmland virtually useless until the return of rain. Consequently, Americans became homeless as thousands were forced out of their land (McElreath, Jensen, et al., 2014). The years of drought significantly changed our Nation by stimulating the most intense period of internal migration in its history, and by 1940, an estimated 2.5 million people had left the states impacted by the Dust Bowl.

The outcome of the disaster prompted the federal government to take the necessary steps to limit the impact of the disaster, and in doing so, changed agriculture and the business of agriculture in the United States. Farming techniques changed and with those changes, the land would become even more productive.

LESSONS LEARNED AND POLICY IMPACT

The Dust Bowl was caused by several economic and agricultural factors, including existing federal land policies, unregulated farming techniques, changes in regional weather, farm economics, and other cultural factors. After the War Between the States, generous opportunities to obtain a homestead attracted thousands from the east toward the west in the hope of achieving the 1800s version of the American dream.

As a result of the Dust Bowl disaster, the farming practices in use at the time were reexamined, and a new focus centered on the relationship between farming practices and soil conservation. President Roosevelt initiated progressive measures that made relief operations a priority of the national government. Such steps included the establishment of the CCC, which provided employment opportunities to thousands seeking work.

In addition, the establishment of the Soil Erosion Service, later known as the Soil Conservation Service, and now the Natural Resources Conservation Service, continues to this day to provide support and technical expertise to our Nation's farmers. The establishment of the Soil Erosion Service also marked the first major federal commitment to the preservation of natural resources in private hands.

In 1935, Roosevelt initiated the Prairie States Forestry Project, which created a "shelter belt" of timber from the Texas Panhandle to the Canadian border. Over the course of 7 years, the U.S. Forestry Service worked in conjunction with the CCC, the WPA, and local farmers by planting nearly 220 million trees that created over 18,000 miles of windbreaks on some 30,000 farms (Woolner, 2010).

Furthermore, the aid from the Agricultural Adjustment Administration became the most significant. Without the financial aid of the Agricultural Adjustment Administration, many farmers in the Dust Bowl would have suffered bankruptcy and lost their lands. The Agricultural Adjustment Administration paid farmers nationwide to reduce production by withdrawing a specific acreage from production. During the Dust Bowl, the Agricultural Adjustment Administration actually paid farmers to reduce production of wheat and cotton. With fewer acres planted with these crops, agency officials believed that the surplus of these commodities nationwide would disappear, and agricultural prices would rise to benefit the farmers.

Moreover, farmers themselves learned valuable lessons from the aid of our government through education programs, and were taught practices that helped with soil conservation and prevented erosion. Included among these notable new practices were:

- **strip farming**—alternating strips of planted and fallow land in certain directions to help prevent wind erosion.
- **crop rotation**—change crop production on various pieces of land, and to make sure pieces of land had fallow seasons.
- **contour plowing**—keeps the soil from going downhill by plowing in the direction of the contour lines of the land.
- **terracing**—the act of developing the land on rolling terrain to be sequentially stepped with level surfaces to enable grow crops, and to keep the soil from going downhill.

These practices, along with other farming practices, such as wetting the soil and mulching the crop stubble to work as a covering to open soil, have helped to reduce the possibility of another major dust storm, like that which occurred during the Dust Bowl era.

In 1936, the Great Plains Committee that was formed by the U.S. government issued its report in an effort to identify the causes, impacts, and necessary remedies to address the crisis in the region. The Committee made numerous recommendations for action, and placed a heavy emphasis on federal and state government intervention in land use management and soil erosion control. Four key areas were addressed by the report, including soil erosion, the loss of forage cover for grazing, the inefficient use of water, and highly variable farm incomes along with high rates of farm indebtedness. The report identified the root causes of each key area and provided suggested government and farm-level actions.

Soil erosion was also a significant problem that contributed to the drought. Recommendations were made for local governments to conduct extensive surveys of land, soil, and water resources, along with state governments creating erosion control districts and zoning regulations that regulated land use based on local conditions. It was suggested that farmers plow along contours and leave behind stubble after harvesting while rotating their cover crops during windbreaks (McElreath, Doss, et al., 2014).

The loss of forage cover for grazing was a direct result of overstocking range lands and expansion of farming into marginal areas. Recommendations from the report included having the federal government acquire range lands with central control, and state governments organizing grazing associations. Farmers needed to reduce their herd sizes or keep herds off fragile lands (McLeman et al., 2014).

Poor farming technologies and practices coupled with inadequate capacity for irrigation failed to conserve soil moisture leading to inefficient use of water. The report recommended the development of systematic irrigation policies at the federal and state levels as well as laws to protect and conserve groundwater. Farmers needed to create deeper and better water ponds for livestock and supplement irrigation in places where it was effective to do so (McLeman et al., 2014).

Finally, undue dependence on wheat as a cash crop led to highly variable incomes coupled with high rates of farm indebtedness. Recommendations included that federal and state governments should establish public finance programs to increase farm size to make farming more viable and funding greater research into pest control. Farmers needed to maintain a higher ratio of fodder and livestock to cash crops, the creation of diversified operational plans, and keeping larger feed and seed reserves (McLeman et al., 2014).

In recent years, an interest in the Dust Bowl, and subsequent droughts, has prompted advances in the atmospheric sciences. Scientists today employ modeling, climatological data analysis, and paleoclimate studies to better understand droughts. These tools have been used to answer some critical questions including what the causes and atmospheric dynamics of the Dust Bowl were, and other droughts of the recent past; and what is the return interval, intensity, and extent of past droughts (McLeman et al., 2014).

IMPLICATIONS FOR MODERN EMERGENCY MANAGEMENT

For modern emergency management, lessons learned include the realization that a major disaster requires a multidimensional response. The causes of this disaster included, but were not limited to, overplanting and overgrazing of the Midwest, combined with severe climate fluctuation. Soil erosion and major dust storms resulted in a multi human disaster, wrecking individual health and economic destruction of individuals, families, and communities. It also resulted in the expansion of federal disaster relief practices in the agricultural sector of the economy. The Dust Bowl stimulated change in the Nation economically, socially, and politically (McElreath, Doss, et al., 2014) and resulted in the expansion of federal disaster relief practices in the agricultural sector of the economy.

SUMMARY

The multiyear drought of the Dust Bowl directly contributed to large-scale soil erosion, dust storms, farm abandonments, personal hardships, and distress migration on scales not previously seen (McLeman et al., 2014). The lessons learned from this historical environmental disaster, and the impact upon our Nation was extensive and multidimensional. The outcome resulted in significant and much needed changes that have made the critical infrastructure of our Nation stronger. An increased focus upon land usage, and the benefit of having protective measures to ensure resiliency of the land, truly captured the attention of policy makers. Subsequently, a considerable number of federal agencies became involved in land management and soil conservation as a result of this disaster. Among the many programs implemented to protect the land, the federal government took the necessary steps to remove and prohibit agricultural production on the land that had been used for farming purpose on the Great Plains, and returned that acreage to prairie grassland. Additionally, in an effort to eliminate unsustainable practices of farm land usage, steps were also taken to educate farmers to reduce and prevent poor farming practices.

REFERENCES

Amadeo, K. (2017, December 6). *How the Dust Bowl environmental disaster impacted the US economy*. The Balance. Retrieved from https://www.thebalance.com/what-was-the-dust-bowl-causes-and-effects-3305689

Ganzel, B. (2003). *The Dust Bowl*. Farming in the 1930s. Retrieved from https://livinghistoryfarm.org/farminginthe30s/water_02.html

McElreath, D. H., Doss, D. A., Jensen, C. J., Wigginton, M. P., Nations, R., Van Slyke, J., & Nations, J. (2014). *Foundations of emergency management*. Debuque: Kendall Hunt.

McElreath, D. H., Jensen, C. J., Wigginton, M. P., Doss, D. A., Nations, R., & Van Slyke, J. (2014). *Introduction to homeland security*. Boca Raton: CRC Press.

McLeman, R. A., Dupre, J., Ford, L. B., Ford, J., Gajewski, K., & Marchildon, G. (2014). What we learned from the Dust Bowl: lessons in science, policy, and adaptation. *Population and Environment*, 417–440. Retrieved from https://www.ncbi.nlm.nih.gov/pmc/articles/PMC4015056

Wilhite, D. A. (2009). *Dust Bowl.* Oklahoma Historical Society. Retrieved from http://www.okhistory.org/publications/enc/entry.php?entryname=DUST%20BOWL

Woolner, D. B. (2010, June 3). *FDR and the new deal response to an environmental catastrophe.* Roosevelt Institute. Retrieved from http://rooseveltinstitute.org/fdr-and-new-deal-response-environmental-catastrophe

7

The 1995 Chicago Heat Wave

Contributed by Jeffrey M. Pavlacic, Emily A. Gawlik, and Stefan E. Schulenberg.

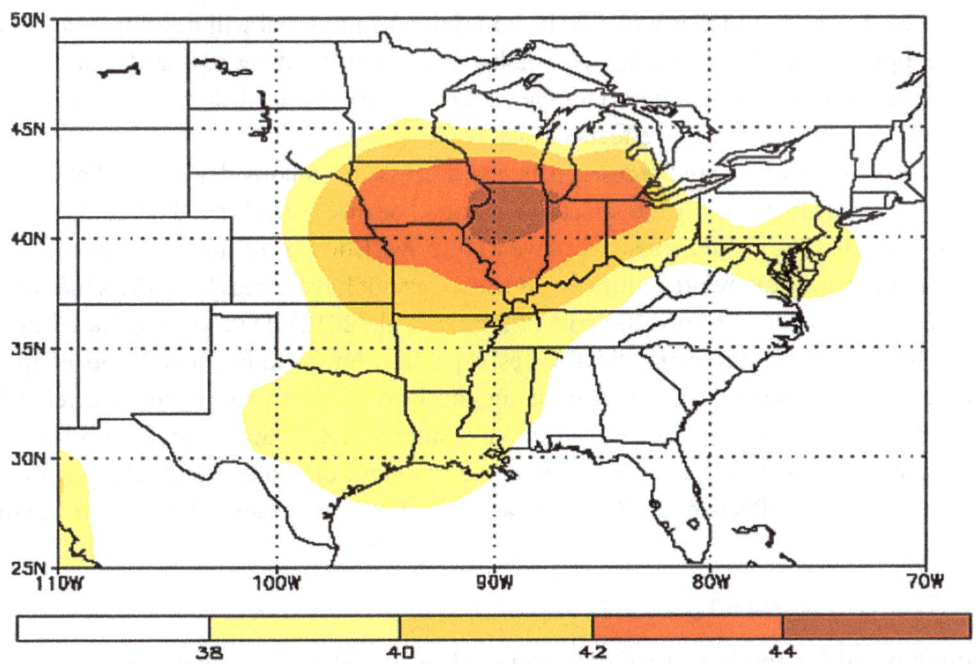

Figure 7.1 Maximum Apparent Temperature (Celsius) During July 13–14, 1995
Source: NOAA

BRIEF OVERVIEW

The July 1995 Chicago heat wave was the deadliest in U.S. history up to that point, lasting 5 days and claiming the lives of over 700 people (Clements & Casani, 2016; Klinenberg, 1999, 2002; Thomas, 2015). Temperatures reached as high as 106 degrees Fahrenheit, pushing the heat index to temperatures over 120 degrees (Klinenberg, 1999; see also Figure 7.1). Generally, a multitude of risk factors contribute to the harsh effects of heat waves, including social isolation of elderly individuals, insufficient emergency preparedness measures, a lack of necessary resources such as air-conditioning and water, and impoverished societal conditions (Gamble et al., 2012). The 1995 Chicago heat wave has implications for future heat waves, especially considering that future heat waves may last longer in duration and be more intense due to climate change (Hayhoe, Sheridan, Kalkstein, & Greene, 2010; Karl & Knight, 1997). Thus, it is necessary to have an understanding of how best to prepare for heat waves, with the circumstances surrounding the 1995 Chicago heat wave serving as a touchstone for awareness and planning. This chapter provides an overview of the 1995 Chicago heat wave, and discusses policy impact, lessons learned from the event, and implications for modern emergency management.

INTRODUCTION

A heat wave is a heat spell that exposes the environment and individuals in a specific area to consistent, hot temperatures. These temperatures may negatively impact the economy, the environment, and human health and well-being (Meehl & Tebaldi, 2004). High temperatures over the summer months put individuals at risk for developing heat-related illnesses such as heat exhaustion, heat stroke, and cramps. Additionally, individuals who do not possess adequate resources (e.g., water and air-conditioning) to combat the heat are more susceptible to the extreme temperatures (Semenza et al., 1999). Elderly individuals who are socially isolated and individuals unable to care for themselves, too, are at risk for experiencing the aforementioned consequences of heat waves (Gamble et al., 2012; Klinenberg, 2002; Naughton et al., 2002; Semenza et al., 1999). Elderly individuals are also more likely to suffer from preexisting conditions like diabetes or cardiovascular disease, further enhancing vulnerability (Changnon, Kunkel, & Reinke, 1996). On a similar note, certain medications can lower tolerance for heat and increase the likelihood that an individual will develop heat-related problems after prolonged exposure (Gamble et al., 2012). Finally, alcohol, illicit drug use, and place of residence have been identified as risk factors for heat waves (Duneier, 2006). Clearly, many different kinds of personal/medical risk factors contribute to the devastating effects of heat waves.

Beyond these sorts of health-related factors, it is also necessary to consider societal and economic influences, such as socioeconomic status, with respect to enhanced or minimized risk. For example, those individuals of lower socioeconomic status (i.e., those who live in impoverished conditions with limited resources) find themselves at greater risk for heat-related illnesses, as they may not have access to utilities like air-conditioning, or they may simply not be able to afford the cost (Gamble et al., 2012). Densely populated urban areas, too, are related to greater mortality rates during heat waves, especially on the uppermost floors of apartment buildings (Semenza et al., 1996). Such suboptimal housing conditions often provide limited relief from extreme temperatures. Further, some structures can even make residents hotter than they would be outdoors due to trapped and rising heat. As an aside, individuals who live in neighborhoods with higher crime rates are more vulnerable to the debilitating effects of stifling heat, as they often have reservations about opening a window (Gamble et al., 2012).

The Setting

Chicago was unprepared for the heat wave that occurred in July 1995, despite experiencing significant heat waves in the past, like those in 1955, 1983, and 1986 (Klinenberg, 1999; Thomas, 2015). So much so that the efforts of first responders proved futile in many instances, as irrevocable heat wave effects had already occurred prior to an organized emergency response (Thomas, 2015). While proficient in responding to instances of extreme cold, Chicago still struggled in its response to episodes of extreme heat. The establishment of designated cooling centers, areas where individuals have access to cool temperatures, was a good step. However, this intervention was not effectively implemented before the event. Without timely advisories and emergency recommendations, Chicagoans were not made aware of the availability of these resources and therefore would not think to utilize them (Changnon et al., 1996). Ultimately, not enough was done in the years leading up to the event to provide adequate support services and a comprehensive plan for protecting the populations most vulnerable to extreme heat, and it was these variables that would result in massive but preventable casualties when Chicago was overwhelmed by severe heat in the week of July 11, 1995.

The Disaster

In the days leading up to the 1995 Chicago heat wave, the National Weather Service warned of an approaching high temperature system that would sweep the Midwest in the weeks to come, recommending that those in urban areas especially prepare for sweltering heat (Changnon et al., 1996). These forecasts came as early as July 5, but the response of Chicagoans varied. Some took precautionary measures, such as purchasing box fans, buying cases of bottled water, and taking to the beach and public pools (Klinenberg, 1999). However, additional preparations that may have mitigated the full impact of the following week of intense heat were not taken despite the warnings (Thomas, 2015). For example, city officials could have encouraged first responders to bring water bottles and provide residents with resources available to help counter the extreme temperatures. Despite insufficient efforts to prepare, no casualties were reported the week preceding the heat wave (Clements & Casani, 2016).

On July 12, what came to be known as the first day of the event, temperatures rose to around 97 degrees Fahrenheit with a heat index of 102 (Klinenberg, 1999). This date marked the beginning of a record-breaking period of intense heat in Chicago that would last five stifling days, with the highest consistency of temperatures recorded in years (Clements & Casani, 2016). On the same day, the National Weather Service issued an alert of impending heat, claiming that the temperatures were only going to increase (Thomas, 2015). The day after, July 13, two toddlers suffocated due to being exposed to extremely high temperatures while in a pickup truck (Klinenberg, 1999). The air was thick, and Chicagoans were subject to the heat as well as dense humidity. A lack of wind and cloud coverage to alleviate the heat made the weather experienced even hotter than it would have been under other conditions (Klinenberg, 1999). Even when night fell, there was little relief, as temperatures continued to hover around the high 80s (Hayhoe et al., 2010).

As the heat rolled in, residents whose homes were equipped with air-conditioning units cranked up the air. Unfortunately, the utilities provider Commonwealth Edison was not ready to handle such a drastic uptick in usage, and generators started failing (Klinenberg, 1999). The heat also caused the train rails to change their shape, speaking to economic impact (Clements & Casani, 2016). This issue was compacted by periodic breakdowns of other crucial pieces of equipment and a general feeling of disarray at Commonwealth Edison. More than 40,000 residents on the North Side ended up losing power for at least a full day, and some remained without electricity through the end of the week (Changnon et al., 1996). Air-conditioning, fans, refrigerators, televisions, and radios, all manner of electronics necessary for keeping cool and keeping current with the heat advisories, were disabled across parts of the city.

Even as temperatures climbed ever higher, the situation for many Chicagoans grew even more unbearable after the unexpected power outages. Public officials provided little in the way of answers. City Hall's reaction was muted. Despite the way that things were rapidly deteriorating as the temperatures skyrocketed, no heat advisories were issued until a full day into the event (Changnon et al., 1996). Moreover, in response to increased concerns at the extremity of the weather and the relative lack of assistance and guidance from emergency services, Mayor Richard Daley dismissed criticisms of his office, mentioning the hot temperature yet encouraging citizens not to overreact (Changnon et al., 1996).

Some residents in parts of Chicago who were affected by the Commonwealth Edison outages or had no air-conditioning to begin with resorted to other means of staying cool. The practice of tapping fire hydrants was prevalent, with approximately 3,000 opened on Thursday alone, when some reports listed the heat index as high as 126 degrees Fahrenheit (Klinenberg, 1999). With no other way to relieve the heat, for many this appeared to be the only option. The result, unsurprisingly, was a reduction in water pressure and water shortages across parts of the city, which caused further problems. Conflict between residents and police and fire departments (so-called "water wars") also arose, with the heat itself serving to ratchet up tensions (Klinenberg, 1999).

It was not until Friday, July 15, that City Hall finally declared a citywide emergency, but at that point, the situation was past dire. As early as July 12, it had become clear that there were not enough ambulances to handle the number of people who needed to be taken to the hospital, forcing emergency services to improvise and use several fire trucks to transport the afflicted (Changnon et al., 1996). Cook County hospitals themselves struggled with the scores of people needing medical attention. As the temperatures had risen, so too had the number of individuals who required assistance. Between Wednesday and the following Tuesday, there were almost 1,100 more hospital admissions and around 3,000 more visits to emergency departments in Cook County than the norm (Hayhoe et al., 2010; Semenza et al., 1999). Most individuals were admitted to the hospital for symptoms of dehydration, heat stroke, or heat exhaustion, with one-fifth of those admitted for heat stroke not surviving hospitalization (Hayhoe et al., 2010).

Although there was disagreement as to just how many deaths in Chicago were a result of the heat wave, even the narrowest estimates at the end of the week placed the death toll in the 500s, with estimates applying more "generous" measures of what constitutes "heat-related" exceeding 700 (Klinenberg, 1999). Regardless of the debate over the death toll, the devastation was widespread and unprecedented, and in its wake, Chicago was left to grapple with both the incredible loss of life and the failures of city officials to properly prepare for the disaster, protect vulnerable populations, or initiate a plan to handle the disaster as it unfolded.

AFTERMATH AND INVESTIGATION

The 1995 Chicago heat wave was a rather unexpected tragedy in American history. However, disaster preparedness actions could have been increased after the warning of impending heat in order to combat the risk factors surrounding the event. Considering these and other specific aspects comprising the event helps to make some sense of the high death rate, as well as factors that exacerbated risk. For instance, Semenza et al. (1999) suggested that the greatest risk factors were a lack of an air conditioner and instances of dehydration related to a lack of water, as the majority of individuals treated from the heat wave suffered from heat-related illnesses. Another risk factor relates to a perceived complacency on the part of Chicago citizens. By way of example, Jerry Taft, Chief meteorologist for ABC-7, indicated that many individuals failed to take action to protect themselves despite the warnings (Thomas, 2015). Thus, the motivation of citizens to take appropriate action in response to the extreme heat was called into question. These factors, coupled with additional risk factors identified below, created a context conducive to extensive heat-related illness and in many cases, death.

Social isolation and age have also been identified as contributing factors with respect to the impact of the event. Arguments along these lines were presented by Klinenberg (1999) and Semenza et al. (1996). In the latter instance, Semenza et al. (1996) conducted a case–control study, concluding that individuals who lived alone or did not leave home on a daily basis were at a higher risk of dying from the heat wave. The median age in this case–control study was 76 years, suggesting age as a potentially important variable to consider. More generally, Klinenberg (2002) found that social support predisaster was crucial in limiting the damaging effects caused by the heat wave. These findings are consistent with a wide range of literature that points to social support as being essential to human health and well-being, in general and during times of extreme adversity (Schulenberg et al., 2008; Southwick & Charney, 2012; Southwick, Litz, Charney, & Friedman, 2011; Tedeschi & Calhoun, 1996, 2004).

Interestingly, issues of social isolation and age appear to be more complex in the current instance than at first examination, however. For instance, Duneier (2006) collected data on demographic characteristics of those who perished during the heat wave, data that challenged the claims made by Klinenberg (1999) and Semenza et al. (1996) as to the importance of social isolation with respect to the elderly. As an example, among the 16 individuals who died in North Lawndale, two were over the age of 65. Further, 12 of the 16 individuals who died in this area lived with families (Duneier, 2006). Thus, the majority of those who died in this area were not elderly and were not living alone. According to Duneier (2006), Klinenberg (1999, 2002) did not collect adequate data that would allow him to make the assumptions he did regarding social isolation and age as contributing factors. Rather, Duneier (2006) asserted that eight of the 16 deaths in North Lawndale were associated with drugs and alcohol, identifying these substances as risk factors. In addition, 15 of the 16 individuals who died in North Lawndale were also living on the top floor of a building, another contributing factor.

In summary, several major risk factors were identified following the 1995 Chicago heat wave that warrant attention and consideration with respect to the occurrence of future heat waves. These factors include a lack of access to air-conditioning and water, apathy with respect to taking action, social isolation and age, the use of drugs and alcohol, and location of residence, demographically speaking and within the context of a particular building. Such factors should be considered holistically in preparing for future heat waves. Moreover, they should be included as focal points in subsequent research studies as a means of reconciling discrepant interpretations.

POLICY IMPACT AND LESSONS LEARNED

After the occurrence of the 1995 Chicago heat wave, the city's mayor, Richard Daley, developed the Commission on Extreme Weather Conditions (CEWC; Naughton et al., 2002) to examine the factors that contributed to heat-related illnesses and death caused by the event. The city of Chicago came to realize that social isolation and age contributed to the enhanced damage caused by the heat wave (Klinenberg, 1999; Semenza et al., 1996). Further, emergency preparedness efforts should have focused on initiating and maintaining contact with the public prior to and throughout the period of extreme heat (Clements & Casani, 2016). Had these efforts been undertaken, and with effectiveness, individuals at risk could have been provided with essential resources and information on how best to protect themselves, galvanizing action.

After the disaster, journalists and media professionals alike attempted to explain it as a natural occurrence, as opposed to uncovering the social constraints that contributed to its effects (Klinenberg, 1999). Klinenberg (1999) discovered that the committee put together by Mayor Daley neglected to include a sociologist. In doing so, Mayor Daley made it possible for the CEWC to attribute the disaster to more natural causes as opposed to the sociological factors that inevitably contributed to the negative consequences (Klinenberg, 1999; Naughton et al., 2002; Thomas, 2015). Attributing the disaster to natural causes is problematic because it passes responsibility from people to natural causes (Klinenberg, 1999; Naughton et al., 2002; Thomas, 2015). In other words, while the impending heat was inevitable, there was some warning prior to the event. Preparations could have been made, protections put in place, and city officials failed to carry out an effective response strategy during the event, or adjust appropriately when resources failed (e.g., electricity, water pressure; see, for example, Clements & Casani, 2016). Portraying the heat wave as a natural occurrence overshadows the more crucial risk factors that lead to heat-related illnesses and death. An understanding of these factors is essential in order to mitigate the effects of future events.

IMPLICATIONS FOR MODERN EMERGENCY MANAGEMENT

The *Morbidity and Mortality Weekly Journal* asserts that deaths from heat are preventable (Centers for Disease Control and Prevention, 1995). Although extremely hot temperatures can have detrimental effects on many aspects of the environment, humans are especially at risk (Easterling, Evans, Groisman, & Karl, 2000). In general, and taking into consideration the many risk factors mentioned above, it is first and foremost necessary to take precautionary measures, as future heat waves are likely to occur that will be more intense and longer in duration (Hayhoe et al., 2010; Karl & Knight, 1997). Cities are at an increased risk for heat waves due to climate change (Revi et al., 2014).

Efforts to enhance the awareness, education, and motivation of people are key to preparation. Such efforts should be made to communicate with citizens, before an impending event, throughout the duration of the event, and with consideration for following up after the event. A comprehensive approach is necessary to ensure that people's needs are being met (McElreath, Doss, et al., 2014). For example, it is critical that individuals have access to resources such as air-conditioning, water, and electricity. Vulnerable populations need to be identified and supported. Individuals who are socially isolated, use alcohol and/or illicit drugs, and/or live on the top floors of buildings appear to be at heightened risk. Elderly adults are particularly susceptible to high temperatures (Gamble et al., 2012), and the literature suggests that age was a factor in the 1995 Chicago heat wave (Klinenberg, 1999, 2002; Semenza et al., 1996). People need to be aware of their social support, willing and able to reach out to others as needed. Social support is essential with respect to adaptive coping (Schulenberg et al., 2008; Southwick & Charney, 2012; Southwick et al., 2011; Vranceanu, Hobfoll, & Johnson, 2007).

SUMMARY

The 1995 Chicago heat wave was a 5-day heat spell that claimed the lives of over 700 people, with temperatures surpassing 100 degrees Fahrenheit (Klinenberg, 1999, 2002). While the extreme temperatures were inevitable, several major risk factors were identified as being related to heat-related illnesses and death from the event.

These factors include a lack of access to air-conditioning and water, a lack of timely communication from city officials, people not being motivated to prepare, social isolation and age, the use of drugs and alcohol, power and water pressure failures, and living context. These and related factors should be considered in future preparation, response, and research efforts. Individuals should have access to essential resources during the summer months as a means of countering the heat (e.g., air-conditioning, cooling centers, water, electricity). People should be aware of their social support, and encouraged to reach out to others as needed. Vulnerable populations need to be identified, informed, and supported (McElreath, Jensen, et al., 2013). Communication and response efforts must be better coordinated, broadly speaking, as extreme temperatures will continue to be problematic in the future.

REFERENCES

Centers for Disease Control and Prevention. (1995). *Extreme heat: A prevention guide to promote your personal health and safety*. Atlanta, GA: U.S. Department of Health and Human Services, Centers for Disease Control and Prevention.

Changnon, S. A., Kunkel, K. E., & Reinke, B. C. (1996). Impacts and responses to the 1995 heat wave: A call to action. *Bulletin of the American Meteorological Society*, *77*(7), 1497–1506. doi:10.1175/1520-0477(1996)077<1497:IARTTH>2.0.CO;2

Clements, B. W., & Casani, J. A. P. (2016). *Disasters and public health: Planning and response* (2nd ed.). Cambridge, MA: Butterworth-Heinemann.

Duneier, K. (2006). Ethnography, the ecological fallacy and the 1995 Chicago heat wave. *American Sociological Review*, *71*(4), 679–688. doi:10.1177/000312240607100040

Easterling, D. R., Evans, J., Groisman, P. Y., & Karl, T. R. (2000). Observed variability and trends in extreme climate events: A brief review. *Bulletin of the American Meteorological Society*, *81*, 417–425.

Gamble, J. L., Hurley, B. J., Schultz, P. A., Jaglom, W. S., Krishnan, N., & Harris, M. (2012). Climate change and older Americans: State of the science. *Environmental Health Perspectives*, *121*(1), 15–22. doi:10.1289/ehp.1205223

Hayhoe, K., Sheridan, S., Kalkstein, L., & Greene, S. (2010). Climate change, heat waves, and mortality projections for Chicago. *Journal of Great Lakes Research*, *36*, 65–73. doi:10.1016/j.jglr.2009.12.009

Karl, T. R., & Knight, R. W. (1997). The 1995 Chicago heat wave: How likely is a recurrence? *Bulletin of the American Meteorological Society*, *78*(6), 1107–1119. doi:10.1175/1520-0477(1997)078<1107:TCHWHL>2.0.CO;2

Klinenberg, E. (1999). Denaturalizing disaster: A social autopsy of the 1995 Chicago heat wave. *Theory and Society*, *28*(2), 239–295. doi:10.1023/A:1006995507723

Klinenberg, E. (2002). *Heat wave: A social autopsy of disaster in Chicago*. Chicago, IL: University of Chicago Press.

McElreath, D., Doss, D., Jensen, C., Wigginton, M., Nations, R., Van Slyke, J., & Nations, J. (2014). *Foundations of emergency management*. Dubuque, IA: Kendall-Hunt.

McElreath, D., Jensen, C., Wigginton, M., Doss, D., Nations, R., & Van Slyke, J. (2013). *Introduction to homeland security* (2nd ed.). Boca Raton, FL: CRC Press.

Meehl, G. A., & Tebaldi, C. (2004). More intense, more frequent, and longer lasting heat waves in the 21st century. *Science*, *305*, 994–997. doi:10.1126/science.1098704

Naughton, M. P., Henderson, A., Mirabelli, M. C., Kaiser, R., Wilhelm, J. L., Kieszak, S. M., Rubin, C. H., & McGeehin, M. A. (2002). Heat-related mortality during a 1999 heat wave in Chicago. *American Journal of Preventive Medicine*, *22*(4), 221–227. doi:10.1016/S0749-3797(02)00421-X

Revi, A., Satterthwaite, D. E., Aragon-Durand, F., Corfee-Morlot, J., Kiunsi, R. B. R., Pelling, M., Roberts, D. C., & Solecki, W. (2014). Urban areas. In C. B. Field, V. R. Barros, D. J. Dokken, K. J. Mach, M. D. Mastrandrea, T. E. Bilir, M. Chatterjee, K. L. Ebi, Y. O. Estrada, R. C. Genova, B. Girma, E. S. Kissel, A. N. Levy, S. MacCracken, P. R. Mastrandrea, & L. L. White (Eds.). *Climate change 2014: Impacts, adaptation, and vulnerability* (pp. 535–562). Cambridge, MA: Cambridge University Press.

Schulenberg, S. E., Dellinger, K. A., Koestler, A. J., Kinnell, A. M. K., Swanson, D. A., Van Boening, M. V., & Forgette, R. G. (2008). Psychologists and Hurricane Katrina: Natural disaster response through training, public education, and research. *Training and Education in Professional Psychology*, *2*(2), 83–88. doi:10.1037/1931-3918.2.2.83

Semenza, J. C., McCullough, J. E., Flanders, W. D., McGeehin, M. A., & Lumpkin, J. R. (1999). Excess hospital admissions during the 1995 heat wave in Chicago. *American Journal of Preventive Medicine*, *16*(4), 269–277. doi:10.1016/S0749-3797(99)00025-2

Semenza, J. C., Rubin, C. H., Falter, K. H., Selanikio, J. D., Flanders, D., Howe, H. L., & Wilhelm, J. L. (1996). Heat-related deaths during the July 1995 heat wave in Chicago. *The New England Journal of Medicine*, *335*(2), 84–90.

Southwick, S. M., & Charney, D. S. (2012). *Resilience: The science of mastering life's greatest challenges*. Cambridge, United Kingdom: Cambridge University Press.

Southwick, S. M., Litz, B. T., Charney, D., & Friedman, M. J. (Eds.). (2011). *Resilience and mental health: Challenges across the lifespan*. Cambridge, United Kingdom: Cambridge University Press.

Tedeschi, R. G., & Calhoun, L. G. (1996). The Posttraumatic Growth Inventory: Measuring the positive legacy of trauma. *Journal of Traumatic Stress*, *9*(3), 455–471.

Tedeschi, R. G., & Calhoun, L. G. (2004). Posttraumatic growth: Conceptual foundations and empirical evidence. *Psychological Inquiry: An International Journal for the Advancement of Psychological Theory*, *15*(1), 1–18. doi:10.1207/s15327965pli1501_01

Thomas, M. (2015). Heat wave. *Chicago Magazine*. Retrieved from http://www.chicagomag.com/Chicago-Magazine/July-2015/1995-Chicago-heat-wave/

Vranceanu, A. M., Hobfoll, S. E., & Johnson, R. J. (2007). Child multi-type maltreatment and associated depression and PTSD symptoms: The role of social support and stress. *Child Abuse & Neglect*, *31*(1), 71–84. doi:10.1016/j.chiabu.2006.04.010

8 Hurricane Sandy (2012)

Contributed by Marcela C. Weber, Emily A. Gawlik, and Stefan E. Schulenberg.

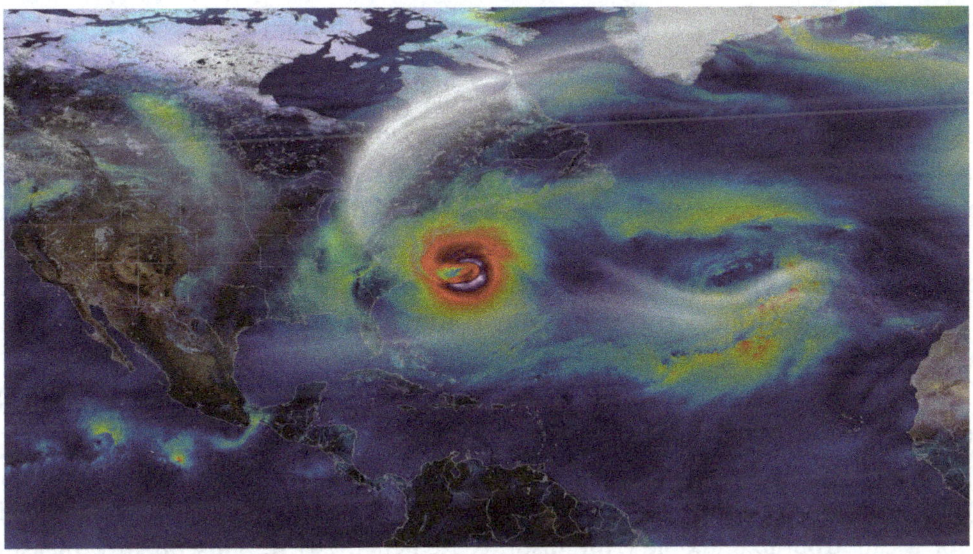

Figure 8.1 October 29, 2012—A day Before Landfall, Sandy Intensified into a Category 2 Superstorm Nearly 1,000 Miles Wide

Source: Video and images courtesy of NASA/GSFC/William Putman

BRIEF OVERVIEW

Hurricane Sandy began developing as a tropical storm on October 22, 2012, made landfall in the Caribbean, and then struck the New York-New Jersey metropolitan area on October 29 (Blake, Kimberlain, Berg, Cangialosi, & Beven, 2013). It was one of the deadliest hurricanes on record, resulting in almost 100 fatalities in the United States alone (Abramson & Redlener, 2013), and over 200 in total (Diakakis, Deligiannakis, Katsetsiadou, & Lekkas, 2015). Beyond fatalities, thousands of individuals were injured (Brackbill et al., 2014). About half of the residents did not evacuate from areas where evacuation orders were mandated (Brown, Parton, Driver, & Norman, 2016). The hurricane knocked out power and caused energy failure in over half of New York City (Tran, Heller, Berger, & Habboushe, 2014), the most densely populated area of the United States.

Between 10% and 20% of residents in the inundation zones experienced posttraumatic stress disorder (PTSD) from Hurricane Sandy (Boscarino, Hoffman, Adams, Figley, & Solhkhah, 2014; Caramanica, Brackbill, Stellman, & Farfel, 2015). PTSD prevalence was as high as 30% for individuals who sustained an injury, lacked social support, or who previously experienced PTSD due to the terrorist attacks of September 11, 2001 (Caramanica et al., 2015). In this way, rather than equipping the city for another crisis, the terrorist attacks on New York City 11 years prior to Hurricane Sandy increased the toll on mental health.

Hurricane Sandy affected modern emergency management in many ways. One major contribution was that it exposed flaws in the measurement and categorization of hurricanes, such that research into alternative measurements is now underway (Kantha, 2013). Hurricane Sandy also demonstrated the importance of redundancy planning, in the event that current technologies fail. For example, hospitals that have switched to electronic records should make hard copies of critical patient information prior to predicted disasters (Adalja et al., 2014). Residents of the U.S. East Coast turned to social media for information on the hurricane, yet the relevant agencies and emergency services did not tend to post emergency information to the sources where residents were looking, such as Twitter or Facebook (Hughes, St Denis, Palen, & Anderson, 2014). In this way, Hurricane Sandy demonstrated the importance of emergency services staying current in the medium used to communicate with the public during a disaster.

INTRODUCTION

Prior to Hurricane Sandy, multiple factors placed New York City at increased risk for damage, injuries, and casualties due to a hurricane. Contrary to public perception that the prospect of a tropical storm striking the region was inconsequential, urban planners had long warned that the region was exceptionally vulnerable to the hazards associated with such an event due to unique infrastructural deficits and population density (Wagner, Chhetri, & Sturm, 2014). The New York City mass transit system presented a particular quandary. With a power supply housed underground and only a few elevated entrances, the subway tunnels were susceptible to flooding. Electrical grid substations were also placed at heights and in locations that had the potential to be problematic. In terms of population distribution, some of the boroughs—Staten Island, in particular—had expanded into areas that could see increased exposure to storm surge. Beachside communities along the New Jersey coastline, too, faced the same danger. Considering that the New York City-Newark area is one of the most densely populated in the United States, this would elevate the potential for extensive damages if the area was directly affected by a hurricane (Blake et al., 2013).

Findings from an investigation conducted by the U.S. Army Corps of Engineers and the New York City Office of Emergency Management after Hurricane Katrina to assess New York City's level of preparedness in the event of a similar disaster affirmed the validity of these fears. The inquiry determined that the population's inexperience with severe storm events and the city's extreme vulnerability to storm surge and winds indicated the potential for a catastrophic outcome (Abramson & Redlener, 2013). Subsequently, measures like storm surge barriers were suggested to address existing structural weaknesses and mitigate the risk of continued urban development into vulnerable coastal areas. However, little headway was made in this endeavor as public officials, citing budgeting concerns, tabled these proposals (Wagner et al., 2014).

The Setting

Before Hurricane Sandy, only three tropical storms had directly affected the Northeast in the previous 100 years; these were the Great New England Hurricane of 1938, Hurricane Agnes (1972), and Hurricane Irene (2011; Magnusson et al., 2014). Tropical cyclones whose progress led them into the North Atlantic usually experienced re-curvature upon tracking northwest, soon swerving away from the mainland and back toward Europe (Kunz et al., 2013).

While systems in the New York City area were not well prepared for a storm of the magnitude of Hurricane Sandy, the inhabitants were not strangers to a major citywide crisis; 11 years prior, they experienced the terrorist attacks of September 11, 2001. A study of the survivors found that approximately 1 in 10 were also living in the inundation area when Hurricane Sandy struck (Brackbill et al., 2014).

The Disaster

Sandy was first recognized as a tropical storm by the National Hurricane Center on October 22, 2012 (Blake et al., 2013). Originating in the Caribbean north of Panama, Sandy built in strength as it slowly, yet surely, progressed north. By the time it reached the southeastern coast of Jamaica on October 24, the intensity of its winds exceeded 75 knots and it had been upgraded to a Category 1 hurricane. Sandy plowed through the island, destroying thousands of homes and displacing countless residents before continuing along its forecasted track through the Caribbean, affecting both the Dominican Republic and Haiti as it steered north (Blake et al., 2013).

While the eye of the hurricane never crossed Haiti, the forceful winds and torrential rains that Sandy brought with it wreaked havoc on a weakened infrastructure still recovering from the major earthquake of 2010. The storm claimed more lives in Haiti than anywhere else in the Caribbean and laid waste to a great portion of the island, including nearly 40% of Haiti's agricultural sector (Blake et al., 2013).

Sandy impacted Cuba soon after as a Category 3 hurricane, with winds reaching peak speeds of 100 knots (Blake et al., 2013). It did, however, grow measurably weaker during its passage over the eastern part of the island. This general trend of weakening and restrengthening of winds would continue as Sandy passed through the Bahamas on October 26 and interacted with more low-pressure systems. Sandy was relegated to tropical storm status again but was gradually becoming extraordinarily large, approximately two times the size of Hurricane Irene (Kantha, 2013). It also took a turn northwest, setting on a new and unusual trajectory. Just as had been predicted by forecast services, Sandy was set on a course to make landfall in New Jersey (Magnusson et al., 2014).

Approaching the Eastern Seaboard, Sandy was heralded by voluminous rainfall and high winds. Interaction with a body of cold air near Appalachia caused blizzard-like conditions to envelop parts of Virginia, Tennessee, and North Carolina and yielded up to a meter of snow in places (Kunz et al., 2013). Sandy continued to swell in size as it interacted with other weather systems in the Atlantic like these—with meteorologists taking note of its massive size, 870 nautical miles wide (Blake et al., 2013). A mandatory citywide evacuation was ordered by officials in New York City, though few residents abided with most having chosen to remain in their homes (Brown et al., 2016). These residents witnessed a storm of incredible magnitude and unprecedented destruction. While it had been relegated to tropical cyclone status, Sandy would make up for what it lacked in wind speed with intensity, pummeling the coast with a force that far exceeded most forecasts (Kantha, 2013).

On the night of October 29, Sandy made landfall as a post-tropical cyclone on the coast of New Jersey near Atlantic City. At the time of impact, wind speeds were calculated at 85 mph and storm surge heights were charted as high as 14 ft (Wagner et al., 2014). New York City transit systems were downed by record-breaking levels of water—streets, major tunnels, ferry, and rail systems all experiencing extensive flooding. Eight subway stations were submerged, and the tarmac at La Guardia and John F. Kennedy International Airports was engulfed by the deluge (Blake et al., 2013). The storm surge continually battered the New Jersey Shoreline, devastating the vulnerable barrier islands and beach communities. While the worst of the flooding was seen in New York City and in New Jersey, where the tide soon rose to 12.65 ft and 8.57 ft past the normal level, respectively, the rest of the Eastern Seaboard fared no better (Blake et al., 2013). Huge swathes of coastline stretching from Maine to Florida—approximately 1,000 km—were affected by the storm surge (Brandon et al., 2014). Some instruments indicated that the magnitude of flooding in New York City following Hurricane Sandy was as rare as return periods every 1,000 years; however, additional research has indicated similar levels of flooding due to other hurricanes in the past few centuries, such as a more compact but intense hurricane that landed in New York City in 1821 (Brandon, Woodruff, Donnelly, & Sullivan, 2014).

The combination of flooding and high winds brought by Sandy also compromised the electrical grid, knocking power out all over the directly affected areas and beyond. Approximately 2.5 million people in New Jersey and 2.3 million people in New York soon found themselves without electricity as the storm raged on (Tran, Heller, Berger, & Habboushe, 2014). Thousands of patients at 37 different healthcare facilities in New York City were forced to evacuate, as hospitals and assisted living centers, too, fought back rising tides and saw their backup generators go out (Adalja et al., 2014). After hours of turmoil, the hurricane finally progressed onward out of the Northeast. From October 30 to October 31, the remnants of the post-tropical system continued to move further north toward the Great Lakes, eventually dissolving near Lake Erie (Kunz et al., 2013).

AFTERMATH AND INVESTIGATION

Evacuation

Unfortunately, many people in Sandy-affected areas defied the evacuation mandate (Abramson & Redlener, 2013). A survey of over 400 residents of Staten Island, the Rockaways, and Brooklyn found that only about half of the residents complied with evacuation orders in Zone A, the most vulnerable zone, where evacuation was mandatory (Brown et al., 2016). Residents who had experienced the terrorist attacks of September 11, 2001 were more likely to evacuate than those who had not (Nejat, Cong, & Liang, 2016). Residents who reported more home damage following Sandy were also more likely to evacuate (Brown et al., 2016), which suggests that the evacuation order was taken more seriously by the residents most at risk. Likewise, residents who lived on higher floors of buildings were less likely to evacuate (Brown et al., 2016).

Fatalities

Although evacuation removed many residents from the dangers and risks of the hurricane, Sandy was still particularly deadly. According to one estimate, Hurricane Sandy resulted in 233 deaths across the United States, Canada, Cuba, Jamaica, Haiti, the Bahamas, Puerto Rico, and the Dominican Republic (Diakakis et al., 2015). The fatalities due to Hurricane Sandy were especially high because it struck a densely populated urban area; 97 of these deaths, or about 40%, occurred in the New York City metropolitan area (Abramson & Redlener, 2013). Seventy-one deaths occurred in New York City, 43 deaths occurred in New Jersey, and 15 deaths occurred in Pennsylvania (Diakakis et al., 2015). Deaths in the New York City area tended to be among older adults, both male and female, while deaths in the Caribbean tended to be among females and the young (Diakakis et al., 2015). The most prevalent cause of death was physical trauma (37% of deaths), followed by drowning (22% of deaths; Diakakis et al., 2015). When examining deaths by activity, 40% of deaths occurred indoors, 20% occurred during cleanup of the disaster, 16% occurred while using a motor vehicle, and 14% were pedestrians (Diakakis et al., 2015). Of direct fatalities, the most common cause was seawater surge (17% of direct fatalities); of indirect fatalities, the most common causes were related to power failure (20%; Diakakis et al., 2015).

Injuries

Much of the data on Hurricane Sandy survivors in the New York City area was gleaned from a study of survivors of the terrorist attacks of September 11, 2001 (i.e., "9/11"), as this study was ongoing at the time Hurricane Sandy struck; it included tens of thousands of New York City residents (Brackbill et al., 2014). Of these survivors, 8,000 were living in a Sandy inundation area and had recently responded to a survey before the hurricane hit. Researchers compared the health data prior to Hurricane Sandy with health data following the event. Ten percent of the survivors were injured as a result of Hurricane Sandy, and 75% of those injured had sustained more than one injury. While this was not a sample of the general population, but of 9/11 survivors, these data suggest that as many as 10% of people living in an inundation area were injured due to Hurricane Sandy.

Economic Impact

Hurricane Sandy was one of the costliest hurricanes ever to hit the United States (Lackmann 2015). Estimated costs of damages range from 50 billion (Abramson & Redlener, 2013) to 65 billion U.S. dollars (Hughes et al., 2014). By some calculations, it was the second costliest hurricane to strike the United States, following Hurricane Katrina (Kunz et al., 2013).

Energy Failures

A particular problem that stemmed from the hurricane was energy failure throughout the dense, urban New York City area. Over 400 hundred carbon monoxide exposures in the aftermath of Sandy were reported to the NYC Poison Control Center (Chen et al., 2013), most of them due to people grilling indoors, using generators, and an apartment building fire. Furthermore, energy failures and gas shortages affected the ability of the emergency response workforce to respond, and eight New York City deaths were due to carbon monoxide poisoning in homes with power failure (Abramson & Redlener, 2013).

Mental Health Impact

In addition to causing fatalities and physical injuries, Hurricane Sandy also adversely affected people's mental health. For example, the study of 9/11 survivors surveyed those affected by Hurricane Sandy at 6 months and 1 year following the storm (Caramanica et al., 2015). For people who survived both 9/11 and Hurricane Sandy, posttraumatic stress disorder (PTSD) prevalence was 11% in the Federal Emergency Management Agency defined inundation zones and 4% in non-inundation zones. PTSD rates were higher for individuals with increased stressors—29% for people who reported low social support prior to Sandy and 32% for people who sustained an injury due to Sandy (Lowe, Sampson, Gruebner, & Galea, 2015). Similarly, PTSD from Hurricane Sandy was more prevalent for individuals who had PTSD following 9/11, at 29% for these survivors (Neria & Shultz, 2012). In a study of primarily middle-aged and elderly adults in the New Jersey Shoreline community, 14% screened positive for PTSD (Boscarino et al., 2014). Another study of the mental health impact of Hurricane Sandy in New York City found that women were more likely than men to have symptoms of posttraumatic stress (Hamama-Raz et al., 2015).

LESSONS LEARNED AND POLICY IMPACT

Hurricane Sandy has informed disaster management practices in the years since the event, including informing flood policy. Prior to Hurricane Sandy, the magnitude of flooding it caused in New York City was considered rare, estimated to occur once every 1,000 years (Brandon et al., 2014). As a result of the high flood levels due to Sandy, environmental scientists examined historical flood levels and concluded that a similar storm surge occurred in 1821 (Brandon et al., 2014). They concluded that flooding comparable to the flooding from Sandy is not nearly as rare as previous estimates predicted. Preparedness for rare, yet extreme disasters is not always a priority for the public, but public support is needed to enact flood preparedness policies (McElreath, Jensen, et al., 2013).

In the case of Hurricane Sandy, the public appeared to support such policies following the event (Schmeltz et al., 2013). A survey conducted in New Jersey 4 months after Hurricane Sandy found that about half of the residents surveyed were in support of the majority of policies described in the survey, which would restrict use of land in flood zones and require houses to be built to resist stormwaters (Greenberg et al., 2014). The residents who supported these policies tended to be in favor of using taxes, bonds, and fees in order to do so.

Specific flood policies were enacted in New York City after Hurricane Katrina, but changes to these policies were needed. For instance, following Hurricane Katrina, New York City hospitals moved their backup generators to higher floors or built 12-ft concrete walls around them to protect against flooding. For some hospitals, these measures were sufficient to maintain operations during Hurricane Sandy (VanDevanter, Kovner, Raveis, McCollum, & Keller, 2014). For others, however, the floodwater levels reached 14 ft and the concrete walls were insufficient (Powell, Hanfling, & Gostin, 2012).

Several additional failures in the emergency medical response following Hurricane Sandy provided valuable lessons. For instance, after Hurricane Katrina, accreditation standards for hospitals were raised to 96 hour of independent functioning, but hospitals in the New York City area delayed meeting these standards and were unable to maintain operations in the hours immediately following Hurricane Sandy (Powell et al., 2012).

IMPLICATIONS FOR MODERN EMERGENCY MANAGEMENT

Hurricane Sandy has had a wide range of implications for modern emergency management, including implications for hurricane classification and measurement, redundancy planning, and the use of social media to communicate about disasters. In 2013, Lakshmi Kantha examined the five major hurricanes Katrina, Isaac, Ike, Irene, and Sandy, concluding that "the size of the hurricane matters, not just the intensity" (p. 124).

At this time, the Saffir–Simpson scale is the main scale used to measure hurricanes. The scale measures potential wind intensity and potential storm surge, but the storm surge component was removed after Hurricane Katrina because it was misleading (McElreath, Doss, et al., 2014. However, after Hurricane Sandy, coastal engineers asserted that using potential wind intensity remains misleading because it is a measure of hurricane intensity, but not hurricane *size*. Hurricanes Katrina, Ike, and Sandy had lower intensities than other storms that hit the same regions, but inflicted less damage. Katrina, Ike, and Sandy inflicted greater damage because of their large size, despite lower intensity, relative to other storms. Analysis of Sandy and the other four hurricanes showed that the Saffir–Simpson scale may be insufficient for effectively informing disaster workers and the public. Because Hurricane Sandy revealed these inaccuracies, disaster scientists are working to develop alternative hurricane classification systems (Rosenzweig & Solecki, 2014).

Redundancy planning for major disasters must be up to date with current technological advancements, as Sandy showed (Parris 2014). For example, in 2012, New York City hospitals had generally transitioned to electronic medical records, while paper files were stored at separate facilities (Adalja et al., 2014). Temporary loss of access to records after Sandy struck demonstrates how essential medical information should be retained in paper form. Even if detailed records are electronic and/or stored elsewhere, essential summary information should be printed before a predicted disaster, such as a hurricane (Adalja et al., 2014). Similarly, hospitals should have two-way radios in case telephone systems fail. During Hurricane Sandy, hospital staff primarily relied on their personal cell phones (Adalja et al., 2014). Beyond hospitals, redundancy planning is needed in other areas of critical infrastructure. For example, gas fuel pumps need a backup power source (Abramson & Redlener, 2013).

As a major disaster that occurred at a time and place with heavy social media use, Hurricane Sandy has major implications for the use of social media to communicate information about disasters. Although the majority of residents in disaster-affected areas looked to social media for information on Hurricane Sandy, most police and fire departments, at the time, were not utilizing social media to provide information to residents of the disaster-affected area (Hughes et al., 2014). Two-thirds of fire and police departments in the New England areas affected by Hurricane Sandy had Facebook accounts at the time the hurricane struck, but only a quarter of departments posted information to their pages during the hurricane (Hughes et al., 2014). Similarly, 13% of fire and police departments had a Twitter account at the time, but only half of those tweeted information about the storm while it was in progress (Hughes et al., 2014). When they did use social media, the primary purpose was to communicate about closures and to provide links to safety information on other webpages (Hughes et al., 2014). At the time of Hurricane Sandy, the police departments in the area were generally more active on social media than the fire departments, but growth in this area is critical to modern emergency management (Hughes et al., 2014). As the general public increasingly looks to social media for information, effective modern emergency management has come to mean that emergency service providers, such as police and fire departments, must use social media to communicate with residents of disaster-affected areas. Social media platforms will likely evolve and their popularity will fluctuate, so it is crucial that police and fire departments stay current.

SUMMARY

When Hurricane Sandy struck the Eastern Seaboard of the United States in October 2012, it had a catastrophic impact. In part because of a lack of experience, proper preparation, or measured assessment of the threat faced, the East Coast of the United States was in poor shape to endure the event. The metropolitan areas it hit lacked the capacity to withstand the levels of flooding that ensued. High winds in tandem with flooding led to power outages that left millions of East Coast residents without electricity. Over 200 lives were lost, either as a direct or as an indirect result of the storm. A large share of these fatalities (97) occurred in the New York City metropolitan area alone. Many more people were injured, and at heightened risk for experiencing symptoms of post-traumatic stress. At the time of occurrence, Hurricane Sandy was the second costliest hurricane in U.S. history, resulting in tens of billions of dollars in damages. Because of Sandy's exceptional scale and the circumstances surrounding its especially devastating impact on the East Coast, the aftermath holds implications for policy decisions and emergency management strategies. For example, Sandy illuminated the need for flood policy reform, highlighted the shortcomings of existing response protocols, exposed the need for hurricane categorization that encompasses both size and intensity, and demonstrated the ways social media must be harnessed in order to effectively warn and inform the public. Such lessons come at great cost, with changes necessary in order to mitigate the impact of future disasters and improve emergency response efforts nationwide, especially as the development and expansion of urban areas in the United States continues over time.

REFERENCES

Abramson, D. M., & Redlener, I. E. (2013). Hurricane Sandy: Lessons learned, again. *Disaster Medicine and Public Health Preparedness, 6*(4), 328–329. doi:10.1001/dmp.2012.76

Adalja, A. A., Watson, M., Bouri, N., Minton, K., Morhard, R. C., & Toner, E. S. (2014). Absorbing citywide patient surge during Hurricane Sandy: A case study in accommodating multiple hospital evacuations. *Annals of Emergency Medicine, 64*(1), 66–73. doi:10.1016/j.annemergmed.2013.12.010

Blake, E. S., Kimberlain, T. B., Berg, R. J., Cangialosi, J. P., & Beven, J. L. (2013). Tropical cyclone report: Hurricane Sandy. *National Hurricane Center.* Retrieved from http://www.nhc.noaa.gov/data/tcr/AL182012_Sandy.pdf

Boscarino, J. A., Hoffman, S. N., Adams, R. E., Figley, C. R., & Solhkhah, R. (2014). Mental health outcomes among vulnerable residents after Hurricane Sandy: Implications for disaster research and planning. *American Journal of Disaster Medicine, 9*(2), 107–120. doi:10.5055/ajdm.2014.0147

Brackbill, R. M., Caramanica, K., Maliniak, M., Stellman, S. D., Fairclough, M. A., Farfel, M. R., . . . & Yu, S. (2014). Nonfatal injuries 1 week after Hurricane Sandy—New York City metropolitan area, October 2012. *Morbidity and Mortality Weekly Report, 63*(42), 950–954. Retrieved from http://www.jstor.org/stable/24855408

Brandon, C. M., Woodruff, J. D., Donnelly, J. P., & Sullivan, R. M. (2014). How unique was Hurricane Sandy? Sedimentary reconstructions of extreme flooding from New York harbor. *Scientific Reports, 4,* 1–9. doi:10.1038/srep07366

Brown, S., Parton, H., Driver, C., & Norman, C. (2016). Evacuation during Hurricane Sandy: Data from a rapid community assessment. *Public Library of Science Currents, 1,* 1–10. doi:10.1371/currents.dis.692664b92af52a3b506483b8550d6368

Caramanica, K., Brackbill, R. M., Stellman, S. D., & Farfel, M. R. (2015). Posttraumatic stress disorder after Hurricane Sandy among persons exposed to the 9/11 disaster. *International Journal of Emergency Mental Health, 17*(1), 356–362. doi:10.4172/1522-4821.1000173

Chen, B. C., Shawn, L. K., Connors, N. J., Wheeler, K., Williams, N., Hoffman, R. S., . . . & Smith, S. W. (2013). Carbon monoxide exposures in New York City following Hurricane Sandy in 2012. *Clinical Toxicology, 51*(9), 879–885. doi:10.3109/15563650.2013.839030

Diakakis, M., Deligiannakis, G., Katsetsiadou, K., & Lekkas, E. (2015). Hurricane Sandy mortality in the Caribbean and continental North America. *Disaster Prevention and Management: An International Journal, 24*(1), 132–148. doi:10.1108/DPM-05-2014-0082

Greenberg, M. R., Weiner, M. D., Noland, R., Herb, J., Kaplan, M., & Broccoli, A. J. (2014). Public support for policies to reduce risk after Hurricane Sandy. *Risk Analysis: An International Journal, 34*(6), 997–1012. doi:10.1111/risa.12203

Hamama-Raz, Y., Palgi, Y., Shrira, A., Goodwin, R., Kaniasty, K., & Ben-Ezra, M. (2015). Gender differences in psychological reactions to Hurricane Sandy among New York metropolitan area residents. *Psychiatric Quarterly, 86*(2), 285–296. doi:10.1007/s11126-014-9333-3

Hughes, A. L., St Denis, L. A., Palen, L., & Anderson, K. M. (2014, April). Online public communications by police & fire services during the 2012 Hurricane Sandy. *Proceedings of the SIG-CHI Conference on Human Factors in Computing Systems,* 1505–1514. doi:10.1145/2556288.2557227

Kantha, L. (2013). Classification of hurricanes: Lessons from Katrina, Ike, Irene, Isaac and Sandy. *Ocean Engineering, 70,* 124–128. doi:10.1016/j.oceaneng.2013.06.007

Kunz, M., Mühr, B., Kunz-Plapp, T., Daniell, J. E., Khazai, B., Wenzel, F., . . . & Zschau, J. (2013). Investigation of superstorm Sandy 2012 in a multi-disciplinary approach. *Natural Hazards and Earth System Sciences, 13,* 2579–2598. doi:10.5194/nhess-13-2579-2013

Lackmann, G. M. (2015). Hurricane Sandy before 1900 and after 2100. *Bulletin of the American Meteorological Society, 96*(4), 547–560. doi:10.1175/BAMS-D-14-00123.1

Lowe, S. R., Sampson, L., Gruebner, O., & Galea, S. (2015). Psychological resilience after Hurricane Sandy: The influence of individual- and community-level factors on mental health after a large-scale natural disaster. *Public Library of Science One, 10*, e0125761. doi:10.1371/journal.pone.0125761

Magnusson, L., Bidlot, J., Lang, S. K., Thorpe, A., Wedi, N., & Yamaguchi, M. (2014). Evaluation of medium-range forecasts for Hurricane Sandy. *Monthly Weather Review, 142*(5), 1962–1981. doi:10.1175/MWR-D-13-00228.1

McElreath, D., Doss, D., Jensen, C., Wigginton, M., Nations, R., Van Slyke, J., & Nations, J. (2014). *Foundations of emergency management*. Dubuque, IA: Kendall-Hunt.

McElreath, D., Jensen, C., Wigginton, M., Doss, D., Nations, R., & Van Slyke, J. (2013). *Introduction to homeland security* (2nd ed.). Boca Raton, FL: CRC Press.

Nejat, A., Cong, Z., & Liang, D. (2016). Family structures, relationships, and housing recovery decisions after Hurricane Sandy. *Buildings, 6*(12), 14–29. doi:10.3390/buildings6020014

Neria, Y., & Shultz, J. M. (2012). Mental health effects of Hurricane Sandy: Characteristics, potential aftermath, and response. *The Journal of the American Medical Association, 308*(24), 2571–2572. doi:10.1001/jama.2012.110700

Parris, A. (2014). How Hurricane Sandy tamed the bureaucracy. *Issues in Science & Technology, 30*(4), 83–90. Retrieved from http://issues.org/30-4/how-hurricane-sandy-tamed-the-bureaucracy

Powell, T., Hanfling, D., & Gostin, L. O. (2012). Emergency preparedness and public health: The lessons of Hurricane Sandy. *The Journal of the American Medical Association, 308*(24), 2569–2570. doi:10.1001/jama.2012.108940

Rosenzweig, C., & Solecki, W. (2014). Hurricane Sandy and adaptation pathways in New York: Lessons from a first-responder city. *Global Environmental Change, 28*, 395–408. doi:10.1016/j.gloenvcha.2014.05.003

Schmeltz, M. T., González, S. K., Fuentes, L., Kwan, A., Ortega-Williams, A., & Pilar Cowan, L. (2013). Lessons from Hurricane Sandy: A community response in Brooklyn, New York. *Journal of Urban Health: Bulletin of the New York Academy of Medicine, 90*(5), 799–809. doi:10.1007/s11524-013-9832-9

Tran, C. N. T., Heller, M., Berger, A., & Habboushe, J. (2014). Hurricane Sandy: How did we do? Assessing a Manhattan hospital's response. *Frontiers in Public Health, 2*(90), 1–2. doi:10.3389/fpubh.2014.00090

VanDevanter, N., Kovner, C. T., Raveis, V. H., McCollum, M., & Keller, R. (2014). Challenges of nurses' deployment to other New York City hospitals in the aftermath of Hurricane Sandy. *Journal of Urban Health, 91*(4), 603–614. doi:10.1007/s11524-014-9889-0

Wagner, M., Chhetri, N., & Sturm, M. (2014). Adaptive capacity in light of Hurricane Sandy: The need for policy engagement. *Applied Geography, 50*, 15–23. doi:10.1016/j.apgeog.2014.01.009

The Yarnell Hill Fire (2013)

9

Figure 9.1 Prescott, Arizona/USA-July 2, 2013: An Extreme Fire Danger Sign is Pictured Outside of the Prescott Fire Station 1 in Prescott, Arizona. The Ongoing Yarnell Hill Fire Killed 19 Firefighting Hotshots
Source: Krista Kennell/Shutterstock.com

BRIEF OVERVIEW

Unless one has seen a wildfire, it is hard to imagine the destructive power it can unleash. Wildfires are quickly becoming commonplace throughout the country as evidenced by the increasing numbers of fires and loss of lives and property. Large fires have demonstrated the power of such events and often the human struggle involved in responding (McElreath, Jensen, et al., 2014). In this case study, we will focus on one such fire, which gripped the country on June 30, 2013, when the Yarnell Fire overran and killed 16 City of Prescott firefighters, who were members of the Granite Mountain Hotshots—an elite team of wildland firefighters. While the Great Peshtigo Fire of 1871 remains the single worst wildfire in history in both size and number of fatalities (see Chapter 1), fires of modern times have devastated areas of California, Texas, and Tennessee.

Throughout U.S. history, fires have been devastating. In 2017, about 9,000 wildfires burned an estimated 1,200,000 acres, destroying almost 11,000 buildings and killing over 46 people in California. In 2008, similar fires burned throughout Northern and Central California, which resulted in 23 deaths and burned over

1.15 million acres. In late November and early December 2016, the resort town of Gatlinburg, Tennessee and surrounding areas were consumed by a firestorm. The fire inflicted significant damage, claiming at least 14 lives, and damaging or destroying over 2,500 homes and businesses. Gatlinburg is dependent on tourists and seasonal homeowners. The fire devastated the local economy. The Yellowstone Fires of 1988 shut down the Yellowstone National Park for several months and destroyed more than a third of the park. At its peak, over 9,000 firefighters struggled to control the blaze to no avail. It eventually burned itself out when a winter snowstorm hit the area.

INTRODUCTION

The Disaster Unfolds

At 5:36 p.m., on June 28, 2013, lightning ignited a wildfire on the Bureau of Land Management land outside of Yarnell, Arizona. Situated about 80 miles northwest of Phoenix, Yarnell was an old mining town and home to approximately 700 residents. Due to the time of the initial lightning strike and size of the fire (less than an acre at the time), officials decided to hold off on sending out any manpower, believing the fire to be minimal and a low risk to the community. Also, safety concerns associated with putting firefighters into steep terrain on a hill at night led the Incident Commander to prepare for full suppression activities the following morning (Mockenhaupt, B., 2014).

Then, on the morning of June 29, firefighters deployed and soon believed the fire was in check. In the afternoon, however, winds significantly increased to about 22 mph, adding to the danger. Soon, the fire began to jump well-established containment lines. In response, the Incident Commander called for reinforcements, and requested a Type 2 Incident Management Team (IMT) to be deployed early the next morning. The IMT consisted of the Granite Mountain Hotshots based out of nearby Prescott. Overnight the fire grew and by morning, it was estimated to have affected 300–500 acres (Mockenhaupt, B., 2014).

On the morning of June 30, 20 members of the Granite Mountain Hotshots arrived on scene for a briefing. Within a few hours the Hotshots were traveling up the ridge on the southwest perimeter of the fire. For most of the day the fire continued to spread to the northeast, threatening the Model Creek and Peeples Valley areas where evacuations were under way. Around 3:50 p.m., the winds suddenly shifted and the fire began to aggressively push toward Yarnell in the southeast and on path toward the location of the Granite Mountain Hotshots team (Mockenhaupt, B., 2014).

When the winds shifted, the Granite Mountain Hotshots were the only crew remaining on the ridge, as others had already withdrawn to assist with the evacuation and protection of property in Model Creek and Peeples Valley. The field command on the ridge was aware that the team was located in "the black," an area that had already been burned with no fuel available to support additional fire response, and as a result, it was assumed they would stay there for safety. No one realized the crew left the black and began heading southeast just after 4 p.m. Then, at 4:30 p.m., the thunderstorm outflows rolled into the area and significantly increased the winds, further increasing the danger of the fire.

Eventually, the fire turned south and overran the Granite Mountain Hotshots around 4:42 p.m., killing 19 members of the team. The lone survivor, 21-year-old Brendan McDonough, had been serving as a lookout for the crew and at the time of the tragedy, was away from the team, hiking out of the area to move the crew's vehicles away from the impinging fire (Mockenhaupt, 2014).

Impact

By July 1, the fire had grown to more than 8,300 acres, and mandatory evacuation orders were issued for the Peeples Valley area. Additional firefights arrived bringing the force to about 400. The teams of firefighters struggled to establish some degree of containment. On July 2, first responders believed a degree of control had been achieved and the situation had begun to stabilize. It was believed the fire was about 8% contained with no additional growth over the course of the next 24 hours.

By the end of the day on July 5, the fire was almost 50% contained and no longer spreading. Residents of Peeples Valley were allowed to return to their homes on July 4. Four more days passed before residents of

Yarnell were able to return to their homes. On July 10, the fire was declared 100% contained. In the end, 127 buildings in Yarnell were lost and two in Peeples Valley. No lives other than those of the Granite Mountain Hotshots were claimed during the event (Wang, Hensley, Collom, Lee, & Clancy, 2013).

RELIEF AND FIRST RESPONSE EFFORTS

The Yarnell Hill Fire in west-central Arizona, just west of the town of Yarnell. The impacted area ran along a north–south mountainous ridge that featured elevations between 4,500 ft and 6,000 ft. The terrain was comprised of very steep ridges along with flat bottom areas in the valley comprised of boulders and other rock outcroppings. Chaparral brush was the dominant vegetation in the area and ranged from 1 ft to 10 ft in height, and the landscape was very dense in many areas. The Yarnell Hill Fire area had not experienced a wildland fire since 1966, making it very ripe and susceptible to wildland fire due to the significant vegetative growth (Nelson, 2017).

Conditions leading up to the fire included severe drought, coupled with very high to extreme fire danger as the area prepared to head into the summer monsoon season. Temperatures were very hot, and the humidity was very low. Storms moving into the area brought lightning, (the cause of the fire) along with unpredictable and quickly shifting winds. Such intense storms drastically affect fire activity.

Statewide, there were 37 active fires on the day the Yarnell Hill Fire began. Resource availability at the time of the fire was average for that time of year and planning level. Available on June 28 were resources that included two Type 1 IMTs, four Type 2 IMTs, three Type 3 IMTs, two Type 1 air tankers, two smokejumper aircraft, approximately 10 single-engine air tankers (SEATs), two Type 1 helicopters, two Type 2 helicopters, eight Type 1 interagency hotshot crews (IHCs), and Type 2 crew capability. Many of these resources would eventually be deployed to both the Yarnell Hill Fire and the Dean Peak Fire a few days later (Yarnell Hill Fire Report, 2013).

The Yarnell Hill Fire incident began on Friday, June 28, with several calls made to the Arizona Dispatch Center around 5:30 p.m., regarding a lightning-ignited fire west of Yarnell. Then, at 5:36 p.m., the dispatch center began documenting the fire at 5:36 p.m., and assigned the name *Yarnell Hill Fire* to the event. An Arizona State Forestry Division (AZSF) firefighter is responsible for coordinating the response to any lightning strikes in the area. The firefighter requested that Air Attack fly to the area around Yarnell, and evaluate the fire and to confirm there are no additional ones in the area. Air Attack had difficulty locating the fire, as there was very little smoke visible. But once it was located, Air Attack reported back that the fire was located in a boulder field with no vehicle access, and that it was less than half an acre in size. Another reconnaissance flight was also scheduled for the next morning (Yarnell Hill Fire Report, 2013).

At 7:40 p.m., the AZSF firefighter officially assumed command of the incident, and reported that the fire was less than half an acre, 80% contained, active in just one corner, and had a low spread potential with no homes or people in jeopardy at the time. With the risk of additional lightning throughout the night and the dangerous terrain, the Incident Commander opted to begin full suppression efforts the next morning. He also requested that two Type 2 Department of Corrections crews, a light helicopter, and an AZFS engine be ready for the following morning (Yarnell Hill Fire Report, 2013).

On the morning of June 29, first responders were preparing to address multiple fires, but the Yarnell Hill Fire was the only one burning as a result of the lightning strikes. The Incident Commander requested that two SEATs drop fire retardant until crews are able to access the area later in the morning. An order was placed for the SEATs, and drops began by mid-morning. At 11 a.m., the requested helicopter had transported seven firefighters to within a quarter of a mile of the ridge. The Department of Corrections firefighters hiked the rest of the way up to the ridge to establish the attack line. The SEATs and Air Attack made a couple of drops throughout the morning. Eventually, firefighters and air resources gained control of the fire for the time being. As a result, the Incident Commander released additional resources staged on site between 2 and 4 p.m., believing that the fire was well managed and under control (Yarnell Hill Fire Report, 2013).

Unfortunately, conditions rapidly changed within half an hour, and the Incident Commander called for the two SEATs and the Air Attack to return to operations at 4:10 p.m. The fire then jumped a road and quickly spread to 6 acres. The Incident Commander then called for additional resources at 5:30 p.m., which included a Type 1 heavy helitanker and large air tanker, along with an additional commander and IMT, fearing the fire would grow over the next 24–48 hours. High winds and incoming severe weather grounded the additional air support requested, leaving just the SEATs to continue dropping fire retardant (Yarnell Hill Fire Report, 2013).

Throughout the evening, the fire grew to more than 100 acres, and elevated the Incident Commanders request for a Type 2 IMT, rather than the previously requested Type 3 IRT. Dispatch sent the Type 2 IMT short team to the scene, and the Incident Commander notified the Yavapai Sheriff's Office to prepare to activate the reverse-911 system in the event that evacuations become necessary. The Incident Commander also requested additional resources for the next morning that included 14 engines, six water tenders, two Type 2 crews, two bulldozers, and numerous aircraft. He also requested that two structure group specialists be assigned to the north and south ends, along with three IHCs. Dispatch also ordered that the Blue Ridge IHC and Arroyo Grande IHC report to the scene, and became the third team deployed from the locally available Granite Mountain IHC based out of Prescott. That evening the fire continued to grow and spread, and 13 firefighters were left on the hill overnight while additional resources remain staged in the Yarnell and Peeples Valley areas (Yarnell Hill Fire Report, 2013).

During a 7 a.m. briefing on June 30, the Incident Commander and other command staff met at the Yarnell Fire Station to discuss the day's strategy, and to share the situational awareness report. The group noted that the Boulder Springs Ranch would make an excellent safety zone, and the command channel and communications plan were then finalized. The Granite Mountain IHC received its mission to establish an anchor point at the heel of the fire, using direct or indirect tactics as necessary. The Incident Command Post (ICP) was established at Model Creek School in Peeples Valley, and command was officially transferred to the incoming Type 2 Incident Commander (ICT2), at 10:22 a.m., upon receiving his briefing, which began at 9:30 a.m., in the ICP. The fire continued to rage and initial evacuation orders were given by the Sheriff's Office to those in the Model Creek and Double Bar A Ranch areas beginning around 10:45 a.m. By 1 p.m., the ICT2 had requested a Type 1 IMT as the fire transitioned from a Type 4 to a Type 1 incident in less than 20 hours (Yarnell Hill Fire Report, 2013).

In the meanwhile, the IHCs headed to their assigned areas. Immediately following a 7 a.m. briefing, the Granite Mountain IHC drove to Yarnell and stationed their vehicles in a clearing that was three quarters of the way down Sesame Street. The crew then convened and discussed a game plan to establish the anchor point and follow the black, or indirectly burn off the two-track road to halt the fire's advancement. Around 8 a.m., the Blue Ridge IHC arrived at the ICP and cloned their radios to frequencies in use at the fire and then moved to the Yarnell Fire Department for staging, and eventually were ordered to drive into the fire area around 10 a.m., and advised to use tactical frequency 3 (Tac 3) on their radios. The crew parked their carriers next to the Granite Mountain IHC vehicles on Sesame Street, and were able to stage themselves to observe the Granite Mountain IHC, while also working along the ridge and cutting lines for the two-track road (Yarnell Hill Fire Report, 2013).

A little before noon, leaders from the Blue Ridge IHC and Granite Mountain IHC met near the anchor point to discuss tactics, and agreed to use a Granite Mountain crewmember as a lookout. Subsequently, they also agreed on a spot at the bottom of the slope, and also discussed some radio communication issues, noting that some of the radios did not have appropriate tone guards, which was adversely affecting radio communications. Eventually, the Granite Mountain Lookout was dropped off at 12:39 p.m., where he situated himself in a good spot where the Granite Mountain IHC could clearly see him along with the advancement of the fire. The Blue Ridge IHC departed toward the Shrine of St. Joseph at a youth camp, while the Granite Mountain IHC resumed working a line along the fire's edge near the good black area. The Granite Mountain IHC then took a short lunch break, but shortly after 2 p.m., severe weather began to approach the area that included additional lightning strikes (Yarnell Hill Fire Report, 2013).

At approximately 3:50 p.m., the Air Attack radioed that the wind had significantly changed direction, and was heading toward Yarnell, and could possibly reach the town within 1–2 hr. The Granite Mountain Lookout also noticed the rapidly changing wind conditions, and radioed that he must reposition himself. He also determined that vehicles were in danger of being overturned, and radioed that he was returning to help move them from harm's way. At that same time, the Granite Mountain IHC team leader responded back that they were closely monitoring the fire and were in the black area, which meant that they were aware of their surroundings and located in a safe area. In addition, it was anticipated that they would stay in this area, but instead they began moving toward the ranch, which inevitably placed them in an area that became the point of no return, and would result in their deaths (Yarnell Hill Fire Report, 2013).

At approximately 4:40 p.m., circumstances dramatically changed for the worse for the Granite Mountain IHC. The team had left the black area, and headed toward the Boulder Springs Ranch without knowing that they were actually in close proximity to the fire. Shortly thereafter, the fire overtook them and the entire 19-member crew was killed. The Granite Mountain IHC was traveling through an unburned area of rough terrain loaded with chaparral brush and a fire moving between 10 and 12 mph. The crew had less than 2 min to deploy their shelters. Temperatures exceeded over 2,000 degrees as the fire swept through the area, and no one stood a chance to survive. Then, at 5:16 p.m., a Department of Public Safety helicopter was deployed along with ground crews to locate the Granite Mountain IHC. Due to heavy smoke, searchers were unable to locate the crew until 6:10 p.m. Upon further observation, the pilot/paramedic crew communicated back at 6:35 p.m., that no one had survived, and that the Granite Mountain IHC members had been in various stages of shelter deployment before being overrun by the fire (Yarnell Hill Fire Report, 2013).

Aftermath and Investigation into the Disaster

The loss of the 19 Granite Mountain Hotshots was the greatest loss of firefighters since 9/11, when 343 firefighters were killed in the collapse of the Twin Towers. A nine-member team of experts in wildland fires and safety arrived in Yarnell on July 2, with a mission to "understand what happened as completely as possible," to prevent a similar situation from ever happening again. The team conducted an investigation that lasted just

Figure 9.2 Prescott, Arizona/USA—July 3, 2013: A Memorial for the 19 Fallen Firefighters Stands Outside of Granite Mountain Hotshots Fire Station 7. Nineteen Firemen Died During the Yarnell Hill Fire

Source: Krista Kennell/Shutterstock.com

3 months, and the Forestry Division released its findings in a report on September 28, 2013. Sadly, there was no evidence of negligence or recklessness in the deaths of the firefighters. In fact, it was determined that an air tanker carrying flame retardant was directly overhead as the firefighters were engulfed in flames. The only significant findings were related to radio communications. For example, some of the radios were not programmed with the appropriate tone guards, and there was excessive radio traffic at the time the fire was encroaching on the team (Yarnell Hill Fire Report, 2013).

A separate investigation and subsequent report produced by the Industrial Commission of Arizona (which oversees workplace safety) pinned the deaths of the 19 firefighters on the state's Forestry Division after an investigation was completed by the state's Division of Occupational Safety and Health. The Commission levied a $559,000 fine against the AZSF on the grounds that they knowingly put the preservation of property ahead of safety instead of life, and should have pulled all crews out earlier (Pearce and Carcamo, 2013).

LESSONS LEARNED AND POLICY IMPACT

Among the lessons learned from the disaster is that it is difficult to undo a century of forestry mismanagement and lax building codes. Over the past few decades, funding to the U.S. Forestry Service, as well as state and local agencies and departments, has continued to be reduced, which has unfortunately laid the groundwork for the perfect storm. Despite repeated investigations, lawsuits, books, and even a movie, many questions have yet to be answered because no one knows for sure what the Granite Mountain Hotshots were thinking or doing after their last radio communication. All we do know is that many of the Hotshots who were killed did not even have the time to fully deploy their shelters. An elite and well-trained team should not have fallen victim to such a circumstance.

The real question lies in the responsibility to prevent something similar from happening again. Should we be asking our firefighters to brave 100-ft flames to protect ill-prepared communities? The National Incident Fire Command (NIFC) reported that over 70,000 communities across the United States, with around 40 million homes housing 140 million people, sit directly in the path of wildfires. According to Chuck Womack with the NIFC, "Wildfires are the only natural disaster we think we can control. If the federal government gave us all the money in the world, we'd still fail to control all the blazes" (Nelson, 2017). The harsh reality is that most homes today do not meet Wildland–Urban Interface (WUI) building codes, and fewer than 2% of America's fire-prone areas have completed defensible space work geared toward resiliency against such fires. When homes are built in forest areas that contain a myriad of vegetation, it poses danger for the next fire disaster. The fire destruction in Gatlinburg in 2016 clearly supports this notion.

Several key recommendations came out of the tragedy that were identified on page 122 of the 2013 Yarnell Fire Report, which was provided by the investigation committee. Among the most significant recommendations were:

1. The Team recommends that the State of Arizona review and possibly update its approach to mitigating wildfire threat to Arizona life and property. This could be modeled after communities such as Prescott, AZ; Santa Fe, NM; or communities in other states. This process could be a cooperative effort to reduce hazardous fuels and improve overall suppression efforts for communities that are at a high risk from wildfire.
2. The Team recommends that the State of Arizona review the statewide wildfire communications plan and program, as well as similar programs and plans in other states, for possible improvements (Fonseca & Dreier, 2013).
3. The Team recommends that the State of Arizona work cooperatively with its fire cooperators to develop a wildland fire staff ride for the Yarnell Hill Fire incident. The staff ride is a process of conveying the lessons learned from this incident for future fire leaders.
4. The Team recommends that the State of Arizona request the National Wildfire Coordination Group (NWCG) to review current technology that could increase resource tracking, communications, real-time weather, etc. For example, this may include GPS units and weather applications.

5. The Team recommends that the State of Arizona request the NWCG and/or Wildland Fire Leadership Council (WFLC) to charter a team of interagency wildland fire and human factors experts to conduct further analysis of this event and the wildland fire communications environment.
6. The Team recommends that the State of Arizona request the NWCG to develop guidance to identify at what point is it necessary to separate the Aerial Supervision Module (ASM) and Air Attack roles to carry out required responsibilities for each platform.
7. The Team recommends that the State of Arizona request the WFLC/NWCG to develop a brief technical tip for fire supervisors/agency administrators on the effective use of Very Large Air Tankers (VLATs). These are new, emerging fire suppression tools that the ground-based fire supervisors may be utilizing regularly in the future (Yarnell Hill Fire Report, 2013).

IMPLICATIONS FOR MODERN EMERGENCY MANAGEMENT

Today's emergency managers are required to address several issues related to wildfires. First and foremost, as part of the risk assessment and Comprehensive Emergency Management Plan—wildland fire potential must be addressed (McElreath, Doss, et al., 2014). Such proactive measures should include identifying fire-prone areas, as well as establishing evacuation protocols and communication measures. The processes of evacuation and effectively distributing warnings and orders in a timely manner are paramount in terms of risk management and the response phase of emergency management. As seen with communication failures in Gatlinburg, breakdowns will cause loss of life—property can always be replaced. Secondly, emergency managers must advocate for stronger codes in areas affected by the WUI. More stringent building codes that take into account the environment can help lessen the impact of fires if they do occur. Attempting to mitigate firestorms by reducing fuel is one of the only means of reducing risk in light of the decades of neglect by the forestry services.

SUMMARY

Perhaps Don Falk, a forest-and-fire ecologist at the University of Arizona, said it best: "Fire is inevitable. You can defer it, but it's a pay-me-now-or-pay-me-later scenario" (Mockenhaupt, 2014). We are paying for our neglect to forests and their proper maintenance and have been avoiding the inevitable for many decades. Built-up fuel is waiting to burn and it is not a matter of "if," but "when" the propensity of such a man-made disaster will occur. Compounding the problem is climate change—years of drought turn that fuel into tinder; fire season is starting sooner and going longer; and forests destroyed by fire are not returning as they once were leaving grass and shrubs in their place ready to burn. It simply means that once fires start, they burn hotter and longer than they ever have before, leaving our wildland firefighters stretched thin season after season.

In an effort to combat this growing threat, we must address the strategies we use to maintain our forests and invest more funding in the forestry service to help mitigate the wildland fire problem facing the United States. Additionally, codes must be developed to restrict housing development that encroaches upon the WUI. As a result of the Yarnell Hill Fire, we have fortunately not yet seen another significant loss of wildland firefighters (Burt, 2013). However, history can and will repeat itself if we do not address the biggest issue at hand—the fuel load available for fires to feed on. You can have all the technology in the world at your disposal but Mother Nature can be very unpredictable in a wildfire environment.

REFERENCES

Burt, C. (2013, August 27). The worst wildfires in U.S. history. *Weather Underground*. Retrieved from https://www.wunderground.com/blog/weatherhistorian/the-worst-wild-fires-in-us-history.html

Fonseca, F. & Dreier, H. (2013, September 28). Communications problems cited in deaths of 19 Arizona Hotshot firefighters. *Associated Press*. Retrieved from http://usnews.nbcnews.com/_news/2013/09/28/20734496-communications-problems-cited-in-deaths-of-19-arizona-hotshot-firefighters?lite

McElreath, D. H., Doss, D. A., Jensen, C. J., Wigginton, M. P., Nations, R., Van Slyke, J., & Nations, J. (2014). *Foundations of emergency management.* Dubuque: Kendall Hunt.

McElreath, D. H., Jensen, C. J., Wigginton, M. P., Doss, D. A., Nations, R., & Van Slyke, J. (2014). *Introduction to homeland security.* Boca Raton: CRC Press.

Mockenhaupt, B. (2014, June). Fire on the mountain. *The Atlantic.* Retrieved from https://www.theatlantic.com/magazine/archive/2014/06/fire-on-the-mountain/361613

Nelson, M. (2017, August 29). Chapter 15—The Yarnell Hill Fire. Retrieved from http://www.paysonroundup.com/chapter---the-yarnell-hill-fire/article_d68c2ef1-56d1-5d2b-84f7-77829832f71b.html

Pearce, M., & Carcamo, C. (2013, September 28). No negligence in deaths of 19 Arizona firefighters, report says. *LA Times.* Retrieved from http://www.latimes.com/nation/nationnow/la-na-nn-yarnell-fire-report-20130928-story.html

Wang, A., Hensley, J., Collom, L., Lee, M., & M. Clancy. (2013, July 1). Yarnell Hill Fire continues to burn out of control. *The Republic.* Retrieved from http://archive.azcentral.com/news/arizona/free/20130701yarnell-hill-firefighters-die.html

Yarnell Hill Fire Report. (2013). Completed by the Serious Accident Investigation Team. Retrieved from http://wildfiretoday.com/documents/Yarnell_Hill_Fire_report.pdf

Yellow Fever Epidemic (1800s)

10

Figure 10.1 Yellow Fever Mosquito
Source: USDA

BRIEF OVERVIEW

Proving much more deadly than cholera in the United States, yellow fever was one of the most feared diseases of the 19th century, frequently resulting in the death of those affected. Before 1900, the cause of the disease was generally a mystery and the treatments, at best, were uncertain. In July 1878, an outbreak of yellow fever was reported in New Orleans, and over the next several weeks, it moved up the Mississippi River infecting Natchez, Vicksburg, Greenville, Memphis, and ultimately extended past Cairo, Illinois. This chapter reviews the origin, transmission, and impact of the yellow fever epidemics with specific attention to the epidemic of 1878 among the communities along the Mississippi River. A commentary is provided regarding emergency management and policy considerations for contemporary times.

INTRODUCTION

Natural and man-made calamities often occur with speed, surprise, and violence. In some cases, infrastructure may be damaged or obliterated with low numbers of casualties. Conversely, some emergencies may exhibit a reverse scenario in which infrastructure is undamaged, but many lives may be lost. These notions introduce the concept of pandemic disease. Modern medicine has a variety of remedies, cures, and preventives whereby a variety of formerly dangerous ailments may be avoided or cured. However, despite medical advancements, some ailments are incurable and deadly, such as the Ebola virus. Dealing with such hazards poses a variety of risks and dangers, ranging from the extreme of death to the potential of incapacitation for some period.

Dealing with modern infectious agents incorporates the learnings of history. Modern society incorporates a variety of preventive means for avoiding or mitigating infectious calamities. Some methods may be intangible, such as policy and regulation. Others are tangible, such as inoculations. Regardless of modality, emergency managers who implement such methods must acknowledge that ailments may spread through contact directly with other people or indirectly through some intermediary, such as encountering insects or animals that may be carriers of infectious agents.

The modern means of addressing infectious agents emerged and matured through time. However, such emergence and maturation originated from necessity and societal duress wherein much suffering occurred before advancements developed within the medical arts and sciences. After all, the polio vaccine resulted from much experimentation, trial, and error. Before its introduction in 1955, thousands suffered annually within American society (Department of Health and Human Services, 2017a). Thus, societal need and suffering influenced the motivation toward crafting an appropriate vaccine.

Many ailments have been practically eradicated during modern times. People in contemporary society, on average, live approximately 70 years. However, before the developing and inventing of modern vaccines, many infectious agents unimpededly ravaged society. Over a century ago, experiencing such illnesses was often deadly for not only victims, but also for first responders and medical personnel. A classic example was yellow fever.

The yellow fever virus decimated societies throughout the 18th and 19th centuries. Although a vaccine now exists to counter the yellow fever virus during modern times, its experiences influenced historical emergency management in a variety of contexts. For instance, it necessitated the crafting of government organizations, regulations, and policies as protective measures to safeguard society from the perils of yellow fever. One such example includes the earlier considerations of American societal quarantine which is traced to the fear of yellow fever. Specifically, in 1878, the Quarantine Act was implemented as a response to the pervasiveness and deadliness of yellow fever. Quarantine is still used during modern times to protect society against the risks of an array of communicable illnesses. Given these notions, this chapter examines the history of yellow fever, and its modern considerations within the context of emergency management.

The Setting

Yellow fever outbreaks instilled a special fear into the Americans of the 1700s and 1800s. It was a disease, shrouded in mystery, whose cause, method of transmission, and cure had yet to be discovered. When contracted, yellow fever typically resulted in a very high mortality rate. From its first appearance in the Americas in the early 1600s, until an effective response to the disease was available, yellow fever wiped out families and devastated communities.

Yellow fever virus originated in Africa and for centuries, it remained on that continent where generations were exposed to, and struggled with, the disease. There, after generations struggled with the disease, the population eventually developed some degree of resistance to the virus. By the early 1600s, yellow fever would be on the move from Africa, transported on the ships involved in the expanding international slave trade. By 1635, yellow fever arrived in the Caribbean where it became endemic (Arden, 2005). From the Caribbean, it was a matter of time before yellow fever moved from the islands into what is now the United States. For approximately 200 years, from 1635 until the early 1900, yellow fever endangered various parts of North America.

Ships and infected crews and passengers ensured the spread of the disease and the growing American ports would become the scene of outbreaks. Before 1900, the cause of the disease was generally a mystery and the treatments, at best, were uncertain (University of Illinois, 2014).

The first recorded outbreak of yellow fever in British North America occurred in Boston, Massachusetts in 1693. In 1699, the disease reemerged in Charleston, South Carolina and Philadelphia, Pennsylvania. In 1702, yellow fever appeared in New York City resulting in the deaths of an estimated 570 people. In 1793, Philadelphia again was the site for an outbreak of yellow fever resulting in the deaths of approximately 4,000 people. Outbreaks reoccurred in New York in both 1790 and 1798, and also at New Orleans in 1796 (Arden, 2005).

The Disaster Unfolds

New Orleans played a unique role in yellow fever and its spread throughout the South and the ports along the Mississippi River. From its first appearance in the Crescent City in 1796, yellow fever would visit New Orleans on an almost annual basis until 1905 resulting in as many as 30 major epidemics and in over 41,000 deaths. Yellow fever proved to be increasingly erratic, mysterious, and terrifying. In the great New Orleans epidemics of 1853 through 1855, an estimated 13,570 people out of a population of 157,000 died from the disease (Carrigan, 1994). Though an accurate count of those who fell victim to yellow fever in New Orleans and the surrounding area may never be known, it is believed that between 50,000 and 100,000 fatalities resulted from the disease (New Orleans Public Library, 2014).

The New Orleans 1853 epidemic would prove exceptionally deadly. The City's Charity Hospital, whose resources were stretched to the limit, was admitting over 100 patients per day. When beds were no longer available, patients were placed on the floor. Thousands of residents began to flee the city. By the time the epidemic ended in December 1853, approximately 7,000–11,000 had died, an estimated 10% of the city's population (McKiven, 2007). New Orleans would not just prove to be the scene of yellow fever outbreaks; it would also be the unintended port from which it would move up the Mississippi River and spread into the populations of the river ports.

In 1878, yellow fever proved especially deadly. Though the actual origin of the 1878 epidemic will remain under speculation, one of the theories is that refugees escaping the ongoing violence in Cuba brought the disease into the U.S. ports along the Gulf coast (Suddath, 2010). Other theories point to the arrival into New Orleans from Havana in late May 1878, from an infected ship and crew. At the time, it was accepted protocol for ships entering U.S. ports to be inspected and if necessary quarantined. For the owners and captains of ships, quarantine could well turn a profitable voyage into a financial disaster if the goods carried were ruined as the ship awaited clearance. Upon entry into the Mississippi River, the ship was inspected by a quarantine physician, and one crew member was diagnosed as having intermittent fever (malaria) and placed in an infirmary. Yellow fever was present in New Orleans and it would soon move up the River, and with it death to thousands.

Impact

Sitting on the east bank of the Mississippi River just north of the Tennessee and Mississippi State line, Memphis, Tennessee would experience the effects of yellow fever throughout the 1800s. Yellow Jack visited Memphis with major outbreaks in 1855, 1867, and 1873, and with each outbreak, the disease took an increasingly greater toll. The 1873 epidemic infected over 5,000 and resulted in 2,000 deaths in Bluff City.

The city of Memphis lies in the perfect setting for yellow fever—the natural warm climate and abundant fresh water allowed mosquitoes to thrive. But other factors contributed to making Memphis the perfect location for yellow fever. At the time, much of the area around the city was swampy and the city itself had the reputation as one of the filthiest and mostly foul-smelling cities in the nation, with open sewers that further provided breeding grounds for mosquitoes.

In 1878, as in the past when the disease appeared down the Mississippi River, reports of the yellow fever outbreak quickly moved up the river. Memphis City officials realized from previous epidemics that if the disease appeared in New Orleans, Natchez, and Vicksburg, then Memphis would soon be in danger, and that steps must be taken to protect the city. Soon thereafter, Memphis officials blocked all cargo from New Orleans, but trade fueled the economy of the city and business leaders quickly pressured for the reopening of trade. Through the lens of history, beginning its journey in New Orleans, yellow fever worked its way upward along the Mississippi River, and wreaked havoc and death throughout communities along the river's route.

On July 18, 1878, the steam towboat *John D. Porter*, which newspapers later referred to as a floating epidemic and the Ark of Death, departed New Orleans moving up the Mississippi River with several cases of yellow fever onboard (Crosby, 2006). When the boat arrived at Vicksburg, two ill crewmen were taken to the local marine hospital. Soon, the *Porter* continued upriver toward Memphis. About 50 miles north of Vicksburg, a crewman died, prompting the boat to return to Vicksburg where it was learned that the two sick crewmen who had been left at the Marine Hospital in Vicksburg had died from yellow fever. The *Porter* left Vicksburg, continuing its journey toward Memphis. In an effort to stop the outbreak, the captain ordered all of the bedding used by the dead crew thrown overboard and the boat disinfected.

The *Porter* appeared little different from the hundreds of boats that annually docked in Memphis when it arrived on July 30, 1878. Prior to docking, the boat was boarded by the port physician who questioned the captain about yellow fever, prior to clearing the boat and crew for docking in Memphis. The captain denied that yellow fever was, or had been, on board, but that some crewmen had become sick and had been left at ports down river due to other causes such as heat prostration and malaria. The physician conducted his inspection and recognized some of the signs of yellow fever, stains of black vomit on the wooden deck. He refused to permit the boat to land and soon the *Porter* continued up river but the damage had been done, Memphis was again exposed to yellow fever. Soon, the first cases of the 1878 epidemic would appear. Over the next 2 months, it was estimated that about 25,000 would flee the city, approximately 17,000 became infected, and over 5,100 died in Memphis alone. Of the thousands who would flee, many took the infection with them to towns inland, effectively spreading the epidemic further (Suddath, 2010).

In Memphis, as in many towns, many who contracted yellow fever were quarantined in an effort to prevent the disease from spreading. Often, they were made to wear yellow jackets or specific marking as a means of identification (History Channel, 2014). Soon, thousands would fall ill and hundreds would also lose their lives. It was estimated that 200 people died every day through September. Disposing of the dead proved a challenge because they were often abandoned. In some cases, entire families were lost because of the overwhelming rapidity of the spread and progress of the disease. The dead were found throughout the city, in abandoned buildings and even public parks. Even responders who provided aid were in danger, often contracting the illness and dying (History Channel, 2014). Thousands died and even more fled the city, some permanently. Declining property values, a generation of poor fiscal management, and another yellow fever outbreak in 1879, which infected 2,000 and caused an estimated 600 deaths, led the city into bankruptcy and the loss of the city's charter in 1879 (Tennessee Encyclopedia of History and Culture, 2014).

Containing the Disaster

In 1878, yellow fever was thought to be caused by bad air in hot and humid lowland areas. As a result, many of those who had the means to escape areas affected by the disease moved to areas they believed "safer," not realizing many took the virus with them. Holly Springs, Mississippi is just one example how the disease traveled from the river areas into communities of North Mississippi. Holly Springs is located 30 miles southeast of Memphis in the hills of North Mississippi and many believed to be immune from the dangers of the disease. Many medical authorities believed yellow fever would not occur in locations 500 ft or more above sea level. As a result, many towns away from the river delta and in the hills were not alarmed as their town filled with refugees from the disease. In 1878, it would prove to be an act of kindness for which they were to pay very dearly. Since there was no hospital in the city, the existing Marshall County Courthouse, at the center of Holly Springs' square, was used as a hospital during the epidemic (City of Holly Springs, 2014). Similarly, churches were also used as makeshift hospitals (City of Holly Springs, 2018). Figure 10.2 shows a church converted into a hospital.

After the epidemic had taken over surrounding cities, a Board of Health was assembled in Holly Springs. The board recommended that Holly Springs be quarantined in order to prevent further spread of the disease. However, it was subsequently voted down by the mayor and alderman who believed that the altitude

Figure 10.2 Display at the Yellow Fever Martyrs Church and Museum
Source: Library of Congres

was too high for a contagious disease to spread. They also based their decision on the sentiment that it would be cruel to turn away those who were trying to flee from infected areas. As a result, the disease spread into Holly Springs, and 304 deaths were reported out of 1,440 cases. The fatalities included a local townsman and "chairman of the relief committee," W. J. L. Holland, who sacrificed his living quarters to those stricken with the disease. His home is now a historical marker called the Yellow Fever House (Haines, 1997).

For most communities affected by the disease, local resources were quickly exhausted or overwhelmed. Even as Memphis struggled with the epidemic, on August 9, 1878, Dr. D.R. Brown, secretary of the Memphis Board of Health, went to Grenada to investigate reports that yellow fever had reached the city arriving with those running from areas already stricken. The small town of Grenada would be another of many Mississippi inland communities to experience death. Before the epidemic would leave the town, Grenada, with a population of 2,000, recorded 367 deaths (Leverette, 1999).

Across the river from Memphis, in Arkansas, yellow fever reached epidemic stages by October. One of the hardest hit communities was Helena, Arkansas where hundreds fled the city in an effort to escape. City officials of Little Rock, Arkansas placed health inspectors at the train stations in an attempt to prevent ill passengers from departing their trains into the city. It soon became apparent that the disease was taking a deadly toll in towns that had traditionally been spared previous epidemics. In response, many communities established quarantine lines staffed with armed citizens who guarded their communities with shotguns, rifles, and pistols. Commerce in the states affected was disrupted. Business deteriorated to the point in Mississippi that the already fragile economy, still trying to recover from the War Between the States in many parts of the state, collapsed. In the impacted towns and cities, the number of dead quickly increased. As a result, individual burials were in some cases abandoned and bodies were placed in mass graves. A wide variety of aid and charity was desperately needed.

In the midst of the death, medical volunteers risked their lives, and in some cases lost their lives in an attempt to render aid and assistance to the victims of the epidemic. The Howard Society, formed in Norfolk, Virginia during the 1855 yellow fever epidemic, provided funds, supplies, and medical personnel across the stricken area (The New York Times, 1878). Many of the Howard Society volunteers fell victim to the disease, some of whom died.

As the communities along the lower Mississippi River Delta struggled with the disease, the *Porter* continued its journey, docking at river ports, to include Cairo, Illinois. The location of Cairo at the confluence of the Mississippi and Ohio Rivers made the city a major port during the era of riverboats. Upon the arrival of *Porter*, the boat was inspected by the mayor of Cairo, the city's quarantine physician and its health officer for signs of yellow fever. The captain denied that the presence of yellow fever was on board, and claimed no one had been sick after leaving Vicksburg. Upon inspection, the crew appeared to be in good health, the boat was in good sanitary condition, and the vessel was given permission to land (Kohn, 2008). Not long after the arrival of the *Porter*, the first cases of yellow fever appeared in the city.

Though the *Porter* is seen as a major factor in the initial spread of the epidemic, once the yellow fever outbreak occurred, it spread rapidly by people fleeing from infected areas. Cairo was such an important transportation center; steamboats and railroads arrived and departed daily, many carrying the virus. In an effort to stop the disease, the Cairo Board of Health intensified quarantine measures.

Local newspapers such as the *Cairo Daily Bulletin* and the *Cairo Evening Sun* published conflicting reports on the likelihood of the spread of the disease. On September 8, 1878, Thomas Nally, the editor of the *Cairo Daily Bulletin,* fell victim to the disease. Nally died 4 days later and with his death, fear swept the city. Thousands fled Cairo, many leaving their belongings behind in the rush to seek safety. Eventually, about one-third of the population of Cairo evacuated the city. The spread of the disease continued until the first frost (University of Illinois, 2014).

Assessing the Disaster

By the end of the epidemic, the disease had inflicted a major toll along the Mississippi River, and to a lesser extent, the Ohio River. From New Orleans past Memphis and Cairo, east to Alabama, and west to Louisiana, approximately 20,000 people in the southeast had died and another 80,000 had survived infection. Over 16,000 cases of yellow fever were reported in Mississippi; there were 4,118 known dead with 1,149 deaths in Vicksburg alone (Nuwer, 2009). At least 2,000 died in Memphis; New Orleans lost more than 4,000 people. Hickman, Kentucky, a community of 1,200 people at the northern end of the epidemic, lost 150 of its citizens, and five of the town's six physicians.

In Memphis, over half of the residents of the city fled eventually in the hopes of escaping the illness (Shapiro, 2008). Within 3 days, between approximately 20,000 and 25,000 people departed Memphis (Shapiro, 2008). Not everyone had the means to flee quickly. Poor residents, many of whom were black or immigrant, remained in Memphis (Shapiro, 2008). Approximately 19,000 individuals remained within the city (Shapiro, 2008). Within this total, approximately 17,000 people became ill, and about 5,000 perished (Shapiro, 2008).

The yellow fever epidemic also created an economic disaster for the areas it touched. Much of the Mississippi River region relied on the production, sale, and distribution of cotton to both domestic and foreign consumers. The quarantine of communities and the restrictions upon travel resulted in significant delays in the transportation of cotton to its various markets. Additionally, the impact of the disease incapacitated a large number of the region's labor force.

Realizing the danger of the disease, some members of the international medical community dedicated themselves in attacking the disease itself. In 1881, Cuban physician Dr. Carlos J. Finlay set forth the hypothesis that was to be later confirmed that the *Aedes aegypti* (Linnaeus) mosquito transmitted the yellow fever virus. Five years later in 1886, his research and findings were published in *The American Journal of the Medical Sciences*. Though at the time many questioned his findings, his work would ultimately be proven correct, thereby eventually leading to removal of yellow fever as one of the deadliest illnesses in the United States.

Figure 10.3 Yellow Fever Vaccination
Source: Antonio Scorza/Shutterstock.com

Within two decades of Dr. Finlay's work, Dr. Walter Reed of the United States Army obtained the experimental results that confirmed Finlay's theory (Carrigan, 1994). The initial research by Finlay, and further work under Reed concerning yellow fever, also proved to be a key element in making the construction of a canal across Panama practical. Renowned Harvard scientist, Max Thieler developed the most effective vaccine against yellow fever in 1937, and his vaccine became the world's standard. This achievement won him the Nobel Prize in Medicine in 1951 (Norrby, 2007).

Because of substantial medical advances, the days of yellow fever were numbered in the United States. The last yellow fever epidemic in the United States occurred in New Orleans in 1905 (Patterson, 2009; Savitt & Young, 1991). By the late 1930s, an effective yellow fever vaccine was developed and distributed, thereby diminishing the threat of illness. Despite the successfulness of a vaccine, yellow fever was never completely eradicated. It exists in various locations globally. Figure 10.3 shows modern yellow fever vaccination methods in developing nations.

RELIEF AND FIRST RESPONSE EFFORTS

The yellow fever epidemic ended with the first frost in October 1878. The deaths in the lower Mississippi Valley, and throughout the southeast, made the yellow fever epidemic of 1878 one of the worst medical disasters in the history of the United States. In July 1879, another siege of yellow fever occurred, but it was incomparable to the impact of the disease the year before (Kohn, 2008).

Attempts to mitigate the effects of the disease were varied. Within hours of the death of the first reported victim in Memphis, her body was burned (Shapiro, 2008). After the illness permeated the populace, house to house searches revealed "abandoned children amid the rotting corpses of their parents" (Shapiro, 2008). Because of ignorance regarding the origin of the disease, its cause was attributed to "bad air" (Shapiro, 2008). As a safeguard against the perceived "bad air," Memphis residents continuously maintained burning fires, and boarded their windows (Shapiro, 2008). Following the death of an infected individual, their beds and clothing were burned among the streets of Memphis (Shapiro, 2008). Despite such efforts, yellow fever continued to ravage the populace until the coldness of the next winter abated its spreading (Shapiro, 2008).

From the epidemic, steps were taken to combat conditions that were suspected of contributing to the spread of the disease. In cities such as Memphis, open sewers and privies were cleaned, thereby destroying the breeding grounds for mosquitoes. Although it was unrealized at the time, such actions contributed to deterring further major epidemics (History Channel, 2014). Changes occurred within the states bordering the city of Memphis. For instance, the 1878 and 1879 yellow fever epidemics prompted Arkansas to create the Arkansas State Board of Health (Gann, 2011).

In Mississippi, the state mandated that discovering and identifying the initial case of any contagious or infectious illness was of the "highest importance," especially instances of "yellow fever, dengue, smallpox, cholera, diphtheria, and scarlet fever" (State of Mississippi, 1900, p. 126). State polices incorporated the use of medical inspectors to achieve these goals (State of Mississippi, 1900, p. 126). These policies also mandated the isolating and observing of infected individuals (State of Mississippi, 1900, p. 126). They also mandated that "no common carrier shall accept for transportation" any infected individuals or "any infected article of any character whatsoever" thereby constraining travel and mitigating the transmission of disease (State of Mississippi, 1900, p. 126). Even the apartments of infected individuals and any surrounding apartments, including "articles therein," were also to be disinfected (State of Mississippi, 1900, p. 126).

AFTERMATH AND INVESTIGATION

The effects of yellow fever were also observed and acknowledged nationally. In 1879, Congress established a National Board of Health, which functioned from 1879 until 1883, to set rules regarding quarantine (Duffy, 1993). A decade later Congress passed the Immigration Act of 1891 which barred the entry into the United States of persons with contagious diseases (Tyner, 2006). On April 26, 1878, President Rutherford B. Hayes signed into law the Quarantine Act of 1878. The purpose of this act was to prevent the introduction of infectious diseases into the United States from foreign ports.

This legislation was enacted in response to the yellow fever outbreak in the Caribbean earlier in the year and the concern that individuals fleeing the infected areas would attempt to enter the United States. This act charged the Marine Hospital Service with the responsibility to stop disease from coming ashore via sailors from ships. The Marine Hospital Service was the first publicly funded healthcare and disease prevention federal agency in the United States created in 1798. It was tasked, along with providing healthcare for maritime personnel, to monitor infectious diseases threatening the nation. With the 1878 legislation, the Marine Hospital Service gained increased power to quarantine ocean and river traffic.

LESSONS LEARNED AND POLICY IMPACT

During modern times, yellow fever exists primarily among various regions in Africa and South America. The yellow fever virus substantively exists within subtropical and tropical environments (Centers for Disease Control and Prevention, 2014). Humans are infected via bites from infected mosquitoes (Centers for Disease Control and Prevention, 2014). The severity of the illness varies from instances of "liver disease with bleeding" to instances of a "self-limited, febrile illness" (Centers for Disease Control and Prevention, 2014).

Typically, during modern times, hospitalization occurs in conjunction with yellow fever (Department of Health and Human Services, 2017b). Certainly, an instance of yellow fever may result in death. Approximately 25% of infected individuals will most likely perish from the disease (Western Cape Government, 2014). No certain treatment methodology exists for yellow fever (Western Cape Government, 2014). Instead, the treatment of patients who have the illness is based on the exhibited symptoms one demonstrates when ill (Western Cape Government, 2014). Diagnosing yellow fever involves symptomology, laboratory tests, findings physically, and travel histories of infected individuals (including the chance of having been exposed to mosquitoes that were possibly infected; Western Cape Government, 2014). Methods to prevent, quash, or diminish yellow fever infections involve the use of insect repellants, vaccinations, and the wearing of "protective clothing" (Western Cape Government, 2014).

As of today, yellow fever still remains a consideration of American health. The City of Houston, Texas represents a potential nexus for introducing yellow fever within the United States. The city exhibits numerous airline flights between the United States and Nigeria, and is home to the world's largest quantity of "Nigerian expats". Their presence within the United States provides an expert skill set within the Houston energy and oil industries. Many people travel between American and Nigerian destinations. Although many Americans refuse yellow fever vaccination when traveling to Nigeria, the African nation has urban locations that represent substantially higher risks for yellow fever. The combination of travelers who return to the Houston area from such African areas of risk, the high amounts of mosquitoes in Houston, and regions of poverty-stricken housing among Houston communities endangers the city and forms a basis for it being considered a risk for any emerging instances of yellow fever.

Although yellow fever in American society is rare, it still represents a potential hazard among various geographic areas. It is spread through the *A. aegypti* mosquito (University of Florida, 2014). This mosquito was introduced among American environments through shipping that occurred during the "exploration and colonization" of the new world by Europeans (University of Florida, 2014). These insects flourish among "unused flowerpots, spare tires, untreated swimming pools, and drainage ditches," and are abundant among urban areas given the close proximity of humans (University of Florida, 2014). This insect population represents a biological network whereby yellow fever may be transmitted geographically and interjected within the human populace, especially among urban areas.

IMPLICATIONS FOR MODERN EMERGENCY MANAGEMENT POLICY

Given the reality of the existence of yellow fever during modern times, societies cannot take its potential lightly. Although vaccines exist to counter the potential yellow fever threat, one death is too many. Thus, within the context of emergency management, modern organizations and governments may consider policy as a method of diminishing the possibilities of yellow fever incidents.

Although the cause of yellow fever is known, and preventive measures exist, the virus remains dangerous. Individuals traveling overseas, to or from destinations that pose a high yellow fever risk, may become carriers of the virus. Economic activities, such as international shipping and trade, present opportunities for infected mosquitoes to be transited and introduced within American society. Yellow fever represents an influential factor for crafting various policies involving macro and micro perspectives. Such policies exist both domestically and internationally. Both macro and micro perspectives of policy exist to address considerations of yellow fever and similar threats. Macro-level policies affect the entirety of nations and regions whereas micro-level policies affect the lowest levels of organizational or societal structuring.

Macro-Level Policy

Various policies exist among nations where yellow fever endangers the populace. For instance, yellow fever risk influences travel policies among African governments. In South Africa, before entering the country, certain travelers must present evidence of vaccination (Republic of South Africa, 2014). A "valid yellow fever certificate" is required by "all citizens and non-citizens over one year of age" who have either an origin from a nation that exhibits a high risk for yellow fever or who transited through an intermediary location within a nation that exhibits a high risk for yellow fever (Republic of South Africa, 2014). The Western Cape Government enforces similar policies. Although it requires the presenting of a yellow fever certificate, exemptions are made for infants less than 1 year of age and for individuals who are allergic to eggs (Western Cape Government, 2014). Otherwise, compliance with governmental policy is mandatory. Failure to comply may result in the denial of entry into the nation (Western Cape Government, 2014). Being uncompliant may also result in being quarantined for a maximum of 6 days or until a yellow fever certificate is validated (Western Cape Government, 2014).

Micro-Level Policy

Micro-level environments also exhibit yellow fever policies. A stringent example is found within the policy documents of the University of California, San Francisco (UCSF) campus. The UCSF policy mandates that all laboratory personnel working directly with and individuals working indirectly near environments housing the yellow fever virus either receive a vaccination or "sign a declination form" (University of California, 2014). The policy is also applicable for personnel who risk "exposure to animal or animal waste" produced by laboratories that work with the yellow fever virus (University of California, 2014). Monitoring occurs to ensure compliance with these policy tenets (University of California, 2014). Any compliance failure may result in individuals having their laboratory privileges either withdrawn or suspended (University of California, 2014).

EMERGENCY MANAGEMENT CYCLE

Both macro-level and micro-level policies are relevant within the phases of the emergency management cycle. The crafting of both forms of policy obviously resulted from localities and regions experiencing the devastating effects of yellow fever. The existence of such policies represents a continuous preventive mechanism that attempts to avert any instances of the yellow fever illness or to quash any spreading of the illness should an outbreak occur. Thus, the necessities of presenting yellow fever certificates when traveling in Africa and the Californian laboratory protections are preventive measures. Certainly, vaccination, the foundation of the African policies, is a preventive measure.

There exists no specific treatment regimen for addressing instances of yellow fever (State of Florida, 2014). Therefore, mitigating the effects of any yellow fever outbreak may be difficult. Individually, at best, the uses of "supportive care and treatment of symptoms" are intended to diminish the effects of the illness (State of Florida, 2014). The African use of quarantine is a mitigation consideration because infected individuals are not exposed to mosquitoes that could become infected parasitically, thereby infecting and endangering other humans. Thus, the quarantine practice also represents a form of prevention.

Based on this realization, when crafting and maintaining policies, government entities and commercial organizations may consider the risks of yellow fever with respect to the emergency management cycle. The phases of the cycle consist of preparedness, mitigation, response, and recovery (McElreath, Doss, et al., 2014). Regarding yellow fever risk, examples of such notions are summarized as follows:

Preparedness—Yellow fever vaccine quantities must be maintained at acceptable levels. Entities whose personnel must travel overseas to high-risk areas should be vaccinated before traveling. If a waiver is medically necessary, then personnel should be examined for symptoms or the presence of the virus within their blood. Among localities, insecticides may be used broadly to control mosquito populations. Potential habitats for mosquitoes should be further neutralized (e.g., removal of standing water, etc.).

Mitigation—Quarantine may be used to ensure that infected individuals are unable to infect others via mosquitoes or human waste. During outbreaks, vaccinations should be given to increase the number of individuals who are insusceptible to the disease.

Response—Emergency action plans or incident response plans should include yellow fever outbreaks as potential hazards. Organizations must specify their specific actions that are to be performed should an outbreak occur. Similar to preparedness, response may also involve the use of measures for controlling insect habitats and populations.

Recovery—As with any disaster, each organization or locale should conduct an after-action evaluation to determine the strengths and weaknesses of its preparedness, mitigation, and response initiatives. Weaknesses should be addressed to ensure that they are neutralized as best as possible. Thus, when some sense of normalcy is restored, the entity is better prepared to either avert or experience similar happenings.

SUMMARY

Modern American society is basically unfamiliar with the dangerousness of yellow fever. Thus, for many, it is an unfamiliar concept and even the term may be unrecognized among individuals. However, yellow fever is a deadly illness whose presence is a viable concern of overseas governments and among American locations.

Given the advent and proliferation of globalism and its created dependencies within the American economy, yellow fever should neither be ignored nor lightly considered by government factions, organizations conducting business internationally, or individuals who travel to, from, or through locations that are high risks for the disease. Thus, some risk for exposure to the disease exists.

Since the yellow fever plagues of the 19th century, many advancements have occurred scientifically, politically, societally, and technologically that contribute toward protecting and insulating American society from the ravages of pestilences. Modern Americans are well conditioned to acts of mass violence that are presented among television news reports and newspapers. Modern Americans are also well conditioned to observing reports about acts of God, ranging from floods and hurricanes to blizzards and earthquakes, which affect mass populations. However, societally, Americans are much less accustomed to reports or experiences of others visibly, painfully, and quickly perishing, in mass quantities, because of infectious diseases.

Despite the unfamiliarity of modern society with imagery of mass deaths or daily fear of illness, the lessons gleaned from the American experiences with yellow fever are timeless. Nature presents a variety of endangerments that are capable of causing mass deaths and impairments. Viruses, such as yellow fever, are certainly within the array of natural hazards that are perilous, and whose outbreaks may grow quickly into epidemics within the populace. Although people can survive and recover from yellow fever, some viruses are currently incurable and fatal.

Historically, the United States has been protected from a variety of threats given its oceans separating it geographically from much of the remainder of the world. However, despite the expansiveness of the Atlantic and the Pacific Oceans, no guarantee exists that some new illness may be introduced within American society. Global trade and shipping present numerous opportunities for transmitting insects and diseases worldwide. Similarly, individuals interacting and traveling to, from, and through locations that are deemed risky also present such an opportunity. Thus, it is possible for new or foreign infectious agents to be distributed among a variety of regions and locales. During modern times, witnessed worldwide, probably the most familiar example is human immunodeficiency virus (HIV).

Viruses are living organisms capable of changing through time. History has demonstrated that viruses are capable of mutating into different strains that wreak havoc societally, that may be incurable despite being manageable, or that are fatal among human populations. Although treatments and vaccines may be appropriate for one strain of a virus, they may be ineffective against a different strain (Government Accountability Office, 2007).

Certainly, one cannot ignore the potential of weaponized viruses among terrorist entities to inflict harm and destruction societally. Around the world, any number of state and non-state groups exist that desire to inflict harm and destruction (Doss, Jones, & Sumrall, 2010; Wigginton, et al., 2015). A large array of resources may be necessary for addressing such incidents. Examples range from the uses of state defense forces throughout affected regions to the use of evacuation (McElreath, Jensen, et al., 2014; McElreath et al., 2013). Governments, organizations, and individuals must recognize the seriousness of viruses to cause mass deaths and unalterably change society. Although some levels of economic recovery may be possible, epidemic outbreaks can interject societal impacts that are unrecoverable. For instance, the historical yellow fever killed entire families, thereby ensuring that no successors existed to pass along the family name and continue the familial lineage. Modern epidemics have an identical capacity for societal damage and destruction.

The 1878 Quarantine Act originated from yellow fever as a policy mechanism whereby government and society dealt with mass calamity. During modern times, one cannot ignore the potential of illnesses to necessitate quarantine. When crafting policy, natural hazards must be considered within the context of emergency management. If a threat assessment reveals that a specific virus represents a viable endangerment, emergency plans should acknowledge its dangerousness throughout all stages of the emergency management cycle. Thus, relevant policies involving every facet of preparedness, mitigation, response, and recovery must incorporate methodologies whereby incidents may be either averted or their effects either neutralized or diminished.

REFERENCES

Arden, W.B. (2005). *Urban yellow fever diffusion patterns and the role of Micro-environmental factors in disease dissemination: A temporal spatial analysis of the Memphis epidemic of 1878.* Retrieved from http://etd.lsu.edu/docs/available/etd-11142005-140210/unrestricted/Arden_thesis.pdf

Carrigan, J.A. (1994). *The Saffron scourge: A history of yellow fever in Louisiana, 1796–1905.* Lafayette, LA: University of Southwestern Louisiana Press.

Centers for Disease Control and Prevention. (2014). *Yellow fever.* Retrieved from http://www.cdc.gov/yellowfever

City of Holly Springs. (2014). *About Holly Springs.* Retrieved from http://www.hollyspringsmsus.com/index.php/live/about-holly-springs

City of Holly Springs. (2018). *A brief history of Holly Springs.* Retrieved from http://www.hollyspringsmsus.com/about-holly-springs

Crosby, M.C. (2006). *The American plague: The untold story of yellow fever, the epidemic that shaped our history.* New York, NY: The Berkley Publishing Group.

Department of Health and Human Services. (2017a). *Polio.* Retrieved from https://www.vaccines.gov/diseases/polio/index.html

Department of Health and Human Services. (2017b). *Yellow fever.* Retrieved from https://www.vaccines.gov/diseases/yellow_fever/index.html

Doss, D., Jones, D., & Sumrall, W. (2010, September). *A quantitative analysis of Animal Liberation Front incidents versus Earth Liberation Front incidents.* Paper presented to the annual meeting of the Southern Criminal Justice Association. Clearwater Beach, FL.

Duffy, J. (1993). *From humors to medical science: A history of American medicine* (2nd ed.). Urbana, IL: University of Illinois Press.

Gann, J. (2011). *Yellow fever.* Retrieved from http://www.encyclopediaofarkansas.net/ encyclopedia/entry-detail.aspx?entryID=3078

Government Accountability Office. (2007). *Influenza pandemic: Efforts under way to address constraints on using antivirals to forestall a pandemic.* Washington, DC: U.S. Government.

Haines, D. (1997). *Yellow fever victims.* Retrieved from http://msgw.org/marshall/locales/victims.php

History Channel. (2014). *August 13, 1878: First victim of Memphis yellow-fever epidemic dies.* Retrieved from http://www.history.com/this-day-in-history/first-victim-of-memphis-yellow-fever-epidemic-dies

Kohn, G.C. (2008). *U.S. yellow fever epidemics.* Retrieved from http://www.fofweb.com/activelink2.asp?ItemID=WE52&iPin=ENPP694&SingleRecord=True

Leverette, J. (1999). The yellow fever epidemic of 1878 in Grenada, Mississippi. Retrieved from http://www.rootsweb.ancestry.com/~msgrenad/grenyfvr.htm

McElreath, D., Doss, D., Jensen, C., Wigginton, M., Kennedy, R., Winter, K., Mongue, R., Bounds, J., & Estis-Sumerel, J. (2013). *Introduction to law enforcement.* Boca Raton, FL: CRC Press.

McElreath, D., Doss, D., Jensen, M., Wigginton, M., Nations, R., Van Slyke, J., & Nations, J. (2014). *Foundations of emergency management.* Dubuque, IA: Kendall-Hunt Publishing.

McElreath, D., Jensen, C., Wigginton, M., Doss, D., Nations, R., & Van Slyke, J. (2014). *Introduction to homeland security* (2nd ed.). Boca Raton, FL: CRC Press.

McKiven, H.M. (2007). The political construction of a natural disaster: The Yellow Fever Epidemic of 1853. *Journal of American History, 94,* 734–742.

New Orleans Public Library. (2014). *Yellow fever deaths in New Orleans, 1817–1905.* Retrieved from http://nutrias.org/facts/feverdeaths.htm

Norrby, E. (2007). Yellow fever and Max Theiler: The only Nobel Prize for a virus vaccine. *Journal of Experimental Medicine, 204*(12), 2779–2784.

Nuwer, D.S. (2009). *Plague among the magnolias: The 1878 yellow fever epidemic in Mississippi.* Tuscaloosa AL: University Press of Alabama.

Patterson, G. (2009). *The mosquito crusades: A history of the American anti-mosquito movement from the Reed commission to the first Earth Day.* Piscataway, NJ: Rutgers University Press.

Republic of South Africa. (2014). *Prevention of yellow fever in South Africa.* Retrieved from http://www.transport.gov.za/Portals/0/Aviation/YELLOW%20FEVER%20COUNTRIES.pdf

Savitt, T.L., & Young, J.H. (1991). *Disease and distinctiveness in the American South.* Knoxville, TN: The University of Tennessee Press.

Shapiro, J. (2008). *Lessons from America's tropical epidemic.* Retrieved from http://www.npr.org/templates/story/story.php?storyId=19241319

State of Florida. (2014). *Yellow fever.* Retrieved from http://www.floridahealth.gov/diseases-and-conditions/mosquito-borne-diseases/yellowfever.html

State of Mississippi. (1900). *Biennial reports of the departments and benevolent institutions for the years 1898-99.* Jacksonville, FL: Vance Printing Co.

Suddath, C. (2010). *Memphis yellow fever.* Retrieved from http://content.time.com/time/specials/packages/article/0,28804,2027479_2027486_2027509,00.html

Tennessee Encyclopedia of History and Culture. (2014). *Memphis.* Retrieved from http://www.tennesseeencyclopedia.net/entry.php?rec=889

The New York Times. (1878). *Horrors of yellow fever: A Mississippi town depopulated.* Retrieved from http://query.nytimes.com/mem/archive-free/pdf?res=9C01EFD6153CE03BBC4D52DFBE668383669FDE

Tyner, J.A. (2006). *Oriental bodies: Discourse and discipline in U.S. immigration policy, 1875-1942.* Lanham, MD: Rowman & Littlefield Publishing.

University of California. (2014). *University of California San Francisco environmental health and safety/biosafety yellow fever immunization policy for the research laboratory setting.* Retrieved from http://or.ucsf.edu/ehs/12288-DSY/version/default/part/4/data

University of Florida. (2014). *Featured creatures.* Retrieved from http://entnemdept.ufl.edu/creatures/aquatic/aedes_aegypti.htm

University of Illinois. (2014). *Yellow fever.* Retrieved from http://www.uic.edu/sph/prepare/courses/ph500/resources/yellowfever.pdf

Western Cape Government. (2014). *Yellow fever certificates.* Retrieved from http://www.westerncape.gov.za/service/yellow-fever-certificates

Wigginton, M., Burton, R., Jensen, C., McElreath, D., Mallory, S., & Doss, D. (2015). *Al-Qods force: Iran's weapon of choice to export terrorism.* Boca Raton, FL: CRC Press.

11 The Cholera Epidemics (1800s)

Figure 11.1 Strong Wooden Coffins Covered with Matting in which the Wealthy Cholera Victims were Buried. Harbin Cholera Epidemic
Source: Library of Congress

BRIEF OVERVIEW

With the advancement in medical science over the last two centuries, many health threats that struck fear in past generations have been eliminated from much of the world. Research has revealed soon after their first contact with Europeans that a very large number of those who lived in what would become known as America in the regions central Mexico and Guatemala, perhaps extending as far south as Peru, were affected by a series of epidemics that devastated the population. A small pox epidemic killed an estimated eight million people. Approximately two decades later as many as 15 million people, the majority of whom were Aztecs, may have contracted and died as a result of the foodborne pathogenic bacterium *Salmonella enterica* (Vonberg, 2018).

Some estimates place the mortality rate by 1600 from various diseases brought to the Americas by European explorers at 80–90% of the Aztec population revealing how fragile a population finds itself when a new disease enters into a vulnerable population (Vonberg, 2018)

In the 1800s and early 1900s, epidemics that included yellow fever, cholera, small pox, and pneumonia wreaked havoc as they spread through parts of the nation leaving death in their wake. Intensifying the fear was the lack of understanding of the cause and method of transfer of the diseases. As with most diseases, the children and the elderly seemed most vulnerable. Medical science struggled to find solutions (McElreath, Jensen, et al., 2014).

The cholera epidemics that struck the United States in the 1830s, 1840s, and 1860s were in no way the most significant epidemics to strike neither the nation nor the deadliest. In fact, yellow fever and the Spanish influenza caused many more deaths than cholera. Polio tragically struck many more Americans and for those who were affected and survived, their lives changed forever.

Cholera holds its own unique place in the nation's history. It is a serious intestinal disease that is spread by the consumption of water and food contaminated with human feces. People who contract cholera generally suffer from diarrhea, vomiting, and cramps. Death from severe dehydration can occur within a few hours or days of the first symptoms. Filth in the nation's cities of the time increased the probability the disease would establish itself and spread through the water and food supply. In the early and mid-1800s, the filthy conditions in the urban setting were right for cholera to strike and strike it did.

INTRODUCTION

Cholera did not originate in the United States. Hippocrates described a cholera-like disease centuries ago. Cholera was found in India as early as the 11th century, but it seems the disease did not become a pandemic until the early 1800s. Most of the outbreaks were traced back to the Ganges River area in India (Schatzki, 2015).

The first pandemic or global epidemic of cholera is believed to have started in 1817 from its endemic area in Southeast Asia and subsequently spread to other parts of the world. Once in Europe, its movement was enhanced by the ever-increasing migration to the United States occurring at the time; it was easy for the disease to be transmitted across the ocean and into the new nation.

During the 19th century, there were several outbreaks of cholera. Once started, cholera, carried by humans, spread. By the 1800s, the demand for goods and the promise of economic prosperity fueled an enormous increase in trade and transportation of people. Traders and travels alike transmitted diseases, often with devastating effect. Cholera was one of those diseases (McElreath, Jensen, et al., 2014).

Moving from India, cholera was reported in Russia by 1831. Soon cases were discovered in Eastern and Central Europe. Once on the move, there was little to stop it from traveling into Western Europe and then the Americas. The movement of the disease was closely followed, yet its method of transmission and cause were unclear (McNamara, 2017). People were terrified by horrific symptoms which seemed to afflict victims instantly—severe diarrhea, vomiting, and cramps. As previously stated, those affected often died from dehydration within a few hours or days after the symptoms first appear.

The Setting

What was unknown at the time was that cholera thrives in waste. Once introduced into the waste of a city, it can easily contaminate an under-protected water system. Once the water system was compromised, all who used the water were at risk of exposure.

Prior to the late 1800s and early 1900s, the majority of the cities, especially in the poorest sections of the cities, were extremely filthy and the perfect setting for the outbreak of disease. For example, residents of New York City's more affluent areas employed private cleaning companies to cleanup and deal with waste management in their neighborhoods. In some of these urban areas, people literally walked through high levels of urban garbage and filth, often a combination of street trash, horse urine and manure, dead animals, food waste and human waste, dumped onto the streets from the buildings (Oatman-Stanford, 2013).

Because of the high levels of unmanaged waste, epidemics of infectious diseases were common. Compounding the problems of waste, cities lacked plumbing and other methods of appropriate disposal common today. Within this setting, outbreaks of smallpox, typhoid, malaria, yellow fever, cholera, and tuberculosis would take a deadly toll (Brackemyre, 2017).

Cholera was the classic epidemic of the 19th century. It first appeared in the United States in the summer of 1832 in the State of New York, apparently arriving from Europe with many of the thousands of newly arrived from outside our nation. This was a time when the new nation seemed open to almost unlimited or unrestricted immigration. Seeking those opportunities in the United States, many of those newly arrived in the nation moved in all directions across the country, taking with them the seeds of an epidemic. With their movement went their skills, cultures, values, and diseases.

Cases of cholera were soon reported in St. Louis, New Orleans, Memphis, and countless other towns and cities. The first cases of cholera were reported in St. Louis in 1832. By the time the first wave of cholera subsided in St. Louis, 300 were reported dead. The disease returned to St. Louis each of the next three summers (O'Neil, 2010).

Within months, cases would be reported in other cities as it traversed through the United States (Snyder, 2015). Over the next several decades, the nation would experience several additional serious waves of cholera, eventually killing thousands.

The Disaster Unfolds

When it first arrived in the United States in 1832, cholera joined other medical challenges of the time that included yellow fever, smallpox, and soon polio, and with a century, Spanish influenza. As is known today, the lack of sanitation was the friend of cholera, but medical science had not yet gained an understanding of the disease (Rosenberg, 1962).

In the 1830s, the social and personal hygiene we so take for granted today was virtually nonexistent. A New England physician was quoted as stating "not one in five of his patients bathed or washed their bodies once a year." For many, basic meals could be extremely dangerous. The lack of refrigeration and an understanding of the need for cleanliness in the preparation and storing of food continually threatened food poisoning. The poor were especially vulnerable, often residing in the most unclean of living situations. City streets, especially in many of the larger urban areas, were filled with various wastes and garbage, also creating a health hazard (Rosenberg, 1962).

The abrupt onset of cholera instilled fear. The disease tended to be most virulent in cities because of their poor sanitation systems. Cholera most commonly struck during spring, summer, and fall. What was not understood at the time was that the cold of winters, at least in the northern parts of the country, routinely killed the cholera germ.

Much of the medical community believed that cholera was transported through the air simply because victims of cholera had trouble breathing. As a result, there was great fear that those affected were highly contagious. Many died with limited, if any, medical assistance.

In some cities, tar was burned in the streets to create a thick smoke many believed would help eradicate the infection from the air. Stray animals were also believed to be vectors of infection, and people would let the dead bodies of the animals decompose in the street instead of removing the bodies for fear of contracting cholera.

In 1848 and 1849, cases of cholera were again reported in the United States, striking many of the cities of the nation much harder that the epidemic of the 1830s. New York City, New Orleans, and St. Louis would each experience the impact of the outbreak. In 1849, New York City reported more than 5,000 cases (Brackemyre, 2017). The same year, approximately one-tenth of the population of St. Louis died from disease. The Western Journal newspaper of St. Louis reported that 8,445 people died, with 4,285 deaths attributed to cholera. Additional deaths may have occurred, since many people were buried outside the city limits.

New Orleans would be hit by an outbreak of cholera in late 1853, which resulted in about 129 deaths, compared with the nearly 8,000 deaths that occurred from yellow fever. Most of those who died in New Orleans, like in other cities, were the poor who often lived in filthy conditions with limited access to clean water.

By the mid-1850s, questions related to the causation and methods of transmission of Cholera were answered. In 1854, British physician John Snow, considered one of the fathers of modern epidemiology, researched cholera cases in the Soho area of London. He concluded that the source of the disease must have originated from the Soho area itself. His investigation led to the identification of the source of the disease—contaminated water from a public well pump. With the source of the disease having been identified in the Soho area of London, an understanding of the relationship of the disease to filth began to develop. It would take almost three decades for a scientific investigation to finally identify the organism that caused the disease. In 1883, Robert Koch, directing a German scientific commission in Egypt, discovered the organism that caused the disease, *Vibrio comma* (Rosenberg, 1962).

Impact

Epidemics created fear and panic. When cities experienced the outbreak of these various diseases, it was not unusual for those who could to flee the danger. The fear of disease became so prevalent that hundreds would seek refuge. In many cases, those seeking safety actually carried the diseases with them. In New York, during the epidemic of 1832, some suggested as many as 100,000 left the city (McElreath, Doss, et al., 2014).

Other epidemics had the same effect on communities, as fear quickly spread among the residents. During the yellow fever epidemic of the 1870s, panicked residents in towns and cities along the Mississippi River attempted to escape the reach of the diseases (McElreath, Doss, et al., 2014).

RELIEF AND FIRST RESPONSE EFFORTS

Unfortunately, for people stricken with cholera, the most common accepted treatment methods, at least before the American War Between the States, were almost as bad as the illness. Doctors routinely prescribed calomel for cholera victims, which contained mercury, and many people died from mercury poisoning or suffered other ill effects from this drug.

From a medical perspective, treatment was very basic and called upon the science of the day. In addition to the use of mercury, it was also common for physicians to treat what they interpreted as the symptoms. In the case of cholera, the diarrhea often resulted in the application of laxative to be used as a purgative for the treatment of bowel illnesses. The purging further compounded the dangers to the patient. As for the use of mercury, the side effects could well include significant poisoning to include the loosening or loss of teeth, loss of hair, and destruction of gums and intestines (Kunesh, 2002).

While many fled, others who were suspected of or exhibited signs of the disease were subjected to the possibility of quarantine. As an example, in 1849, St. Louis established a quarantine station on a small island about 4 miles south of downtown. Arsenal Island, which was renamed Quarantine Island, served as a landing from which boats moving up the Mississippi toward the city were stopped and inspected. Anyone onboard showing symptoms of cholera were required to remain on Quarantine Island in a hastily built hospital, until they either were declared free of the disease or died. Those who died on Quarantine Island were buried in a graveyard that would later be washed away by the river. Memphis would later establish a quarantine station during the yellow fever epidemic.

In 1849, the New York City Board of Health established a hospital to deal with the thousands of cholera cases that occurred during the epidemic when it stuck the city. By the end of the 1849 epidemic, more than 5,000 people died.

AFTERMATH AND INVESTIGATION

Cholera mortality was great in the large cities, especially among the young and the elderly. In the 1832 cholera epidemic, at least 500 residents of St. Louis died. Cincinnati reported the deaths of 732. In Detroit, about 322 died of the disease. When the disease returned in 1849–1851, the death toll was even greater. This time, St. Louis lost 4,557 to cholera. Cincinnati reported the deaths of 5,969. Deaths in Detroit reached 700. In each outbreak, deaths averaged between 5% and 10% of the population. Boards of Health were activated but lapsed after a time. After John Snow reported that drinking water was responsible for cholera in a London outbreak and the 1884 discovery of the bacillus by Koch, sanitation efforts slowly became more sincere and, eventually, more successful (Daly, 2008).

LESSONS LEARNED AND POLICY IMPACT

As the cause of many of these diseases became understood, steps were taken to prevent another outbreak. Cities began to understand more fully the need for clean water and food combined with proper sanitation and waste disposal. On the local, state, and federal levels, legislation was enacted to establish heath- and sanitation-related rules and other legislation to ensure compliance. Even today, the exposure to inappropriate sanitation during a post-disaster event can be significant, establishing the conditions under which illness may spread through those who remain in the impacted areas (McElreath, Doss, et al., 2014).

On the local level, cities began taking health and sanitation conditions seriously within their jurisdictions. In New York City, the Department of Street Cleaning was established in 1881, but it would take the efforts of reform Mayor William Strong to turn the Department of Street Cleaning into an effective and productive organization. When Strong appointed George Waring to head the organization, Waring took the appointment seriously, established performance expectations, and held those in the Department accountable for their actions. Very quickly the conditions on the streets of the city improved (Oatman-Stanford, 2013).

Eventually, the federal government took steps to establish quarantine stations to isolate those suspected of infection. The National Quarantine Act of 1878 gave the federal government control over quarantine measures. In 1891, federal immigration legislation made the Marine Hospital Service responsible for performing medical examinations on all newly arriving immigrants.

In addressing the condition of food supply, the U.S. Meat Inspection Act of 1906 authorized the Secretary of Agriculture to direct the inspection of meat-slaughtering and meat-packaging factories and meat storehouses, another step in protecting the nation's food supply (Tucker, 2017).

As sanitation improved within the United States, including chlorination of water, the illness became less common. As cities outgrew local wells for the supply of water, they began the construction of water reservoir systems which, by the mid and late 1800s, would be supplying the cities with safer water.

With changes of this type, the occurrence of cholera in developed countries declined significantly, but the disease has not gone away. In 2016, the World Health Organization reported 132,121 cholera cases and 2,420 deaths, 54% of the cases occurring in Africa, the majority of which occurred in Yemen (Toy, 2017). The standard treatment for cholera today is to keep the ill person hydrated with germ-free water, other fluids, intravenous hydration plus antibiotics (Schatzki, 2015).

IMPLICATIONS FOR MODERN EMERGENCY MANAGEMENT

Cholera epidemics continued in the United States until the early 1900s. In 1911, a ship with 13 cases of cholera arrived in New York. The passengers were quarantined and the disease did not spread (Schatzki, 2015). By the end of the 19th century, science gained a better understanding of the cause and management of infectious diseases. A large part of this process was the widespread acceptance of germ theory (Brackemyre, 2017).

In 1961, the seventh cholera pandemic wave began in Indonesia and spread rapidly to other countries in Asia, Europe, Africa, and finally in 1991 to Latin America, which had been free of cholera for more than one century. The disease spread rapidly in Latin America, causing nearly 400,000 reported cases and over 4,000 deaths in 16 countries of the Americas that year (The World Health Organization, 2017).

Though rare in the United States today, cases of cholera do occur. In the aftermath of the Rwanda crisis in 1994, outbreaks of cholera caused at least 48,000 cases and 23,800 deaths within 1 month in the refugee camps in Goma, the Congo (The World Health Organization, 2017). Since 2000, there have been outbreaks of cholera across Africa, in Iraq during the war, in India, and in Vietnam.

The lessons are clear as to the importance of proper sanitation in the prevention of disease. Contamination of water and food can well serve as the foundation of the outbreak of a major epidemic even today. It is critical to approach disease from both a proactive and reactive strategy. Proactive approaches require taking the steps to prevent or stop an outbreak, whereas reactive approaches require responding to the crisis as quickly and efficiently as possible. Effective response requires a partnership between science, medicine, government, and the private sector. Many medical situations need a multinational approach, which includes the cooperation, support, and assistance of nongovernmental organizations.

Today's emergency managers include pandemic planning as part of their comprehensive approach to emergency management planning. This became a focus for emergency managers when avian influenza A (H5N1) became a more pronounced threat in the United States. As a result of the fear of avian and swine influenza, the U.S. Department of Health and Human Services (HHS) developed a Pandemic Influenza Plan to assist with the coordination and response to A (H5N1) viruses. The next pandemic remains on the horizon and preparations at the local, state, and federal levels must be made. Three significant tools have been developed to assist emergency managers on all three levels of government which align with the Word Health Organization's (WHO) global framework to include the Pandemic Intervals Framework (PIF), Influenza Risk Assessment Tool (IRAT), and the Pandemic Severity Assessment Framework (PSAF). Other various planning tools can be retrieved through the Department of Homeland Security, HHS, and Centers for Disease Control.

SUMMARY

While modern sewage and water treatment have helped eradicate cholera in some countries, the virus is still present and presents a danger. Cholera is not the only threat to health that challenges the human race. Over the past two decades, Ebola, Middle East respiratory syndrome (MERS), pandemic influenza, and the Zika virus have each demonstrated the extraordinary health, economic, and security risks associated with infectious disease outbreaks. In the future, infectious disease outbreaks may well become ever more complex and challenging. A combination of high population density, poverty, changes in social structures, and a lack of public health infrastructure will create progressively more favorable conditions for communicable diseases (World Economic Forum, 2017).

REFERENCES

Brackemyre, T. (2017). *Immigrants, cities, and disease.* U. S. History Scene. Retrieved from http://ushistoryscene.com/article/immigrants-cities-disease

Daly, W. J. (2008). The Black Cholera comes to the Central Valley of America in the 19th century - 1832, 1849, and later. *Transactions of the American Clinical and Climatological Association,* 143–153. Retrieved from https://www.ncbi.nlm.nih.gov/pmc/articles/PMC2394684

Kunesh, T. (2002, March 9). *Sickness and death in the Old South.* TNGenWeb. Retrieved from http://www.tngenweb.org/darkside/index.html

McElreath, D. H., Doss, D. A., Jensen, C. J., Wigginton, M. P., Nations, R., Van Slyke, J., & Nations, J. (2014). *Foundations of emergency management.* Debuque: Kendall Hunt.

McElreath, D. H., Jensen, C. J., Wigginton, M. P., Doss, D. A., Nations, R., & Van Slyke, J. (2014). *Introduction to homeland security.* Boca Raton: CRC Press.

McNamara, R. (2017, September 14). *The Cholera Epidemic of 1832 as immigrants were blamed, half of New York City fled in panic.* Retrieved from ThoughtCo.: https://www.thoughtco.com/the-cholera-epidemic-1773767

Oatman-Stanford, H. (2013, June 24). *A filthy history: When New Yorkers lived knee-deep in trash.* Collectors Weekly. Retrieved from https://www.collectorsweekly.com/articles/when-new-yorkers-lived-knee-deep-in-trash

O'Neil, T. (2010, July 18). *A look back: Cholera epidemic hit a peak here in 1849.* St. Louis Post Dispatch. Retrieved from http://www.stltoday.com/news/local/metro/a-look-back-cholera-epidemic-hit-a-peak-here-in/article_f50b669f-a4c8-595b-bc6a-d3d9833ffc14.html

Rosenberg, C. E. (1962). *The cholera years: The United States in 1832, 1849 and 1866.* Chicago and London: The University of Chicago Press.

Schatzki, S. C. (2015). Cholera. *American Journal of Roentgenology.* Retrieved from http://www.ajronline.org/doi/full/10.2214/AJR.14.13787

Snyder, C. G. (2015, April 20). *The 1849 cholera epidemic in Illinois.* ILGenWeb. Retrieved from https://illinoisgenweb.org/historyproject/the-1849-cholera-epidemic-in-illinois.html

The World Health Organization. (2017). *Global epidemics and impact of cholera.* The World Health Organization. Retrieved from http://www.who.int/topics/cholera/impact/en

Toy, S. (2017, June 28). *Yemen faces worst cholera outbreak in the world, health authorities say.* USA Today. Retrieved from https://www.usatoday.com/story/news/2017/06/28/yemen-facing-worst-cholera-outbreak-world-health-authorities-say/436858001

Tucker, K. (2017, April 25). *Sanitary conditions in the late 1800s in America.* Sciencing. Retrieved from https://sciencing.com/sanitary-conditions-late-1800s-america-16334.html

Vonberg, J. (2018, January 16). *Mystery over death of 15 million Aztecs may be solved after nearly 500 years, study suggests.* The Independent. Retrieved from https://www.yahoo.com/news/salmonella-epidemic-may-wiped-15-105216789.html

World Economic Forum. (2017). *Global health security.* Retrieved from https://www.weforum.org/projects/managing-the-risk-and-impact-of-future-epidemics

Spanish Influenza (1918) 12

Figure 12.1 U.S. 39th Regiment in Seattle Wearing Masks to Prevent Influenza
Source: Everett Historical/Shutterstock.com

BRIEF OVERVIEW

The 1918 Spanish influenza pandemic was described as the "the greatest medical holocaust in history" (Ellis, 2016, p. 4). Within a period of 18 months, the pestilence spread quickly and globally infecting nearly one-third of the world's population (Natera, 2012). Its death toll exceeded the quantity of deaths wrought in World War I wherein approximately 16 million people perished (National Archives and Records Administration, 2017). Worldwide, during the years 1918 and 1919, the influenza pestilence killed at least 20 million individuals, and may have killed an upward estimate of 50 million individuals (National Archives and Records Administration, 2017; Stanford University, 2017). When compared to the death toll resulting from the bubonic plague, the Spanish flu killed more individuals in a solitary year than those who died between 1347 and 1351 (Stanford University, 2017). Because of Spanish influenza, American life expectancy was diminished by approximately

12 years overall during the early 20th century (National Archives and Records Administration, 2017). Essentially, the virus was heralded as one of the deadliest in recorded history. From a modern perspective, the death toll resulting from 1918 Spanish influenza exceeded that of the modern human immunodeficiency virus (HIV) death rate. Essentially, the 1918 pestilence killed more individuals in 24 weeks than HIV killed in 24 years (Finger, 2006a).

INTRODUCTION

The 1918 pestilence was attributed to a pandemic of Spanish influenza. The etiology of the word "influenza" originated within the Italian language for astrological purposes (Merriam-Webster, 2018). Basically, the word held the literal meaning of "influence" (Merriam-Webster, 2018). However, during the Middle Ages, use of the word was applied to mean disease outbreaks (Merriam-Webster, 2018). Throughout the 1600s and the 1700s, the word "influenza" was applied to instances of the disease that is now known as "flu" or "influenza" (Merriam-Webster, 2018).

Three types of influenza exist: A, B, and C. The A version exhibits a broad array of possible hosts, such as humans, pigs, seals, horses, and other types of animals (Ryan, 2008). Typically, the natural host for this category is the wild bird (Ryan, 2008). The type A variety is generally considered the most virulent strain of the influenza virus, and it encompasses all identified varieties of avian influenza (Ryan, 2008). The B version is nearly exclusive to humans, but may be found in some forms of animals (Ryan, 2008). The B variety has never caused a known pandemic, but was the catalyst for epidemics (Ryan, 2008). The C variety is limited to swine and humans, and causes neither epidemics nor pandemics (Ryan, 2008). The A and B varieties have the potential of causing severe instances of diseases whereas the C variety may cause milder illness (Ryan, 2008).

Influenza is dispersed via droplets in the air when an infected person talks, coughs, or sneezes (Mayo Clinic, 2018). An influenza infection may occur when someone inhales aerosol droplets or encounters the virus from interaction with some form of carrier object (e.g., keyboards) and transits the germs to the nose, eyes, or mouth (Mayo Clinic, 2018). Typically, infected individuals can be contagious from the day before symptoms are exhibited until approximately 5 days after noticeable symptoms begin (Mayo Clinic, 2018). In some cases, they may be contagious for a period of 10 days after the commencement of symptoms (Mayo Clinic, 2018). Although it is typically unfatal and one may recover without medical intervention, influenza can be lethal (Mayo Clinic, 2018).

Influenza may strike anyone. Itemized below are the primary characteristics for influenza risks (Mayo Clinic, 2018; Vincent, 2010):

- Young children under 5 years of age, and especially those who are less than 2 years of age
- Adults 65 years of age or older
- Residents of nursing homes and other long-term care environments
- Pregnant women and women up to 2 weeks postpartum
- People with weakened immune systems
- People who have chronic illnesses, such as asthma, heart disease, kidney disease, diabetes, and so forth
- People who are very obese, with a body mass index of 40 or greater

Some may recover from influenza without impairment whereas others may experience complications or death. Complications associated with influenza include pneumonia, bronchitis, asthma attacks, heart difficulties, and ear infections (Vincent, 2010). Not everyone will experience influenza. However, for those who do, symptoms are typically exhibited as follows (Mayo Clinic, 2018):

- Fever over 100.4 F (38 °C)
- Aching muscles, especially in your back, arms and legs
- Chills and sweats
- Headache
- Dry, persistent cough

- Fatigue and weakness
- Nasal congestion
- Sore throat

The Setting

In many ways the world had turned itself upside down. The major powers of Europe had faltered to a war in 1914, which led to global devastation among nations, and concluded by opening a new chapter in human history. The United States would reframe from entering the conflict until 1917, but quickly took steps to assemble and mobilize an army capable of entering the conflict.

In the process of recruitment, training, and preparation for deployment to Europe, millions of men and women quickly entered the military or joined the nation's industries expanding to meet the new challenges facing the country.

In this setting, the possibility of disease outbreak and spread was significant. In 1918, early reports from Fort Riley, Kansas indicated soldiers were victims of an influenza, which was quickly proving fatal. Compounding the problem was that many soldiers, exposed to the disease at Fort Riley, had been transferred to other duty stations awaiting deployment. With them and their travels, they carried the deadly virus. Figure 12.1 shows the U.S. 39th regiment in Seattle wearing masks to prevent influenza as they prepare to leave for France.

The Disaster Unfolds

American society was devastated by Spanish influenza. Within the United States, an estimated 22 million individuals were infected with the illness (Armstrong, 2015). Approximately 675,000 Americans perished from the illness, especially individuals between the ages of 20 and 40 (Billings, 1997). The quantity of deaths was approximately 10 times the number of Americans who perished in World War I (Billings, 1997). Among adults between 20 and 39 years of age, a minimum of 280,000 deaths resulted from secondary pneumonia during the period encompassing 1918 and 1919 (Glezen, 1996). The quantity of deaths represented approximately one individual per 100 within this age group (Glezen, 1996).

Death rates represented approximately 28% of American society (Billings, 1997). Its pattern of illness and affected population were atypical of the flu viruses of the period because others usually affected children or elderly people (Billings, 1997). Spanish influenza also ravaged American military forces. Among all U.S. servicemen who died in Europe during World War I, around 50% perished from influenza, and not from combat action against the enemy (Billings, 1997).

Spanish influenza occurred in four primary phases (Wever & van Bergen, 2014). The first phase occurred in spring 1918, and exhibited mildness with few deaths (Wever & van Bergen, 2014). It consisted of an initial "three-day-fever" succeeded by an illness of greater severity (National Archives and Records Administration, 2017). The first phase was usually unwarned and unexpected; few deaths occurred from the initial fever with recovery necessitating merely a few days (National Archives and Records Administration, 2017). The summer months were relatively calm regarding progression of the virus (Wever & van Bergen, 2014). The second phase occurred in the autumn of 1918 (Wever & van Bergen, 2014). When the pestilence occurred again during the fall months, its severity occurred with much greater intensity and lethality. Individuals were perishing so quickly that medical personnel and scientists were initially unable to discern the disease itself (National Archives and Records Administration, 2017).

The third phase occurred during the first few months of 1919, wherein a high mortality rate existed (Wever & van Bergen, 2014). The fourth phase occurred during the initial months of 1920 among various locations geographically (Johnson & Mueller, 2002; Wever & van Bergen, 2014). Thereafter, the succeeding annual epidemic patterns of influenza exhibited much less lethality and altered strains (Taubenberger & Morens, 2006).

At its greatest intensity, the illness occurred quickly and viciously; some died within a few hours of showing any initial symptoms of sickness whereas others died after several days (National Archives and Records Administration, 2017). Victims who languished for days eventually died because of suffocation resulting from pneumonia (their lungs filled with fluid) (National Archives and Records Administration, 2017). The influenza virus of 1918 exhibited a magnitude of strong virulence that was seen neither before nor after its emergence within society (Glezen, 1996).

Although the exact location of the initial infection which prompted the pandemic remains unknown during modern times, some speculate that it occurred within the United States (Glezen, 1996). Regardless of its location of inception, practically no location was safe geographically from the illness. It spread throughout trading routes and shipping lanes across the globe (Stanford University, 2017). Reports of Spanish influenza occurred in Africa, Asia, Europe, India, North America, South America, and regions of the South Pacific (Stanford University, 2017). Within India, the rate of mortality was approximately 50 deaths per 1,000 individuals (Stanford University, 2017). The spreading of the influenza outbreak was exacerbated by the activities of World War I with the movements of individuals, armies, and ships worldwide (Stanford University, 2017).

Impact

The illness occurred in conjunction with World War I. Initially, the illness was witnessed in the United States, segments of Asia, and Europe before quickly spreading worldwide (History Channel, 2018). Although the actual origin of the disease was unknown during the period, many speculated that it was the result of some biological warfare campaign wrought by the Germans, trench warfare, mustard gas, or other noxious fumes generated from warfare (Stanford University, 2017). Others believed it to be some unknown or new agent (Stanford University, 2017). Further conjecture blamed a variety of possible causes—plants, "cosmic influence," and even "dusty books" (Ryan, 2008, p. 58). However, the virus responsible for the Spanish influenza pestilence was not definitively identified until 1997 (Byerly, 2005). It was eventually determined that the illness resulted from influenza A virus (Morens & Fauci, 2007).

Virology was in its infancy as Spanish influenza ravaged the world during 1918 and 1919. Although doctors had known that viruses were responsible for some diseases, such knowledge had existed for about 20 years (Ryan, 2008). With their limited understanding, during October 1918, some scientists hypothesized that a virus was the cause of influenza (Ryan, 2008). However, 12 years later, in 1930, the influenza virus was isolated. Later, in 1933, further isolation of the influenza A virus occurred, thereby revealing the lethal cause of the global pandemic (Ryan, 2008). Respectively, during 1940 and 1950, influenza B and C viruses were isolated (Ryan, 2008).

Over 70 years later, samples of the actual Spanish influenza virus were examined empirically. During the 1990s, samples of influenza virus were obtained from an Alaskan who had perished during the 1918 pestilence. Analyzing the sample's genetics revealed much about its lethality and provided insight regarding how to safeguard against possible outbreaks during modern times (Associated Press, 1998). Several analyses showed that its genetic blueprint was unlike that of other influenza viruses, and was related to "swine" influenza (Associated Press, 1998). Essentially, it was suggested that, after evolving in pigs, it appeared as though the virus mutated and adapted to humans (Associated Press, 1998). Afterward, the illness was disseminated by mobilizing American military personnel during World War I (Associated Press, 1998). Such findings supported notions that swine influenza was strongly virulent toward humans (Associated Press, 1998).

However, research endeavors of the 1990s concerning Spanish influenza were neither final nor definitive. During modern times, additional examinations of the 1918 virus further shed light regarding its origin. Taubenberger et al. (2005) and Taubenberger (2006) suggested that Spanish influenza arose from avian origins and later adapted to humans. Taubenberger (2006) also indicated that the 1918 Spanish influenza was the ancestor of classical swine H1N1 influenza and human influenza viruses. During 2009, an H1N1 pandemic exhibited combinations of influenza genes that were previously unseen in humans or animals (Centers for Disease Control and Prevention, 2018). According to the Centers for Disease Control, between the months of April 2009 and April 2010, it was estimated that there were approximately "60.8 million cases (range: 43.3–89.3 million), 274,304 hospitalizations (195,086-402,719), and 12,469 deaths (8868-18,306) in the United States due to the (H1N1) pdm09 virus" (Centers for Disease Control and Prevention, 2018).

Despite much research, mysteries still cling to the 1918 Spanish influenza. For instance, Morens and Fauci (2007) indicated that, during modern times, the identity of the host which was the source of the virus remains unknown. Morens and Fauci (2007) also indicated that the exact specifics of the 1918 viral adaptation among humans remain unclear. Nearly a century of empirical investigation has yielded no final answers regarding all details of the 1918 Spanish influenza virus. Despite its genetic structure resembling avian flu, it remains "genetically distinct from any of the hundreds of avian or mammalian influenza viruses collected and examined between 1917 and 2006, primarily because of greater-than-expected numbers of silent nucleotide changes" (Morens & Fauci, 2007).

RELIEF AND FIRST RESPONSE EFFORTS

The Case of Memphis, Tennessee

The city of Memphis, Tennessee experienced Spanish influenza just as did urban areas throughout the nation. Being an economic hub, maritime port along the Mississippi River, and connection between the Eastern and Western states, the unavoidable emergence of pestilence was only a matter of time. Memphis was no stranger to pestilence. Before Spanish influenza, it experienced a devastating bout with yellow fever in 1878 that killed 5,150 people, infected at least 17,000, and caused at least 25,000 to flee the city (McElreath, et al., 2013). Thus, the city had lingering familiarity with disease that killed quickly. With yellow fever, symptoms became physically obvious with yellowed skin and blackened vomit, and victims died slowly over days. However, the 1918 Spanish influenza was more insidious than yellow fever. The first reported Memphis influenza case occurred on September 25, 1918 (McBride, 2007). Although the Memphis Health Department admitted the pestilence had arrived, it stated that "no cause for alarm existed" (McBride, 2007, p. 77). Although the city encouraged residents to be unalarmed, the first Memphis death, attributed to Spanish influenza, occurred on September 26—the very next day (McBride, 2007).

Despite the presence of deadly pestilence, on October 1, Memphis held a Liberty Bond parade that generated an attendance of approximately 70,000 (Finger, 2006a). A Liberty Bond was a type of war bond that was sold to support financial funding for the Allied forces during World War I. Purchasing the bonds was symbolic of patriotic duty. The event was heralded as the "greatest crowd ever seen on Main Street" (Finger, 2006a). Typical with most parades, the participants included floats, bands, and vehicles. Because of the war in Europe and the advertising of Liberty Bonds for its financing, the participants also included many rows of military vehicles, tanks, and soldiers marching in formation (Finger, 2006a). Many of the marching soldiers were probably already infected with influenza. Those attending the Liberty Bond parade unwittingly subjected themselves to illness while fulfilling their sense of patriotic duty.

The parade was a mass gathering of Memphians wherein the Spanish influenza spread quickly and lethally. Within 72 hr, the *Memphis Commercial Appeal* newspaper indicated that Memphis industries were crippled by influenza, and that thousands of influenza cases were being reported (Finger, 2006a). The newspaper also indicated that city services, including the fire and police departments, city railway, and telephone company, were decimated by influenza (Finger, 2006a). City residents were told to only make only "urgent" telephone calls (Finger, 2006a).

The virus killed indiscriminately. City administrators and leaders died just as quickly as commoners. The newspaper soon reported that prominent judges and attorneys were infected, and that 20 deaths were possibly attributed to influenza (Finger, 2006a). The illness was so widespread and exhibited such magnitude that churches, schools, and amusement halls were all closed in order to protect public safety. Such closures were commensurate with safety measures that existed elsewhere throughout the nation. However, in Memphis, the pestilence became so pervasive and deadly that even funerals were disallowed and prohibited (Finger, 2006a).

Medical services were also impaired within the city. For instance, although hospitals and medical centers necessitated personnel, few were available. Many doctors and nurses had succumbed to the Spanish influenza. Others became ill, and ignored pleas for assistance from the Red Cross simply because they were physically unable to perform duties (Finger, 2006b). On October 12, over 100 cases were reported by the health department, but the city's medical capacity was unable to process them efficiently (Finger, 2006b). Reported cases continued to increase faster than hospitals and physicians could address them (Finger, 2006b). As an emergency measure,

Central High School, which had already been closed because of illness, was converted into a Red Cross hospital solely for treating influenza (Finger, 2006b). Although treatment was provided for free, patients were asked and encouraged to make donations for the services (Finger, 2006b). After the onset of the illness, physicians indicated that they were too occupied with treating illness and countering pestilence to report patient numbers or attend other matters (Garrett, 2007).

After illness receded, the Memphis Health Department reported that the city experienced 5,617 cases of influenza, and that 493 individuals had perished from the illness (Finger, 2006b). Before the onset of illness, Memphis exhibited an influenza death rate of approximately 179.3 per 100,000 individuals in 1915 (Garrett, 2007). Afterward, in 1918, the death rate soared to 666.1 per 100,000 individuals (Garrett, 2007). During 1919, the death rate was 340.6 per 100,000 individuals (Garrett, 2007). Throughout its practically unstoppable rampage, the illness pervaded every fabric of Memphian society. Both the poor and rich died from pestilence. Children, as they skipped rope, sang a metaphorical tune reminiscent of the pestilence's appearance and effects: "I had a little bird, its name was Enza. I opened the window, and in-flu-Enza" (Finger, 2006b). Such a ditty compared influenza to a ravenous animal with a voracious, desultory appetite for human life, seeking victims unceasingly.

Being a port along the Mississippi River and a crossroads between the major halves of the country, much commerce occurred in the area. The economic effects of influenza devastated Memphis and its serviced regions. Industrial resources and factories were handicapped greatly regarding production (Garrett, 2007). Although its transportation infrastructure employed around 400 folks for its railway service, a total of 124 reported illness and became incapacitated in one day alone, thereby impeding public travel within the city (Garrett, 2007). The Cumberland Telephone Company reported the absences of at least 100 operators because of illness, thereby substantially impeding its ability to process telephone calls (Garrett, 2007). In the Memphis area, among coal mines, a 50% decrease in production occurred because of influenza, and some considered absolute closure (Garrett, 2007).

The Case of Philadelphia, Pennsylvania

The city of Philadelphia, Pennsylvania had one of the worst Spanish influenza experiences. Because of World War I, approximately 26% of the city's physicians were serving in the military and approximately 75% of its hospital medical personnel were stationed overseas (Armstrong, 2015). Because of such absences, the city was unprepared for the impending devastation of Spanish influenza. Although the surrounding areas and fairly distant cities (e.g., Boston) were experiencing illness, Philadelphia was unaffected initially. However, its pristine status was soon blemished with an initial report of illness followed quickly by masses of sick individuals.

During 1918, the city's population was approximately 1,700,000 with an additional 300,000 transients who supported the war efforts (Armstrong, 2015). Spanish influenza appeared among the masses on September 18, 1918. Immediately, the following day, 600 sailors reported illness, and local hospitals received infected individuals. Influenza was so excessive that Philadelphia became known as the city with the greatest, most accumulated death rate. At its height, Spanish influenza killed 1,700 individuals in a single day (Armstrong, 2015). The month of October was excessively deadly. At its height, the death rate in Philadelphia itself was approximately 700 times greater than the normal rate (Ryan, 2008).

The spreading of influenza occurred in conjunction with public celebration. On September 28, at least 200,000 folks congregated to see commencing of the Fourth Liberty Loan Drive (Ryan, 2008). After the gathering, by October 4, a total of 636 influenza cases and 139 deaths were reported in Philadelphia (Ryan, 2008). As a result, the Philadelphia Board of Health mandated the closing of all churches, schools, saloons, and theaters (Ryan, 2008). Many of the closed schools were not opened until the spring of 1919 (Ryan, 2008). The city also mandated that all citizens wear masks when in public as a deterrent toward spreading illness (Ryan, 2008). Following Philadelphia's example, the remainder of the state enacted the same order the next day (Ryan, 2008).

Spanish influenza negatively affected Philadelphia's contribution toward the war effort. Among the Philadelphia Navy yards and at the aircraft factory, thousands reported illness, and at least 30 deaths occurred (Ryan, 2008). During the middle of September, a troop ship departed Philadelphia, and sailed to Washington (Ryan, 2008). When it arrived at its destination, it contained 700 cases of influenza, and one person died from illness (Ryan, 2008).

Overall, Philadelphia experienced a plethora of deaths. One account indicated that four ladies gathered to play a bridge game one evening; three died the next day (Finger, 2006b). Between the months of June and November, approximately, and at least 16,000 died from the illness (Williams, 2013). Tragically, within the period of 5 weeks between September and November, approximately 12,000 individuals died from influenza (Williams, 2013). Although dead bodies were accumulated among the city's morgues, an insufficient quantity of personnel existed for burial (Williams, 2013). As a result, about 200 seminaries from the St. Charles Borromeo Seminary volunteered to dig graves and to bury the dead (Williams, 2013).

Although Philadelphia experienced a dismal calamity, much of it may have been avoidable. Before the pestilence arrived, Philadelphia was forewarned of the possible tribulation. The city was aware of the disease's effects in Europe. During July, the Board of Health issued a warning notice that informed of the possible spreading of the disease from Europe to America. Because influenza permeated surrounding areas and distant cities, Philadelphia was aware of its potential to become infected. However, they city's health administration did not view influenza as a "reportable disease" until an epidemic was blatantly apparent. Essentially, despite having knowledge of Spanish influenza, Philadelphia chose to remain unprepared for encountering the disease.

AFTERMATH AND INVESTIGATION OF THE DISASTER

Early 20th-century physicians and hospitals were overwhelmed by influenza victims. Medical treatments were rudimentary because vaccines against the influenza virus had not been invented, and antibiotics were also undiscovered during the period of the pandemic. Essentially, no preventive or defensive medicines existed for either avoiding infection or curing patients. Therefore, no methods existed for addressing the primary viral infection or the secondary bacterial infections (e.g., pneumonia) that could result from influenza (Centers for Disease Control and Prevention, 2018). In other words, few, if any, resources existed for effectively treating patients or avoiding infection.

Some attempts were made to craft vaccines that supposedly guarded against influenza (Ott, Shaw, Danila, & Lynfield, 2007). For instance, in St. Paul, Minnesota, two experimental vaccines were administered in the hope of safeguarding individuals (Ott, Shaw, Danila, & Lynfield, 2007). However, they were both ineffective because neither contained any influenza virus (Ott, Shaw, Danila, & Lynfield, 2007). Regardless, despite the lack of knowledge concerning influenza, such endeavors provided rudimentary means of preventing future infections via vaccination.

Despite the lack of medicinal approaches, physicians of the period attempted to mitigate the effects of the illness via a variety of methods. Examples included quarantining and isolating infected individuals, using disinfectants, and implementing personal hygiene practices (Centers for Disease Control and Prevention, 2018). In some cases, whiskey toddies were administered medicinally (Naval History and Heritage Command, 2018). Patients were encouraged to cover their faces and mouths with handkerchiefs to minimize spreading of the disease (Naval History and Heritage Command, 2018). Used handkerchiefs were to be boiled as a disinfectant method (Naval History and Heritage Command, 2018). Some public assemblies were either limited or disallowed completely (Centers for Disease Control and Prevention, 2018). However, such methods and approaches were not applied uniformly (Centers for Disease Control and Prevention, 2018).

Numerous efforts were made to enhance public safety. Individuals who necessitated public contact, such as postal employees, mass transit personnel, police officers, and so forth, wore masks to avoid illness (National Archives, 2018). Among urban areas in which buildings possessed elevators, some debated closing and disallowing elevator use (Ott, Shaw, Danila, & Lynfield, 2007). Others advocated for the closing of various "public places," such as churches, schools, dance halls, and theaters (Ott, Shaw, Danila, & Lynfield, 2007). In some cases, streetcar windows were mandated to be kept open during travel (Ott, Shaw, Danila, & Lynfield, 2007). Operating hours for businesses were also regulated as a method of minimizing public congestion (Ott, Shaw, Danila, & Lynfield, 2007). In some cases, enforcement of such regulations and mandates was exercised by police forces (Ott, Shaw, Danila, & Lynfield, 2007). Regardless, many individuals, groups, and organizations soon transgressed such dictates, and congregated to pursue various activities (Ott, Shaw, Danila, & Lynfield, 2007).

Even with the use of elementary approaches and practices, Spanish influenza was devastating and overwhelming for medical centers. For instance, the U.S. Navy and Marine Corps hospitals overflowed with patients who were quarantined by medical personnel (Naval History and Heritage Command, 2018). Despite their best efforts, physicians and nurses were unable to provide robust medical care and services. Given the rates of fulminant illness and deaths, some personnel reported that insufficient time existed for treating patients adequately—neither blood pressure nor temperature was taken among individuals reporting to the facilities (Naval History and Heritage Command, 2018). During 1918, a total of 121,255 Navy and Marine patients were admitted to hospitals with Spanish influenza (Naval History and Heritage Command, 2018). Among these individuals, a total of 4,158 died from illness (Naval History and Heritage Command, 2018). Astonishingly, infected Navy and Marine patients expended at least 1 million sick days globally (Naval History and Heritage Command, 2018).

The U.S. Army was also inundated with influenza infections. Army hospitals experienced similar conditions to those of the navy. For instance, one army hospital, which contained 360 beds within its influenza ward, processed 824 sick individuals within 24 hours (Byerly, 2005). Army hospitals dispensed combinations of glycerin, aspirin, quinine, digitalis, and strychnine medicinally to treat patients (Byerly, 2005). Among its soldiers, Spanish influenza affected approximately 26% of the entirety of the U.S. Army, thereby representing a minimum of 1 million soldiers (Wever & van Bergen, 2014). Wever and van Bergen (2014) indicated that the illness killed 45,000 American soldiers among training centers and within the European theater. Approximately 30,000 soldiers among army training centers were killed by the virus before they could deploy to Europe (Wever & van Bergen, 2014). Near Washington, D.C., it was reported that 90 deaths per day occurred from influenza infection (National Archives and Records Administration, 2018). An army epidemiologist surmised that if it had continued uncontrolled and unabated, the Spanish influenza virus would have eradicated civilization itself "from the face of the earth" within mere weeks (Byerly, 2005, p. 5). Figure 12.2 shows influenza patients at Walter Reed Army Hospital.

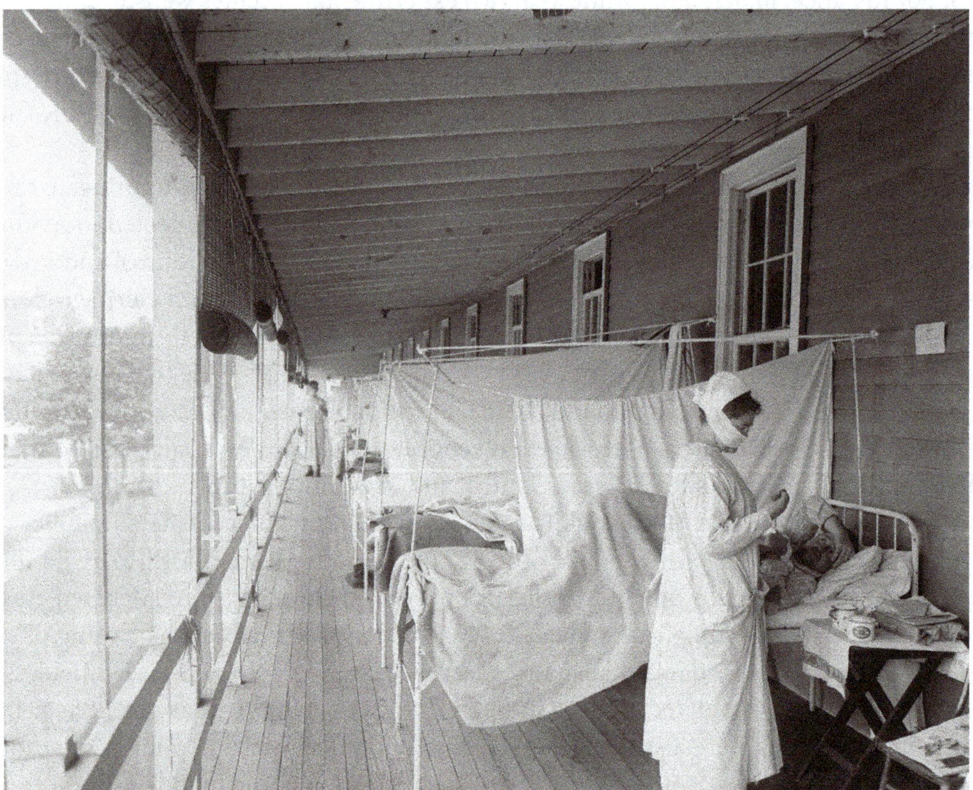

Figure 12.2 Influenza Patients at Walter Reed Army Hospital
Source: Everett Historical/Shutterstock.com

The illness yielded little mercy or grace for medical personnel who unceasingly attempted to care for patients. For instance, Byerly (2005) indicated that an army orderly, who was responsible for 45 influenza patients, fell ill to the virus. However, the orderly did not quit working and obtain bed rest until his temperature became 102 degrees (Byerly, 2005). Upon recovering somewhat, he immediately commenced orderly duties again only to experience additional reinfection and hospitalization (Byerly, 2005). Although the orderly survived influenza, others were unlucky and succumbed fatally to the illness. Among navy hospitals, 32 nurses died from influenza infections while caring for patients who were victims of the pestilence (Naval History and Heritage Command, 2018).

Emergency responses to the influenza pandemic experienced a variety of impediments. Since the pandemic occurred during global war, it provided opportunities to insidiously attack enemy forces. For instance, near Washington, D.C., at an army hospital, two German spies were found posing as doctors, and were transmitting the influenza virus to American soldiers (National Archives and Records Administration, 2018). When discovered, the German soldiers were executed (National Archives and Records Administration, 2018).

Attempts were made to provide information to individuals so they could avoid infection. People were encouraged to avoid the following: individuals who coughed or sneezed; constipation; crowds and crowded forms of transportation; public gatherings and places (e.g., theaters); traveling by railroad; using glasses or cups that were unclean; poorly ventilated areas and rooms; damp clothes; and physical fatigue (Department of the Navy, 1918). Additionally, people were expected to remain in good physical condition; eat nourishing meals; hydrate sufficiently; keep feet dry and remain warm; sleep a minimum of 7 hr daily (Department of the Navy, 1918).

Dissemination of information within the general public involved a variety of methods. Individuals who had much public contact, such as postal employees and teachers, were used to provide information to the general public (Ott, Shaw, Danila, & Lynfield, 2007). When schools closed, as volunteers, many teachers were asked to teach health classes to the general public (Ott, Shaw, Danila, & Lynfield, 2007). The primary goals of the health courses were to abolish the uses of shared drinking cups and to eliminate roller towels used when washing hands (Ott, Shaw, Danila, & Lynfield, 2007).

Generally, influenza required an average of 6 weeks from start to finish. Typically, patients suffered with the illness for about 2 weeks (Richardson, 1918). Afterward, sequelae and deaths occurred over a 2-week period (Richardson, 1918). For those survived, convalescence occurred over a 3-week period (Richardson, 1918). Despite the best efforts toward countering Spanish influenza, the epidemic ravaged the nation. Both urban and rural areas were affected lethally. Although some efforts were made to identify and delineate the characteristics of the influenza virus, initial efforts were unfruitful. Essentially, for many, influenza infection meant a strong possibility of death personally or for a loved one. In many cases, response and management incorporated nothing more than empathetic efforts to make individuals comfortable while they suffered.

LESSONS LEARNED AND POLICY IMPACT

Although tragic, the 1918 Spanish influenza epidemic necessitated the implementing of mass public safety exercises. Through such experiences, both urban and rural areas across the nation gained valuable insight toward managing epidemic illness. Influenza did not discriminate; anyone fell ill to the disease. Thus, administrators had to adapt to constant change among response resources when responders themselves became ill with the disease. In some cases, given the combination of deployed personnel for the war effort and folks become ill with influenza, insufficient quantities of people were available for staffing medical centers or public burial details. Such a scenario has potential implications for modern emergency management policy. During modern times, hospitals and medical centers may craft plans whereby contingencies may be accommodated when pestilence occurs. Such plans may be creative with respect to identifying potential sources of qualified personnel and replacement personnel should illness befall primary responders.

The quickness of death polarized the medical community toward identifying and understanding the germ responsible for the illness. Essentially, the pestilence was a catalyst for medical research and vaccine development throughout the remainder of the 20th century. After identification of the virus and a stronger understanding of its characteristics, influenza viruses were crafted in the hope of lessening the risk of future outbreaks. During modern times, the general public is encouraged to receive influenza vaccination. However, no guarantee exists that the vaccination offered will match the actual strain of influenza that occurs. Thus, if an influenza pandemic or epidemic emerges again, it would be months before supplies of correctly formulated vaccines could be produced and distributed to needy patients.

The 1918 pestilence exhibited masses of sick people overrunning hospitals and medical centers. Such masses often exceeded the quantities of patients for which medical centers were designed and had resource capacities. With respect to modern emergency management, the scenarios provide important lessons for crafting and maintaining various "plans for surge capacity and community containment" (Ott, Shaw, Danila, & Lynfield, 2007). By doing so, better distribution of resources and provision of treatment may be achieved while simultaneously enacting methods for mitigating the spreading of disease within society.

The 1918 virus was also a lesson in human nature. Given that human nature is static, the behaviors of societal predecessors may provide considerations for modern times. When ill, members of the public were supposed to inform a doctor about their condition (Ott, Shaw, Danila, & Lynfield, 2007). Afterward, the physician was supposed to isolate the patient within their personal residence and inform the health department (Ott, Shaw, Danila, & Lynfield, 2007). However, not everyone desired to play by the rules. Many sick individuals did not immediately report their sickness because they had no desire for their house to be publicly known as a location that was infected by influenza (Ott, Shaw, Danila, & Lynfield, 2007). Because of such fear, many sick individuals did not seek medical assistance for their condition or waited until they were "gravely ill" before seeking a physician (Ott, Shaw, Danila, & Lynfield, 2007). As a result, secondary pneumonia had already developed by the time many sought medical treatment (Ott, Shaw, Danila, & Lynfield, 2007). During modern times, should pestilence occur, who is to say that society won't behave differently? Would contemporary families fear the public knowledge that their house was an unsafe location or that their relatives were the carriers of a potentially deadly disease?

Pestilence also brought about changes in sanitation and cleanliness within society. As a result, cities passed ordinances that mandated the sterilizing of cups and dishes among bars and restaurants. Ordinances were also passed that prohibited "roller towels" and the use of "common drinking cups" among "public restrooms" (Ott, Shaw, Danila, & Lynfield, 2007, p. 1). Such actions were precursors to many standards that exist among communities throughout the nation during modern times. For instance, people within contemporary society would believe it to be unthinkable that restaurant dishes were not sanitized between uses. Thus, many modern sanitation, health, and cleanliness policies were derived from attempts to quash the 1918 Spanish influenza.

Examining the 1918 pandemic provides insight regarding coordinated, effective measures toward delaying or avoiding disease. The experience provided opportunities for administrators to craft methods of societal control as methods of mitigating influenza while simultaneously attempting to benefit public health. With enhancement, such concepts permeate modern emergency management policies. For instance, during modern times, one method of addressing potential pandemics involves a three-stage approach: (a) delay, (b) contain, and (c) protect (Thackway & McAnulty, 2010). The initial stage, delay, attempts to thwart pestilence from entering a location (Thackway & McAnulty, 2010). Contain, the second stage, involves identifying infected people and limiting the transmission of disease (Thackway & McAnulty, 2010). Protect, the final stage, involves identifying and protecting individuals who have a high chance of severe illness.

IMPLICATIONS FOR MODERN EMERGENCY MANAGEMENT

Examining the historical case of the 1918 Spanish influenza epidemic provides considerations for modern emergency management. Some straightforward lessons arise from examining the historical records:

- Natural threats to human life may emerge and spread with speed, surprise, and violence. Despite the best technology and knowledge of the time, no guarantees exist that contemporary measures against disease will be effective.

- Mass illness may arise at any time.
- Lethal illness may be spread quickly via mass gatherings (e.g., mass Liberty Bond assemblies in both Memphis and Philadelphia).
- Disease does not discriminate; anyone may become infected, including first responders and medical personnel.
- Sufficient resources may not always exist when pestilence strikes.
- People may not always be compliant with public health mandates.
- Planned strategies, tactics, and operations should exist for addressing mass illness.
- No plan will be implemented flawlessly; unforeseen events will occur to impede or halt progress.
- Contingency planning should always exist for identifiable risks.
- Local infrastructure may be converted into makeshift resources for processing and treating patients.
- Despite much research and investigation, one may never fully know and understand the exact source of a specific illness.

Since 1918, two other massive, pandemic bouts with influenza occurred in the 20th century—1957 and 1968. Respectively, the three pandemics of the 20th century were eventually deemed to be Spanish, Asian, and Hong Kong forms of influenza (Fineberg, 2014). Although not completely pandemic or pandemic in adults, other substantial bouts with influenza included a 1947 "pseudopandemic" that exhibited a low death rate, a 1977 pandemic that affected children, and a 1976 "abortive epidemic of swine influenza" that had the potential of becoming pandemic (Fineberg, 2014, p. 1).

Modern times also have experienced influenza pandemic. During 2009, another influenza pandemic affected societies worldwide. Appearing in Mexico during the months of February and March, the H1N1 influenza virus spread quickly throughout the global community. By the month of June, a total of 73 nations had reported at least 26,000 cases (confirmed by laboratories). This quantity of reported infections officially qualified the illness as a pandemic per standards of the World Health Organization (Fineberg, 2014). Regarding the 2009 pandemic, between 123,000 and 203,000 deaths were estimated worldwide during the last 9 months of the year (Simonsen, et al., 2013). Between approximately 62% and 85% of infected individuals who died were less than 65 years of age (Janukavicius, 2014). Globally, the influenza pandemic killed approximately 284,000 people (Janukavicius, 2014).

Some believe that mass illnesses, such as influenza strains, are recurring, natural events. In other words, they represent a pandemic cycle. A pandemic cycle represents alternating periods of mass illness and overall wellness within society. Within the pandemic cycle, it is assumed that a form of illness will strike again at some point in the future. Some credence for this hypothesis was provided by the recurring instances of influenza pandemics throughout the 20th and 21st centuries. However, some debate the efficacy and accuracy of pandemic cycles. Juckett (2006) indicated that any theories of a "predictable pandemic cycle" should be forsaken. Contrastingly, some indicate that modern times are due for another pandemic, somewhere between the years 2017 and 2020 (Armstrong Economists, 2018). Regardless of one's position concerning the viability of pandemic cycles, one tautology exists, which is that an illness will invariably result in an epidemic.

Given such an observation, one may ponder the potential of another influenza pandemic. Yeung (2006) examined the potential relationship between influenza A and sunspot cycles through the period encompassing the years between 1700 and 2000. Using a binomial testing methodology and sunspot sensitivity exceeding 50 to detect sunspots and potential pandemics, the study showed a sensitivity of approximately 85% with respect to the detection of influenza pandemics (Yeung, 2006). The specificity was approximately 51% (Yeung, 2006). In short, the study suggested that using sunspot cycles may be an inexpensive and easy method for detecting potential influenza pandemics (Yeung, 2006). When it was published in 2006, the study forecasted that a period of great risk existed between the years 2008 and 2013 (Yeung, 2006). Given the occurrence of an influenza pandemic in 2009, Yeung's (2006) study merits additional investigation as a potential predictive tool for pandemic influenza outbreaks.

To date, with full certainty and accuracy, no discernable, absolute pandemic cycle has been observed regarding pandemic influenza throughout the preceding 300 years (Morens, Taubenberger, Harvey, & Memoli, 2010). However, some question whether pandemic influenza exhibits "macrocyclicity" wherein cyclical appearances of H1, H2, and H3 strains are exhibited about every 68 years, and are influenced by "waning population immunity of approximately the same duration as the mean human life span" (Morens, Taubenberger, Harvey, & Memoli, 2010, p. 10). Although such notions represent completely fascinating inquiries and hypotheses, much more data (centuries) will be necessary before any hypothesis may be testing regarding such ideas.

SUMMARY

The potential for public health emergencies due to natural causes, and/or human error, can never be overlooked due to the tremendous potential for mass casualties, as well as significant economic losses when government services and private sector operations are disrupted (Waugh, 2000). Tragically and regrettably, improvements among emergency management methods and resources result from experiencing great calamities and extreme hardships.

Given its horrendous death tolls, the 1918 Spanish influenza pandemic was the catalyst for profound conjecture toward hypothesizing that influenza was caused by a virus. The tragic deaths of many millions motivated scientists and researchers alike to pursue a variety of disciplines with greater vigilance. Certainly, improvements among the disciplines of virology, bacteriology, pathology, and immunology occurred because of the Spanish influenza.

Within the context of emergency management, the pandemic provided a variety of lessons for both modern and future generations. Through the lens of history, it yielded insight regarding what worked well or failed toward quashing the illness or mitigating its negative effects. Regardless of the best efforts toward abating the pestilence, it ravaged society unimpededly. Death tolls and infected quantities were most likely underreported since not all states participated in some type of formal reporting process. Because of its magnitude and lethality, medical researchers feverishly sought the catalyst of the illness and effective treatments. Given the extensiveness of sickness and high death rates that devastated society a century ago, one may contemplate the severity of a potential influenza pandemic during modern times.

Despite much progress technologically, medically, and pharmaceutically, no guarantees exist that modern society could withstand pandemic pestilence. After all, what good are such advancements if they become either inaccessible or unavailable? Although American society has experienced a variety of pestilences, its healthcare infrastructure is limited. Budgetary and financial reductions have weakened societal healthcare to lower levels than those that existed during 1976, when the swine influenza occurred (Juckett, 2006). Quantities of hospital beds were reduced substantially by managed and regulated healthcare (Juckett, 2006). Both the capacity for fostering emergency response and the availability of respirators have been diminished (Juckett, 2006). Nationally, little surplus capacity remains within the healthcare system, and fewer locations are available to provide medical care if pandemic illness occurs (Juckett, 2006).

Another factor that may impact any response to a modern pandemic involves the overuse of pharmaceuticals during modern times. Unlike the period of the early 20th century in which antibiotic and antiviral medicines did not exist, modern society has witnessed a vast overuse of such medicines. As a result, many modern strains of illness are insusceptible to some antibiotics. Similarly, resistance to antiviral medicines is also occurring (Morens, Taubenberger, Harvey, & Memoli, 2010). Thus, treating any pandemic outbreak may prove to be difficult, problematic, and challenging.

An influenza outbreak may be considered from an economic perspective. Doss, Sumrall, McElreath, and Jones (2013) indicated that any economic system must address three key questions societally: what to produce, for whom to produce, and in what quantity to produce? Such questions are central to assessing production and consumption within any society (Doss, Sumrall, McElreath, & Jones, 2013). Within the context of influenza, producing vaccines requires a minimum, static period of 6 months with respect to the availability of fertile chicken eggs (Juckett, 2006). The production process also slows because of essential quality management issues and requirements to ensure the quality of any produced vaccines (Juckett, 2006). Additionally, no machine is perfect; equipment used for producing vaccines may break or need maintenance, thereby impeding production processes (Juckett, 2006).

Such economic perspectives are common within the medical community. After a century of examining influenza, medical sciences and arts have advanced knowledge and understanding of the virus and the resulting illness. However, just as in 1918, transportation modalities, wherein individuals experience close quarters and travel to distant locations quickly, provide opportunity for speedy transmission of illness geographically. Although numerous treatments exist in modern times that were unavailable a century ago during the 1918 pandemic, a primary issue involves the availability and sufficiency of resource quantities necessary for addressing pandemic illness (Morens, Taubenberger, Harvey, & Memoli, 2010). In other words, a prime concern of dealing with modern pandemics effectively and efficiently involves medical capacities and resource availabilities (Morens, Taubenberger, Harvey, & Memoli, 2010).

The future is both unknowable and uncertain. No guarantee exists that an influenza pandemic will appear soon. However, society must remain vigilantly prepared for the emergence of pestilence. During modern times, the potential of quickly spreading any illness, not just influenza, exists via the use of several transportation methods (e.g., airplanes). Pestilence may proliferate anywhere, anytime. Contemporary society possesses better sanitation and hygiene practices, medicines, healthcare, technology, and knowledge of influenza than those that existed in 1918, but also experiences failure risks associated with any emergency mitigation and response initiatives. Key personnel may become ill (or die); equipment may fail or be unavailable; insufficient or tainted medicines may exist; and so forth. Despite such risks, the necessity of preparedness complements the inevitability of pestilence. At some point, pandemic influenza will occur again. Thus, emergency management resources must be as prepared as possible to address the potential of pandemic.

REFERENCES

Armstrong Economists. (2018). *Plague cycle: Moving into peak 2017-2020.* Retrieved from https://www.armstrongeconomics.com/uncategorized/plague-cycle-moving-into-peak-2017-2020

Armstrong, J.F. (2015). *Philadelphia, nurses, and the Spanish Influenza pandemic of 1918.* Retrieved from https://www.history.navy.mil/content/history/nhhc/research/library/online-reading-room/title-list-alphabetically/i/influenza/philadelphia-nurses-and-the-spanish-influenza-pandemic-of-1918.html

Associated Press. (1998). *Alaskan victim of 1918 flu yields sample of killer virus.* Retrieved from http://www.nytimes.com/1998/02/08/us/alaskan-victim-of-1918-flu-yields-sample-of-killer-virus.html

Billings, A. (1997). *The Influenza Panic of 1918.* Retrieved from https://virus.stanford.edu/uda

Byerly, C.R. (2005). *Fever of war: The influenza epidemic in the U.S. Army during World War I.* New York, NY: New York University Press.

Centers for Disease Control and Prevention. (2018). *Past pandemics.* Retrieved from https://www.cdc.gov/flu/pandemic-resources/basics/past-pandemics.html

Department of the Navy. (1918). *Influenza.* Retrieved from https://www.archives.gov/exhibits/influenza-epidemic/records/washington-dc-directive-l.jpg

Doss, D., Sumrall, W., McElreath, D., & Jones, D. (2013). *Economic and financial analysis for criminal justice organizations.* Boca Raton, FL: CRC Press.

Ellis, C. (2016). *Prepping for a pandemic: Lifesaving supplies, skills and plans for surviving an outbreak.* Berkeley, CA: Ulysses Press.

Fineberg, H.V. (2014). Pandemic preparedness and response–Lessons from the H1N1 influenza of 2009. *New England Journal of Medicine, 370*, 1335–1342.

Finger, M. (2006a). *The return of the Spanish Lady.* Retrieved from http://memphismagazine.com/features/the-return-of-the-spanish-lady

Finger, M. (2006b). *The return to the Spanish Lady (Pt. II).* Retrieved from http://memphismagazine.com/features/the-return-of-the-spanish-lady-pt-ii

Garrett, T.A. (2007). *Economic effects of the 1918 Influenza pandemic: Implications for a modern-day pandemic.* St. Louis, MO: Federal Reserve Bank.

Glezen, W.P. (1996). Emerging infections: Pandemic influenza. *Epidemiologic Reviews, 18*(1), 64–76.

History Channel. (2018). *1918 Flu Pandemic*. Retrieved from http://www.history.com/topics/1918-flu-pandemic

Janukavicius, P. (2014). High H1N1 prevalence and mortality rates a concern. *CMAJ, 186*(3), 104.

Johnson, N.P., & Mueller, J. (2002). Updating the accounts: Global mortality of the 1918-1920 "Spanish" influenza pandemic. *Bulletin of the History of Medicine, 76*(1), 105–115.

Juckett, G. (2006). Avian influenza: Preparing for a pandemic. *American Family Physician, 74*(5), 783–790.

Mayo Clinic. (2018). *Influenza*. Retrieved from https://www.mayoclinic.org/diseases-conditions/flu/symptoms-causes/syc-20351719

McBride, C. (2007). *Walking into a new spirituality: Chronicling the life, ministry, and contributions of Elder Robert E. Hart, B.D., LL.B., to the CME church and COGIC: With some additional COGIC history*. Lincoln, NE: Universe Publishing.

McElreath, D., Jensen, C., Wigginton, M., Doss, D., Nations, R., & Van Slyke, J. (2013). *Introduction to homeland security* (2nd ed.). Boca Raton, FL: CRC Press.

Merriam-Webster. (2018). *Influenza*. Retrieved from: https://www.merriam-webster.com/dictionary/influenza

Morens, D.M. & Fauci, A.S. (2007). The 1918 Influenza pandemic: Insights for the 21st century. *The Journal of Infectious Diseases, 195*(7), 1018–1028.

Morens, D.M., Taubenberger, J.K., Harvey, H.A., & Memoli, M.J. (2010). The 1918 influenza pandemic: Lessons for 2009 and the future. *Critical Care Medicine, 38*, 10–20.

Natera, F. (2012). *Coyame: A history of the American settler*. Bloomington, IN: Xlibris Publishing.

National Archives and Records Administration. (2018). *Letter from nurse to her friend at the Haskell Indian Nations University, Kansas, October 17, 1918. Bureau of Indian Affairs*. Retrieved from https://www.archives.gov/exhibits/influenza-epidemic/records/volunteer-nurse-letter.pdf

National Archives and Records Administration. (2017). *The deadly virus: The Influenza epidemic of 1918*. Retrieved from https://www.archives.gov/exhibits/influenza-epidemic

Naval History and Heritage Command. (2018). *Influenza of 1918 (Spanish Flu) and the US Navy*. Retrieved from https://www.history.navy.mil/research/library/online-reading-room/title-list-alphabetically/i/influenza/influenza-of-1918-spanish-flu-and-the-us-navy.html

Ott, M., Shaw, S.F., Danila, R.N., & Lynfield, R. (2007). Lessons learned from the 1918-1919 influenza pandemic in Minneapolis and St. Paul, Minnesota. *Public Health Reports, 122*(6), 803–810.

Richardson, D. A. (1918). *Influenza epidemic at Pueblos of Albuquerque Day School section*. Retrieved from https://www.archives.gov/exhibits/influenza-epidemic/records/visiting-doctor-letter.pdf

Ryan, J. R. (2008). *Pandemic influenza: Emergency planning and community preparedness*. Boca Raton, FL: CRC Press.

Simonsen, L., Spreeuwenberg, P., Lustig, R., Taylor, R., Fleming, D., Kroneman, M., Van Kerkhove, M., Mounts, A., & Paget, W. (2013). Global mortality estimates for the 2009 influenza pandemic from the GlaMOR project: A modeling study. *PLOS Medicine*. Retrieved from https://doi.org/10.1371/journal.pmed.1001558

Stanford University. (2017). *The Influenza Panic of 1918*. Retrieved from https://virus.stanford.edu/uda

Taubenberger, J. K., Reid, A. H., Lourens, R. M., Wang, R., Jin, G., & Fanning, T. G. (2005). Characterization of the 1918 influenza virus polymerase genes. *Nature, 437*, 889–893.

Taubenberger, J. K. (2006). The origin and virulence of the 1918 "Spanish" influenza virus. *Proceedings of the American Philosophical Society, 150*(1), 86–112.

Thackway, S. V., & McAnulty, J. M. (2010). *Pandemic (H1N1) 2009 influenza in NSW*. Retrieved from http://www.publish.csiro.au/NB/pdf/NBv21n2

Vincent, J. (2010). *Yearbook of intensive care and emergency medicine, 2010*. New York, NY: Springer.

Waugh Jr., William L., (2000). *Living with hazards dealing with disasters: An introduction to emergency management*. Armonk, New York: M.E. Sharpe.

Wever, P. C., & van Bergen, L. (2014). Death from 1918 pandemic influenza during the First World War: A perspective from personal and anecdotal evidence. *Influenza and Other Respiratory Viruses, 8*(5), 538–546.

Williams, P. J. (2013). *Philadelphia: The World War I years*. Charleston, SC: Arcadia Publishing.

Yeung, J. W. (2006). A hypothesis: Sunspot cycles may detect pandemic influenza A in 1700–2000 A.D. *Medical Hypotheses, 67*(5), 1016–1022.

The Sinking of the Sultana (1865)

13

Figure 13.1 The Sultana
Source: Library of Congress

"The sight of 2000 ghostly, pallid faces upturned in the chilling waters of the Mississippi, as I looked down on them from the boat, is a picture that haunts in my dreams." (Huffman, 2009)

BRIEF OVERVIEW

It is hard, if not impossible, to truly understand the terror experienced by those who were involved in the disaster of the Sultana. In what would prove to be the greatest maritime disaster in the history of the United States, thousands would die, and those who survived would, for the rest of their lives, sustain both physical and psychological injuries after the complete devastation of the Sultana, due to the swift currents of the mighty Mississippi River, which occurred one night in 1865, just above Memphis, Tennessee.

The American Civil War had just ended, and the South was in ruins. Most of the military personnel from both sides found themselves far from home, and were facing an uncertain future as they began the reconstruction of their lives. The majority had vastly aged by their wartime experiences. For some, those involved in the Sultana disaster, the war would have at least one more chapter.

The Sultana was a worn-out wooden side-wheeler, and was one of many vessels that had traveled the nation's internal series of rivers while moving people and cargo during the conflict. Overworked, and in need of repair, the Sultana was pressed into service for another trip up the flooded Mississippi River, and would soon become the center figure in the tragic series of events. At least 1,500 individuals perished in the disaster (Berry, 1892).

INTRODUCTION

The nation's rivers played a vital role in the evolution of the nation, serving as the nation's early highway system. By the early 1800s, steam had been captured and powered the riverboats, which proved to be an essential resource that was vital to the economic growth of our new nation. In times of peace, the rivers and ports located at St. Louis, Cape Girardeau, Cairo, Memphis, Vicksburg, Natchez, and New Orleans were major centers of economic development. In times of war, these locations also played a significant role in implementing a variety of strategic actions.

The value of the nation's rivers, along with the hundreds of vessels used during the war for combatant and transport purposes, cannot be understated. The numerous vessels would be critical in the movement of soldiers from the battlefields to their home during the discharge process, and their role remained an integral part of the conclusion of the Civil War. In many ways, these ships served as the source of transportation from war to peace.

In 1865, the Sultana was just one of many overworked riverboats, struggling against the currents of the river, now taking over a thousand Union soldiers, many recently released from Confederate prisoner camps, and civilians up the Mississippi and to civilian life. Constructed of wood and iron, the Sultana was designed to be powerful enough to withstand the strain of the nation's great river and large enough to be profitable.

With each trip, the ship pushed its limits of operational safety as maintenance was often deferred to keep the ship in service. For over 2 years, the Sultana moved up and down the river, transporting people and supplies as her cargo, but eventually, the ship's infrastructure suffered from its numerous trips, and it was in dire need of repair.

The Setting

The War Between the States was rapidly coming to a conclusion. The Mississippi and the ports along the river were under the control of federal authorities. Military demobilization would soon be in full swing as hundreds of thousands of soldiers from both sides began their travels home from the various battlefields. The need to rapidly move men and materials created a logistical challenge for military authorities. With many of the railroads in the South destroyed in the struggle and the primitive nature of roads and overland travel, the reliance upon the nation's networks of rivers to move men and cargo was significant.

In this setting, hundreds of boats, in various states of repair, were pressed into service and were believed to be capable of moving men and cargo throughout the river system. Understanding the economic opportunity presented by these federal contracts, many boats, some of which were in need of repair or overhaul, were pushed into service. The Sultana was one of these vessels.

The Sultana was just one of hundreds of coal-burning steamships that moved people and cargo along the nation's rivers in the mid-1800s. Constructed and launched in January 1863, the Sultana was of solid construction and reflected the technology in river boat construction of the time. The Sultana was a coal-burning side-wheeler, 260 ft in length with multiple desks, adequate room for cargo and passengers, which was designed to have the size and power to navigate in the brown waters of the Mississippi River system. Although the Sultana was designed to carry a maximum load of 375 passengers and cargo, as with similar ships, it was generally overloaded with passenger and cargo to increase profit (Madden, 2015).

For the commercial vessels that survived the journey supplying the federal forces, the War Between the States proved profitable. Military officers responsible for contracting cargo and troop transportation served with little oversight. Shipowners found moving people and cargo very lucrative. For this last trip, the government agreed to pay $5 for each enlisted man, and $10 for each officer who made the trip. Overloading of the

Sultana proved appealing to both the Army officials and the shipowner: (1) for the Army contract officer, overloading the Sultana would expedite the soldiers return home; and (2) for the shipowner, more profit.

During the War, contracts were plentiful, and such financial assistance to transport men and materials was intended to supply and replenish the Federal Army as they moved closer to victory. For many shipowners, scheduled and needed maintenance was delayed in favor of profits. As the War drew to a close, many of these boats had reached the end of their operational life, but the desire for another contract kept them in service. Consequently, as in the case of the Sultana, economic opportunity, greed, and a disregard for safety would be the ingredients for a disaster.

When constructed in January 1863 in Cincinnati, the 260-ft Sultana was a state of the-art vessel, and was registered at 1,719 tons (Potter, 2006). The Sultana operated with a crew of about 80, and, when constructed, included the most modern safety equipment available—safety gauges that fused open when the internal boiler pressure reached 150 lb/in.^{2}, three firefighting pumps, a metallic lifeboat and a wooden yawl, 300 ft of fire hose, 30 buckets, five firefighting axes, and 76 life belts (Ambrose, 2001).

By April 1865, though only 27 months old, the Sultana was worn-out beyond its years. A veteran of river service, the Sultana was used extensively to move troops and cargo along the river throughout the second half of the Civil War. Often overloaded, the strain on its boilers had taken its toll. Combined with her route between St. Louis and New Orleans, the life of a river vessel could only endure several months while navigating through the mighty Mississippi River.

For the Sultana, its last trip began in New Orleans on the night of Friday, April 21, 1865 when she started her journey up the river with 75–100 cabin passengers and a cargo, which included a hundred hogsheads of sugar and a hundred head of assorted livestock (Larson, 1955). In charge of the Sultana was Captain Frederick Speed, an experienced river captain who methodically navigated the ship, and sought to obtain one of the many contracts being awarded to transport soldiers north.

The Sultana continued its journey up the river, but there was an issue with one of its boilers. In Vicksburg, the Sultana was inspected, and its boilers were subjected to repair. Repairs took time. Records indicate that the repair work continued for as many as 33 hr until the ship was declared safe again.

Vicksburg had been turned into a great repatriation center, and here were gathered thousands of gaunt, worn-out men in faded blue uniforms—Union prisoners of war, just released from the horrors of prison compounds like Andersonville, waiting in Vicksburg for transportation to their northern homes (Larson, 1955). In Vicksburg, Army contractors booked approximately 2,000 Union solders onto the Sultana, to be transported, many of whom were recently released prisoners of war. Loaded and with a quick boiler repair, the Sultana was declared safe and capable of continuing her journey to St. Louis. Captain J. Cass Mason of St. Louis oversaw the Sultana, and had a reputation as an experienced veteran of the Mississippi River (Larson, 1955). As they boarded, it was obvious that the ship was overloaded, although there is no indication that anyone expressed concern.

The Sultana arrived in Memphis where it received further repairs and unloaded some passengers and cargo. Some of the passengers disembarked. The hogsheads of sugar were unloaded, and some of the stronger ex-prisoners helped in the work. Still others left the ship to enjoy Memphis and failed to return before the ship departed. About midnight, the overloaded Sultana pulled into the main current of the river, straining to move up current in a river swollen by recent rains. The majority of the passengers settled down to sleep. About 7 miles north of Memphis the end of the Sultana was near.

The Disaster

It was a dark April night in 1865 when the Sultana strained and struggled against a cold, swollen and rapidly running Mississippi River, as it moved northward a few miles north of Memphis. The Sultana was overloaded, carrying about 2,600 passengers and cargo. Of the passengers, about 2,400 were Union soldiers, many recently released from Southern Prisoner of War Camps, all traveling north and for the majority, an expectation of military discharge and return to civilian life. In addition to the soldiers, there were also about 180 civilians, and a crew of about 85, all hoping for an uneventful journey. Below deck, crewmen fed coal into the four boilers, which provided power to the Sultana as it struggled within the swollen river.

A few miles above Memphis, the unexpected would soon wreak havoc, which would result in a catastrophic nightmare. As the Sultana pushed up the Mississippi River, its boilers that were overworked and in need, began to strain as they fought against the current. At approximately 2:00 a.m., three of the overworked boilers powering the vessel exploded, which started a series of events that resulted in a burning shipwreck, confusion, and all passengers suddenly fighting for their lives. Unfortunately, the quiet night changed into a scene of chaos, death, and destruction.

Over a brief period, an explosion ensued on the Sultana, which sent an orange-colored flame into the dark sky that was visible for miles. In Memphis, the night watchman on the river gunboat USS *Grosbeak* saw the light and heard the noise. Immediately, the night watchman notified the captain of the *Grosbeak*, who believing a disaster had just occurred, ordered the cast off the mooring lines and movement of the *Grosbeak* up the river. Other steamers on the Memphis waterfront also suspected a disaster had occurred and joined the movement of the *Grosbeak* up the river (Larson, 1955).

As the tragedy unfolded, the subsequent explosion, fire, confusion, and rapid current of the river inflicted a maritime death toll of a level unseen in the United States. Many of the passengers were asleep when the explosion occurred, and the sudden shock, combined with the darkness, intensified the confusion. Ultimately, the hopeless fate of the Sultana and its many passengers was sealed in the crux of the mighty Mississippi River. Figure 13.2 shows a historical depiction of the disaster.

The deck supporting the Sultana's main passenger cabins collapsed at one end into the fire, and its wreckage immediately began to drift downstream. The twin smokestacks, hallmark of every Mississippi packet boat, soon crashed, which resulted in additional fatalities (Larson, 1955). As the Sultana continued to disintegrate, fire continued to spread that was fueled by the timbers and cargo of the ship and gale forced April winds, which further contributed to impending deaths and injuries to passengers.

Of those onboard, over 1,500 were killed and hundreds more injured (Berry, 1892). Almost immediately, the first of over 2,000 entered the water. Many could not swim, and were quickly pulled under the rapidly moving currents, while others struggled to survive amid a hostile river.

The fortunate ones who found themselves in the river clung to anything that would float, which included debris in the river, horses and mules that had escaped the boat, anything hoping to make it to shore. Some tried to swim, others decided to float. The level of confusion and fear can only be imagined. The night was dark, the river swift and almost 5 miles wide at the point of the disaster (Dennis, 2015). The section of the Mississippi

Figure 13.2 The Sinking of the Sultana

Source: Library of Congress.

River above Memphis had now become the scene of the nation's greatest maritime disaster as the burned Sultana drifted and began to sink.

Toward morning, the wreckage of the Sultana struck a small island, and some of those still aboard jumped ashore. Approximately, 20–30 more passengers managed to fabricate a makeshift raft from broken timbers. Eventually, as the flames died down, what was left of the Sultana drifted away and sank.

Along the shoreline of the Mississippi River, the sound of an explosion was heard, and many local residents took immediate steps to quickly rescue those they could. Such rescue efforts resulted in the saving of many who would have otherwise been lost. Throughout the night and into the next day, many, including former enemies, risked their lives to save the victims of the disaster, but the death toll would be great.

Eventually, boats from Memphis arrived and rescued hundreds from the Mississippi River. Survivors struggled in the river, clinging to anything they believed would float—logs, rafts, spars, barrels, sections of railing, and other bits of wood. Others were found on both shores of the Mississippi, clinging to trees or driftwood, many of them badly burned and without clothing (Larson, 1955).

A subsequent investigation of the wreckage determined that three of the Sultana's four boilers had exploded. The steam and scalding water from the ruptured boilers immediately killed those nearby, and injured others, many critically. Flames spread throughout the wooden ship, with the intensity of the fire fueled by cargo and personal items of the passengers.

Survivors of the Sultana wreckage were sent to hospitals in Memphis. Many were naked by the time they were rescued, having shed their clothes to make it easier to swim. In Memphis, most were given red long underwear, which some of them wore as they wandered the streets of Memphis, while they waited for the Army to provide a new set of clothing and shoes. When they were well enough, the survivors were put on other boats and sent north (Huffman, 2009).

AFTERMATH AND INVESTIGATION

Each story of survival reveals one part of a great tragedy. The exact toll of those injured, who would later die of sustained injuries, is unknown. Approximately 700 of Sultana's passengers and crew who sustained injuries arrived in Memphis, where an estimated 200 of them soon died (Madden, 2015). Organized efforts to recover the dead continued for days with bodies eventually found as far south as Vicksburg. In Memphis, the riverfront became a large morgue with more than a 100 coffins lined up as bodies were continually being brought, or washed ashore.

Many passengers of the Sultana remained unaccounted for. For several months after the disaster, passengers and crewmen traveling on the Mississippi River in other boats reported seeing what they believed to be human remains along the river bank, which they assumed to be from the Sultana disaster. Thereafter, many of the bodies recovered were so decomposed that it was impossible for the remains to be identified.

On May 19, 1865, less than a month after the disaster, Brigadier General William Hoffman, Commissary General of Prisoners, released the results of his preliminary investigation of the Sultana disaster. He reported an overall loss of soldiers, passengers, and crew of 1,238, a number that upon further investigation, proved to be low. Such disaster aspects that included the explosion, a fire, and the river, all combined to make true accountability impossible.

A military inquiry convened to investigate the disaster in an effort to determine cause and responsibility. One of the initial findings found that there was notable pressure to quickly move thousands of soldiers north, which created a demand for boat space that far exceeded available capacity. In addition, it was also discovered that General Dana, the Union Commander for the Department of the Mississippi, had ordered that the recently freed federal military prisoners of war be sent northward from Vicksburg on privately owned steamboats, with the vessels' owners receiving 5 dollars per enlisted man carried and 10 dollars for each officer (Potter, 2006).

The captain of the Sultana, J. Cass Mason, met with Brigadier General Morgan L. Smith, and Lieutenant Colonel Reuben B. Hatch in Vicksburg, where he secured a commitment to transport hundreds of men upriver, which was above and beyond what the ship was designed to safely carry (Potter, 2006). With this demand for space, contracting officers willingly approved the requests of the Sultana's captains and owners to compromise its safety limits. It was clear that the Sultana was extremely overloaded, carrying over six times its expected capacity as it departed Memphis (Ambrose, 2001).

The official cause of the Sultana disaster was determined to be a mismanagement of water levels in the boiler, exacerbated by the fact that the vessel was severely overcrowded, and too top heavy. As the Sultana made its way north following the twists and turns of the river, it listed severely to one side then the other. The inner workings of the Sultana consisted of four boilers that were interconnected and mounted side by side, so that if the boat tipped sideways, water would tend to run out of the highest boiler. With the fires still going against the empty boiler, this created hot spots, which contributed to its demise.

When the boat tipped, water rushed into the empty boiler resulting in an instant flash of steam and a pressure surge. This proved to be a critical element in the disaster. The official inquiry found that the boat's boilers exploded due to the combined effects of careening, low water level, and a faulty repair to a leaky boiler made a few days earlier.

LESSONS LEARNED AND POLICY IMPACT

The scope of the Sultana disaster quickly led to an official investigation. General Dana and Brigadier General William Hoffman, the U.S. Army Commissary General of Prisoners, each conducted an investigation. Hoffman's findings were the most critical of the military's involvement in the Sultana tragedy, and he concluded that the "shipment of so large a number of troops (1,866) on one boat was, under the circumstances, unnecessary, unjustifiable, and a great outrage on the troops." Hoffman's report also pointed a finger of guilt at General Smith, noting that although Smith "had nothing officially to do with the shipment of the troops; yet as it was officially reported to him by Captain Kerns that too many men were being put on the Sultana, it was proper that he should have satisfied himself from good authority whether there was sufficient grounds for the report, and if he found it so he should have interfered to have the evil remedied. Had [Smith] done so, the lives of many men would have been saved" (Potter, 2006).

IMPLICATIONS FOR MODERN EMERGENCY MANAGEMENT

Although this disaster occurred over 150 years ago, there are many lessons drawn from such events, which pertain to passenger and cargo limits, vessel inspection, riverway and channel safety, maritime safety regulations, and search and rescue. The significance of the nation's rivers remains a critical element to our economy, but the profile of river traffic has changed over the last century (McElreath, Doss, et al., 2014). The steamboats of the 1800s have been replaced by over 5,500 tugboats and towboats that move more than 880 million tons of domestic cargo each year. Counted annually among this cargo is 60% of U.S. grain intended for export, and approximately 244 million tons of petroleum (Thorsen, 2017).

To keep the rivers safe, the nation relies on, among other agencies, the United States Coast Guard, the National Oceanic and Atmospheric Administration (NOAA) and the United States Army Corps of Engineers (USACE). Maintaining Aids to Navigation is the Coast Guard's oldest mission, tracing its roots to the ninth law passed by Congress, which created the U.S. Lighthouse Establishment in 1789 (McElreath, Jensen, et al., 2014).

The Coast Guard, USACE, and the NOAA have a shared responsibility for navigable waterways in the United States. As such, the NOAA produces nautical charts, and provides tidal and weather information, whereas the USACE conducts hydrographic surveys, maintains locks, and dredges waterways. As the multi-mission maritime service, the Coast Guard is responsible for the safety, security, and stewardship of American waterways, and maintains the Aids to Navigation system, which includes more than 48,000 buoys, beacons, and electronic aids. The system safely guides millions of mariners and trillions of dollars in trade into U.S. ports (Ham, 2017).

SUMMARY

The tragedy of the Sultana would quickly be overshadowed by the death of President Lincoln. By April 1865, the nation was tired of the conflict, and the deaths and injuries related to the Sultana proved to be a footnote in the history of the nation. Many of those recovered could not be identified, and the names of others initially identified were lost in the confusion of the burial relocation process that occurred in the late 1860s, to the Memphis National Veterans Cemetery, which holds the second largest number of unknown soldiers of any national cemetery, and is the final resting place for hundreds of casualties of the **SS** Sultana disaster.

Some of the wreckage of the Sultana was recovered and sold for salvage, the rest of the ship lies buried in a field in Arkansas, which resulted from the movement of the river away from the location of the sinking. One final note of interest involves the unknown whereabouts of the Sultana's safe that reportedly contained over $32,000, much of it in gold, which was mysteriously never recovered.

REFERENCES

Ambrose, S. (2001, May 1). *Remembering Sultana.* National Geographic News. Retrieved from http://news.nationalgeographic.com/news/2001/05/0501_river5.html

Berry, C. D. (1892). *Loss of the Sultana and reminiscences of survivors.* Lansing, MI: Thorp.

Dennis, B. (2015, March 27). The Sinking of the Sultana: A disaster lost in the lingering fog of the Civil War. *The Washington Post.*

Ham, W. (2017, January 10). *Army engineers, Coast Guard partner to chart U.S. water navigation's future.* Department of Defense.Retrievedfromhttps://www.defense.gov/News/Article/Article/1046239/army-engineers-coast-guard-partner-to-chart-us-water-navigations-future

Huffman, A. (2009). *Sultana.* New York: Harper.

Larson, C. A. (1955, October). *DEATH on the DARK RIVER, The Story of the Sultana Disaster in 1865.* American Heritage. Retrieved from http://www.americanheritage.com/content/death-dark-river

Madden, D. (2015, April 27). *The Sinking of the Sultana.* The New York Times. Retrieved from https://opinionator.blogs.nytimes.com/2015/04/27/the-sinking-of-the-sultana/?_r=0

McElreath, D. H., Doss, D. A., Jensen, C. J., Wigginton, M. P., Nations, R., Van Slyke, J., & Nations, J. (2014). *Foundations of emergency management.* Dubuque: Kendall Hunt.

McElreath, D. H., Jensen, C. J., Wigginton, M. P., Doss, D. A., Nations, R., & Van Slyke, J. (2014). *Introduction to homeland security.* Boca Raton: CRC Press.

Potter, J. O. (2006, June 12). *Sultana: A tragic postscript to the Civil War.* HistoryNet. Retrieved from http://www.historynet.com/sultana-a-tragic-postscript-to-the-civil-war.htm

Thorsen, L. (2017, February 6). *River commerce depends on aging, often broken Coast Guard cutters for navigation.* St Louis Post Dispatch. Retrieved from http://www.stltoday.com/business/local/river-commerce-depends-on-aging-often-broken-coast-guard-cutters/article_a89d9b6e-7ab1-54c4-8ad9-7c1c5191388c.html

14 Triangle Shirtwaist Factory Fire (1911)

The World.

154 KILLED IN SKYSCRAPER FACTORY FIRE; SCORES BURN, OTHERS LEAP TO DEATH.

700 WORKERS, MOSTLY GIRLS, TRAPPED; BODIES OF DEAD HEAP THE STREETS; ONLY ONE FIRE ESCAPE FOR ALL.

Employees Caught on Eighth, Ninth and Tenth Floors—The Blaze Spreads with Great Rapidity—Victims Jump from Window Ledges with Clothing Aflame and Pile Up Below, Dead and Dying—Life Nets Either Torn from Grasp of Rescuers or Burst by Force of Numbers—Criminal Negligence May Be Charged for Locked Fireproof Doors Leading to Stairs—Blaze Is in Triangle Waist Co. Rooms, Washington Place and Greene St.

Figure 14.1 Newspaper Reporting Shirtwaist Fire
Source: Everett Historical/Shutterstock.com

BRIEF OVERVIEW

On March 25, 1911, a fire erupted in the Triangle Shirtwaist Factory. In the end, of the 600 women, at least 145 people were killed, and many more injured. When they found the escape routes from the upper floors blocked, many of those killed jumped to their deaths to escape the intense fire; others fell victim to the fire and smoke inhalation. The danger of fire in factories like the Triangle Shirtwaist was well known, but there was little incentive to make safety-related changes from the government or the garment industry. These shops employed a large percentage of young women, the majority of whom were recent immigrants (Staff, 2009b). The Triangle Shirtwaist Fire remains one of the worst industrial disasters in the history of New York City and served in many ways as a "wake-up" call bringing attention to the conditions in the hundreds of small manufacturing shops operating in the late 1800s and early 1900s. This "wake-up" call inspired a series of laws and regulations that better protected the safety of workers in the industrial setting.

INTRODUCTION

The Triangle Shirtwaist Factory was one of numerous small factories located in New York City's garment district in the early 1900s. These shops employed thousands of young, female immigrants who worked long hours at low wages, and often in poor working conditions. The International Ladies Garment Workers Union (ILGWU) made an attempt to organize these workers, but their efforts failed.

The Triangle Shirtwaist Factory manufactured shirtwaists, a garment worn at the time by working women who wanted a stylish but comfortable fashion. The shirtwaist was simple could be mass-produced. The Triangle Shirtwaist Company was the nation's largest manufacturer of shirtwaists at the time of the fire (Lifflander, 2011).

The Triangle Shirtwaist Company occupied the eighth, ninth, and tenth floors of the 10-story Asch Building. By 1911, half of New York City's laborers in the various garment factories worked above the seventh floor in their workplace (Behrens, 1983). The New York City Fire Department's ladder companies could not reach floors at this height. In the early 1900s, factory fires were viewed as a risk of the industry. Deaths in the manufacturing setting were common with some estimates suggesting one life per day was lost in the workplace (New York Factory Investigating Commission, 1912, p. 28).

The Setting

The Triangle Shirtwaist Factory occupied three floors of the 10-story Asch Building, which was located in the Greenwich Village area of New York City. The company employed about 600 workers, most of whom were young, female immigrants aged 13–23. Most of those employed had recently migrated from Germany, Hungary, Russia, and Italy. Many barely understood English. Workers typically put in a 52-hr work week (Pence et al., 2003).

At the time of the fire, there were four elevators in the building with access to the factory floors, but only one was fully operational, and it could hold only 12 people at a time. There were two stairways down to the street, but one was locked from the outside to prevent theft by the workers, and the other only opened inwardly. The fire escape, as all would come to see, was shoddily constructed, and could not support the weight of more than a few women at a time (Staff, 2009b).

Few factory owners felt responsibility for the safety or working conditions of those employed (Behrens, 1983, pp. 5, 29). As Stein (1962) noted: ". . . the women who produced the shirtwaists were both figuratively and literally invisible to the owners who profited from them" (p. 161).

The Disaster Unfolds

The fire erupted around 4:40 p.m., on Saturday afternoon, March 25, 1911 as workers were preparing to end their shift. The fire was believed to have started in a scrap bin located in a corner of the eighth floor. The first alarm made to the fire department came in at 4:45 p.m., when a passerby in the street noticed smoke coming from the eighth floor. Although smoking had been forbidden, some workers did smoke and a cigarette may well have been the cause of the blaze, as the Fire Marshal concluded in his report of the disaster (Chicago Sunday Tribune, 1911).

Figure 14.2 Triangle Shirtwaist Factory Interior Gutted by the Fire that Killed 146 Women Workers on March 25, 1911

Source: Everett Historical/Shutterstock.com

The smoke was initially spotted on the eighth floor. Once the fire started it rapidly spread. Workers on the eighth floor began yelling "FIRE!" as the phone operator on the floor dialed the ninth and tenth floor offices. The tenth floor housed factory offices as well as the pressing and shipping departments. Many on the upper floors safely evacuated the building by climbing to the rooftop and crossing to an adjoining building or by taking the executive elevator down. Those on the eighth floor crowded into stairways or pushed their way into two passenger elevators as the smoke and fire quickly spread. Within minutes, escape routes were cut off. Barrels of machinery oil caught fire near the elevators adding to the disaster. Surprisingly, the majority of those on the eight floors survived the fire.

The majority of fatalities occurred on the ninth floor because the fire warning did not reach the ninth floor quickly enough. As a result, those on the ninth floor were the last to learn of the danger, and remained trapped between two floors completely engulfed by fire because many of the exit doors were locked. The elevators were no longer working. Panic quickly spread as many faced the terrible decision having to choose between death by fire or death by fall.

Those on the street were horrified to see many jump from the upper floors to the sidewalks and streets below rather than to remain in the building and face death from the flames. According to witnesses, it was horrible to see ladies, many wearing clothes burning brightly, plummet to the street below. Battalion Chief Edward Worth would later testify in court, "they hit the pavement just like rain" (Lee, 1988).

Impact

In less than 18 min it was over. A total of 63 bodies were scattered along the sidewalk outside the building, and most had fallen from the ninth floor (Saperstein, 1979). Inside the building were charred bodies. Survivors relayed horror stories to the media and eventually at the trial of the owners. The disaster shocked the nation and served as a major catalyst in the movement to form organized labor.

Containing the Disaster

The reports of the fire quickly reached the New York Fire Department, and fire companies were rapidly dispatched. New York Fire Department Engine 72, the horse-drawn steam pumper and its tender answered the first call and arrived quickly and connected to a high-pressure hydrant. The flames were intense and concentrated in the upper floors, what eluding the reach of the fire equipment. Other units arrived quickly at the scene. The first Battalion Chief at the Triangle fire ordered a stream of water aimed at the cornice over the windows to try to set up a water curtain to provide protection for the people in the windows getting ready to jump. Firefighters began to enter the building.

Relief and First Response Efforts

Firefighters witnessed a horrible scene. The girls who did not make it to the stairwells or the elevator were trapped by the fire inside the factory and began to jump from the windows to escape it. The bodies of the jumpers fell on the fire hoses, making it difficult to begin fighting the fire. Also, the firefighters' ladders reached only seven floors high and the fire was on the eighth floor. In one case, a life net was unfurled to catch jumpers, but three girls jumped at the same time, ripping the net. The nets turned out to be mostly ineffectual (Staff, 2009a).

The firefighters held the fire to the top three floors of Mr. Asch's building in spite of no firebreaks, windows blown out, and elevator shafts that served as giant flues for a 10,000 square foot factory floor that became a furnace where everything between the walls of the allegedly fireproof Asch Building burned (Greer, n.d.). Police attempted to help, working with the firefighters in the rescue efforts. Police wagons brought the bodies to the morgue, where hundreds gathered to discover who were among the dead.

AFTERMATH AND INVESTIGATION

There was no question that exits were purposely locked by the owners, who believed that workers were robbing them blind of shirtwaists—that was the rationale for locking many exits and access to the rooftop. However, an inspection was conducted of the factory just weeks before the fire by P.J. McKeon, a fire prevention expert from Columbia University, who was hired by the insurance company for underwriting purposes. His report highlighted the unsafe working conditions, and in a report filed just 9 days before the fire by Dr. George M. Price, Director of the New York Board of Sanitary Control, those sentiments were reaffirmed. In the report Price filed with the New York City Building, Fire and Police Departments, he noted that garment shops such as the Triangle Shirtwaist Company were the worst offenders in the area of worker safety (Stein, 1962, p.26).

Subsequently, the owners of the Triangle Shirtwaist Company, Isaac Harris and Max Blanck, were indicted on charges of first- and second-degree manslaughter under a New York statute, which mandated that: "factory doors shall be constructed to open outwardly, where practical, and shall not be locked, bolted, or fastened during working hours." After just an hour and 45 min of deliberation, the defendants were acquitted of all charges by a jury of their peers (Lewin, 1986). Members of the jury included New York businessmen, many with ties to the garment industry.

In addition, Edward F. Croker, who had been chief of the Fire Department of New York since 1899, resigned 5 weeks after the Triangle Shirtwaist Factory fire to devote the rest of his life "to preventing, instead of fighting fires." He had had enough.

LESSONS LEARNED AND POLICY IMPACT

The Triangle Shirtwaist Fire brought fire prevention to the attention of the world. There was a lot of finger-pointing after the fire as to who was responsible for the tragedy. The listing of responsible parties not only included the proprietors, but also included city and state inspectors, superintendents, elected officials, architects, licensors, the public, and the victims themselves (New York Times, 1911). Organized labor was also a target of blame—the union should have gone on to strike after the fire against the company's exploiting of its workers but it did not.

According to the *Literary Digest* (1912), "The blood of those victims was on more than two heads; on more than 20 heads, perhaps on more than a million heads. Everyone connected with the actual neglect of the fire and building laws, whether in an official or unofficial capacity, shared in the blame" (p. 6).

Labor leaders as Samuel Gompers worked to organize unions with the intent to provide workers with a voice and strength to influence wages and working conditions. Soon, the movement to form unions would gain

Figure 14.3 Triangle Shirtwaist Fire Victims in Coffins at the Morgue, March 25, 1911
Source: Everett Historical/Shutterstock.com

support and with the unified strength of workers, the unions began to flex their muscles through the mechanism of the strike and refusal to work, for many, bold and radical concepts (Drehle, 2003).

The New York State Legislature created the nine-member Factory Investigating Commission to examine issues of worker health, safety, well-being, physical conditions of work, wages, hours, child labor, and even sanitary conditions within factories. Abram I. Elkus, counsel for the FIC wrote: "it is the duty of the state to safeguard the worker against the incidents which are the ordinary occurrences of industrial life" (Serrin, 1986, p. 2). Between 1911 and 1915 when the committee disbanded, 36 broad reform measures were passed including provisions to protect workers that would prevent a similar event from happening again.

IMPLICATIONS FOR MODERN EMERGENCY MANAGEMENT

Today, fire codes and fire inspections are designed to mitigate such events from occurring. Despite these efforts, tragedies occur as seen not only with the Imperial Foods Plant Fire but as in the case of the Rode Island Night Club Fire which also saw exits blocked and locked along with numerous other fire code violations. Emergency managers partner with fire officials to ensure emergency planning is done to address commercial occupancies as well as any occupancy designated as large assembly structures such as schools, churches, concert venues, movie theaters, and hospitals. Compliance can mean the difference between life and death.

SUMMARY

Sadly, during this era in U.S. history, business owners' profits were more important than the safety of workers. Our country had a very poor reputation for its treatment of workers, especially the most vulnerable—young girls and women who were immigrants to this country. While the labor unions eventually fought to improve working conditions and wages, sweatshops in New York City and other places still exist today. Despite more stringent fire codes, enforcement remains an issue in many cities and towns across the United States. The Imperial Foods Plant Fire, like the Triangle Shirtwaist Fire, produced casualties as a result of locked doors by owners who were concerned about theft. However, change has occurred and in North Carolina the Imperial

Foods Plant Fire is often credited with driving code enforcement to a much higher level in the state. Additionally, criminal convictions were handed down in the Imperial Food Plant case unlike the Triangle Shirtwaist Fire. This sent a message to other business owners who might have used the same methods to try and prevent theft—that worker safety is their responsibility and that they would be held accountable for reckless actions.

REFERENCES

Behrens, E. (1983). The Triangle Shirtwaist Company fire of 1911: A lesson in legislative manipulation. *Texas Law Review, 62*(2), 361–385.

Chicago Sunday Tribune. (1911, 26 March). New York fire kills 148: Girl victims leap to death from factory. *Chicago Sunday Tribune*, p. 1.

Drehle, David von. (2003). *Triangle: The fire that changed America.* New York, NY: Grove Press.

Greer, B. R. (n.d.). *Triangle firefighting.* Remember the Triangle. Retrieved from http://rememberthetrianglefire.org/open-archive/wp-content/uploads/2011/02/Triangle_Essay_OpenArchive.pdf

Lee, H. (1988, April 1). They hit the pavement like rain: The Triangle Shirtwaist Fire, 1911. *Daily News*, New York, NY, p. 29.

Lewin, T. (1986). Archives of business: The Triangle Shirtwaist Fire - 25 March 1911. *New York Times*, pp. 3, 10.

Lifflander, M. L. (2011). The tragedy that changed New York. *New York Archives* (Summer 2011).

Literary Digest. (1912, January 6). 147 dead, nobody guilty. *Literary Digest*, p. 6.

New York Factory Investigating Commission. (1912). *Preliminary report of the Factory Investigating Commission, 1919, vol. 1.* New York, NY: The Argus Company, pp. 13–20.

New York Times. (1911, March 23). Blame shifted on all sides for fire horror. *New York Times*, p. 1.

Pence, P., Carson, P., Carson, K., Hamilton, J., and B. Birkenmeirer. (2003). And all who jumped died: The Triangle Shirtwaist Fire. *Management Decision, 41*(4), 407–421.

Saperstein, S. (1979, February 1). Reliving pain. *Washington Post*, p. C1.

Serrin, W. (1986, March 25). Labor marking 75th anniversary of Triangle Shirtwaist Fire. *New York Times*, p. B2.

Staff, H. (2009a). *Fire kills 145 at Triangle Shirtwaist factory.* History. Retrieved from http://www.history.com/this-day-in-history/fire-kills-145-at-triangle-shirtwaist-factory

Staff, H. (2009b). *Triangle Shirtwaist Factory Fire.* History. Retrieved from http://www.history.com/topics/triangle-shirtwaist-fire

Stein, L. (1962). *The Triangle Fire.* New York, NY: Carroll and Graf/Quicksilver Books.

15

Sinking of the Eastland (1915)

Figure 15.1 Sunken *Eastland*
Source: Library of Congress

BRIEF OVERVIEW

American history records the *Titanic* disaster as one of the worst maritime disasters that spurred changes toward the incorporating of lifeboats and life preservers aboard maritime vessels. Although the death toll from the *Titanic* was high and the experiences of its survivors were traumatic, the incident was not the worst shipwreck in American history. Instead, in Chicago, among the Great Lakes, another shipwreck surpassed the notable *Titanic* disaster. During 1915, the sinking of the *Eastland* surpassed the *Titanic* as the worst American maritime disaster. Despite such a tremendous death toll, modern American society remembers little of the *Eastland* disaster. A number of conjectures offer some explanation for the event's contemporary obscurity. Some believed it was overshadowed by a prominent wedding of the period; others believed it was belittled because many of the *Eastland's* deceased were immigrants and factory workers whereas the *Titanic* ferried aristocratic individuals (Bellows, 2007; Chicago Sun-Times, 2016). Regardless, in terms of emergency management, the Eastland tragedy provides insight for addressing modern disasters and calamities.

INTRODUCTION

During the early morning hours of July 24, 1915, a drizzly morning foreshadowed a depressing maritime cataclysm wherein the losses of human lives exceeded those of the *Titanic*. Despite a joyous congregation of workers who gathered to celebrate gaily the fifth annual picnic of the Western Electric Company, tragedy occurred unceremoniously wherein public shock and emergency response ensued. As part of the festivities, the *Eastland* was detailed to ferry personnel, families, and friends of Western Electric between Chicago, Illinois and Michigan City, Indiana (Eastland Disaster Historical Society, 2017). However, before it departed its berth, the ship listed, capsized, and instigated one of the largest calamities witnessed in the Great Lakes region.

After its construction, the *Eastland* was discovered to have design flaws, making it susceptible to listing. The ship was top-heavy, which became evident when passengers massed upon its upper decks. During July 1903, overcrowding caused the *Eastland* to list tragically with water flowing upward along one of the ship's gangplanks. Although the situation was quickly rectified, it was the first of several events foreshadowing the concealed lethality that loomed while the ship was docked. Later, during the same month, the *Eastland's* stern was damaged when it was backed into a tugboat. Nearly a year later, during August 1906, another incident of listing occurred which resulted in the filing of complaints against the *Eastland's* new owner, the Chicago-South Haven Line. The company had purchased the ship earlier in the year. Despite its defectiveness, the *Eastland* continued its operations among the Great Lakes.

The Eastland was capable of transporting 2,500 passengers and crew. However, the ship's capacity for danger was high given its historical issues. On July 24, 1915, because of structural changes that caused unrecoverable listing to the *Eastland*, the resulting quantity of deceased passengers and personnel surpassed those of both the *Titanic* and the *Lusitania* (Stranahan, 2014). A total of over 900 individuals perished approximately 20 ft away from the dock (Bicknell, 1915). Approximately 70% of the deceased people represented folks under the age of 25 years (Stranahan, 2014). Although the death toll was substantial historically, modern considerations of emergency management are affected by the tragedy.

Contemporary maritime vessel designs are much safer than those which were present during the early 20th century. However, such modern safety enhancements are neither random nor the products of sheer luck. Instead, actions toward diminishing endangerments are often the result of experience through time. Before one can truly safeguard against any certain calamity, it must be imaginable and reasonably expected within the context of an emergency incident. Thus, through the lens of modern emergency management, despite the horrendous loss of life exhibited during the 1915 sinking of the *Eastland*, the decedents perished neither vainly nor senselessly. Instead, their deaths contributed toward the protecting and safeguarding of future lives—not only in the Chicago Great Lakes area, but nationally as well.

The Disaster Unfolds

On July 24, 1915, the *Eastland*, operating in tandem with the steamers *Theodore Roosevelt*, *Petoskey*, *Racine*, and *Rochester*, were chartered along the Great Lakes to ferry personnel from the Western Electric Company in Cicero, Illinois, to a picnic in Michigan City, Indiana (Hilton, 1995). Because many of the employees were unable to have holiday time off from work, the event was prized among the workers. Among the *Eastland's* passengers were Czech immigrants hailing from Cicero. Tragically, a total of 220 of them perished when the ship irrecoverably listed (Hilton, 1995).

During 1915, the federal Seamen's Act was passed because of the ill-fated RMS *Titanic* disaster of 1912. The legislation mandated that a complete array of lifeboats be incorporated among passenger vessels, including vessels such as the *Eastland* (Zavodnyik, 2011). However, the *Eastland* was designed and constructed before the implementation of the legislation, and was never intended to conform to its lifeboat requirements. The additional weight caused by the inclusion of lifeboats atop the vessel may have increased the *Eastland's* dangerousness because it potentially exacerbated the existing severity of the problem of the ship being top-heavy. Because it was known that the *Eastland* was top-heavy, it necessitated special constraints regarding the quantity of passengers that could be ferried with expectations of safety.

Figure 15.2 Port View of the *Eastland*
Source: Library of Congress

Poor design contributed toward the maritime disaster. The *Eastland* was constructed in 1903, long before the fated *Titanic* incident and the corresponding Seaman's Act. Thus, any consideration of sufficient lifeboat necessity was not a primary concern of engineers and designers when crafting the *Eastland's* design. Given the technology of the period, the *Eastland's* design enabled it to traverse waters at the rate of 19 miles per hour. Although its sleekness facilitated maritime speed, it also contributed toward the vessel being thin and top-heavy. Basically, the ship's design made it 300 ft in length (approximately the length of an entire football field), but only the whole of 38 ft in width (Egan, 2017). Figure 15.2 shows the port view of the *Eastland*.

On the day of doom, passengers commenced boarding the *Eastland* along the south bank of the Chicago River, between LaSalle and Clark Streets, at approximately 6:30 a.m. (Egan, 2017). It achieved its maximum passenger constraint of 2,572 passengers by 7:10 a.m. (Egan, 2017). The ship was both full and heavy with numerous passengers standing along its upper decks (Egan, 2017). Because of the uneven weight distribution, it listed slightly toward the port side. Although the crew attempted stabilization by adjusting the water level inside the ballast tanks, the efforts were unsuccessful. Sometime during the next 15 min, as passenger weight shifted atop the vessel, at 7:28 a.m., the *Eastland* listed sharply toward portside at an angle of 45 degrees, rolled completely, and came to rest against the river's bottom (approximately 20 ft below the surface) (Eastland Disaster Historical Society, 2017; Stranahan, 2014). The combination of passenger weight distribution, insufficient keel weight caused by unfilled ballast tanks, and the unsuppressed intake of water during listing contributed to the *Eastland's* sinking (Bicknell, 1915).

Impact

Regardless of the origin, calamity often occurs unexpectedly with speed, surprise, and violence (Doss et al, 2017a, 2017b). When calamities originate from nature or humans, they may also be unwarned, unforeseen, and unstoppable. Because of such attributes, time to react immediately or to escape endangerments may be insufficient. These notions are applicable regarding the *Eastland* incident. There was no time for anyone below decks to escape the rolling hull (Wachholz, 2005). Hundreds of people were trapped inside the *Eastland* by the water and the sudden rollover; some were crushed by heavy furniture, including pianos, bookcases, and tables (Wachholz, 2005). The highest chance of survival existed for individuals who were above decks and were able to crawl over the ship's railings as it capsized (Wachholz, 2005). Individuals who were below decks had practically little to no chance for survival (Wachholz, 2005).

INCIDENT RESPONSE AND MANAGEMENT

Sister vessels responded to the tragedy. The *Kenosha*, a nearby tugboat, provided some measure of security for survivors of the incident. Despite the deaths, some were spared by the actions of the *Kenosha's* captain, John O'Meara (Wachholz, 2005). He ordered that his vessel be moored alongside the adjacent wharf thereby creating a bridge between the capsized *Eastland* and the wharf (Wachholz, 2005). As a result, hundreds of individuals were able to walk to safety after climbing over the *Eastland's* starboard railings while the ship capsized (Wachholz, 2005). Other vessels, the *Ryerson*, the *Indiana*, and the *Rita McDonald*, also ferried many to safety (Wachholz, 2005).

A temporary maritime morgue was established aboard the nearby *Theodore Roosevelt*. Although it was a passenger vessel that was scheduled to depart soon after the *Eastland*, its passengers panicked and fled when they witnessed the capsizing of the *Eastland* (Wachholz, 2005). However, several of its crew and passengers immediately attempted to assist the *Eastland's* survivors by throwing life preservers into the water (Wachholz, 2005). A trio of the *Theodore Roosevelt's* deck hands manned a lifeboat, and repeatedly retrieved survivors from the water and ferried them ashore (Wachholz, 2005).

Local resources were converted quickly to accommodate both the living and the dead. Wholesale outlets were quickly converted into makeshift morgues and medical centers for survivors. The bodies of the victims were held for identification among various temporary morgues established in the area. By the afternoon, the remaining unidentified bodies were amalgamated at the armory of the Second Regiment (Bicknell, 1915). The armory was converted into a temporary medical station that was staffed unceasingly by Red Cross nurses both "day and night" (Bicknell, 1915, p. 307).

Initial responders consisted of different factions. Physicians and nurses swarmed to the incident site. They attempted to resuscitate many individuals who were retrieved from the water (Bicknell, 2015). In the hopes of rescuing entrapped passengers, fire departments were dispatched to cut holes in the *Eastland's* hull to provide access to the interior of the ship (Bicknell, 1915). Law enforcement personnel were dispatched to maintain order and to provide security (Bicknell, 1915). Retail stores quickly supplied numerous supplies, including blankets, cots, and "other supplies" (Bicknell, 1915, p. 307).

Containing the Disaster

During the aftermath of the incident, a financial relief fund was managed by the American Red Cross. A variety of relief efforts occurred among organizations and individuals. The efforts between these factions generated approximately $400,000 (Wachholz, 2005). The Western Electric Company provided $100,000 to relief and recovery efforts of family members of the victims of the disaster (Wachholz, 2005). A streetcar union provided assistance of $5,000 (Wachholz, 2005). In Chicago, the predecessor to Wrigley Field, Weeghman Park, donated all financial proceedings from its baseball event on July 29 toward the benefit of surviving families (Wachholz, 2005). Not only were baseball admissions paid by fans who attended to view the game, but also by anyone at the ballpark—including coaches, team members, and umpires (Wachholz, 2005).

RELIEF AND FIRST RESPONSE EFFORTS

The *Eastland* incident represented the first emergency addressed by the Chicago Red Cross. The Chicago organization was founded nearly 6 weeks before the tragedy). The Red Cross provided leadership that was necessary for enacting emergency operations and ensuring their effectiveness and efficiency). Specifically, the Red

Cross implemented counseling activities, first aid, and cooperated with other relief organizations. Despite the tremendous loss of life, the incident contributed toward solidifying the Red Cross as a viable, reputable relief organization capable of conducting response and recovery operations. Through its benevolence and cooperation with the locality, the Red Cross amassed approximately $200,000 toward relief funding. This effort may well have been the "first organized corporate and individual disaster fundraising drive in Chicago.

AFTERMATH AND INVESTIGATION

Several inquiries commenced within days of the accident. Cook County officials asserted their jurisdiction immediately. Despite the initial hastiness, it was 24 years before the resulting *Eastland* litigation concluded. The disaster was attributed to a condition of instability that resulted from a combination of passenger overload, shifting weight, vessel design and construction, and mismanaged ballast (Hilton, 1995). A total of 10 grand jury indictments occurred when the matter came before the justice system (Bonansinga, 2004). Specifically, manslaughter indictments were brought against the president and three officers representing the maritime company that owned the *Eastland* (Hilton, 1995). Indictments for criminal carelessness were also brought against the *Eastland's* captain and engineer (Hilton, 1995). Overall, the criminal sanctions involved a possible period of imprisonment comprised of 10 years for "criminal carelessness" and "two years for conspiracy" as well as financial fines totaling around $10,000 (Bonansinga, 2004, p. 218). However, insufficient evidence existed to "prove negligence or wrongdoing on the part of the company or crew" (Newton-Matza, 2014, p. 233).

Throughout the court proceedings, Sidney Jenks, a primary witness who was president of the maritime corporation that constructed the *Eastland*, testified that the ship was initially designed as a "fast ship to transport fruit," and was "capable of making 20 mph (32 km/h) and carrying 500 passengers" (New York Times, 1916). Defense counsel Clarence Darrow asked whether Jenks was concerned about the conversion of the ship into a passenger vessel with a capacity of 2,500 or more passengers that greatly exceeded its original expectation (New York Times, 1916). Jenks replied, "I had no way of knowing the quantity of its business after it left our yards . . . No, I did not worry about the *Eastland*" (New York Times, 1916). Jenks testified that an actual stability test of the ship never occurred, and stated that after tilting to an angle of 45 degrees at launching, "it righted itself as straight as a church, satisfactorily demonstrating its stability" (New York Times, 1916).

During the court proceedings, civil litigation continued for about two decades. Per court decree, the *Eastland* was liquidated for cash after its salvage (Dial, 2016). Maritime law of the time limited any compensation for liability to the value of the *Eastland*, which was $46,000 (Dial, 2016; Powers, 2009). After litigation was completed, despite the filing of hundreds of lawsuit, the victims and families were uncompensated because no money remained for reparation payments (Dial, 2016; Powers, 2009). The court held that negligence was absent because the *Eastland* was deemed seaworthy, operators had attempted precautionary measures, and the responsible party was the engineer who improperly "filled the ballast tanks" (Powers, 2009, p. 235). The court proceedings were lengthy and arduous. Erickson, the engineer, died during the course of the legal proceedings, thereby providing an opportunity for others to blame him for the incident (Dial, 2016; Powers, 2009). Although evidence strongly suggested that Pedersen had been negligent, he was not prosecuted. Officers of the ship's company also escaped prosecution (Dial, 2016; Powers, 2009). All criminal charges were dropped, and the owners avoided any legal finding of negligence (Dial, 2016; Powers, 2009). Overall, litigation involving the *Eastland* continued for about 24 years (Stranahan, 2014).

LESSONS LEARNED AND POLICY IMPACT

Few, if any, of the passengers boarding the vessel noticed that the *Eastland* possessed a complete array of lifeboats, rafts, and life preservers. The quantities of safety equipment and lifeboats were necessary for complying with federal law passed after the *Titanic* disaster. However, although the *Eastland* complied with law, the bulk of additional items created a safety hazard because of the increased weight aboard the vessel. Thus, by complying with safety requirements, the *Eastland* became more dangerous because the increased weight contributed to a greater chance of accidents resulting from top-heaviness.

The *Eastland* was constructed in 1902. It was initially designed for ferrying 500 individuals for lake excursions and for hauling produce during the return trips to Chicago (Stranahan, 2014). The *Eastland* possessed no keel, was dependent on poorly designed ballast tanks within its hold to maintain its uprightness; and was top-heavy. As the ship changed owners and experienced various refits through time, the alterations increased the vessel's speed and passenger capacity (Stranahan, 2014). However, the structural changes also contributed toward increased instability. For instance, structural alterations, completed before July 24 decreased its metacentric height to 4 in., thereby exacerbating the vessel's dangerousness (Hendee, 1915; Stranahan, 2014).

IMPLICATIONS FOR MODERN EMERGENCY MANAGEMENT

Although safety regulations and laws existed and were complied with by the *Eastland's* crew and complements, they were meaningless given the structural imbalances and deficiencies posed by the vessel's design and physical constraints. Compliance with the safety regulations caused the vessel to assume additional weight from lifeboats and equipment, thereby contributing to a greater potential of listing and toppling given weight distributions. Thus, regardless of compliance with any safety regulations, the potential of danger was increased, especially when an influx of passengers contributed to greater weights and indiscriminate load-balancing.

The significance of the *Eastland* incident also necessitates a consideration of Red Cross emergency operations among the Great Lakes. The disaster represented the first major maritime incident to which a substantial Chicago Red Cross response was necessary. Although it was a trial by fire, the incident provided opportunity for the organization to exercise its operational capacity, determine its strengths and weaknesses, and improve itself to address future calamities. The Chicago Red Cross certainly did not respond alone to the incident. Instead, it necessitated cooperation, communication, and coordination among multiple response organizations (McElreath, Jensen, et al., 2013). Essentially, for the newfound Chicago Red Cross, the *Eastland* incident tested immediately its ability to interface with response peers, such as government, police, fire, and military organizations. Through such experience, Chicago's emergency response organizations gained insight for improving their respective capacities to plan for and to address maritime incidents.

Although the modern emergency management cycle (EMC) did not emerge until the latter half of the 20th century, the *Eastland* incident may be viewed from its cyclical perspective. Typically, the EMC phases represent activities of preparedness, prevention, mitigation, response, and recovery. The *Eastland* incident provided immediate experience for the latter three categories. However, lessons learned from the response efforts provided a basis for enhancing the capacities of preventing and preparing for similar incidents.

SUMMARY

The *Eastland* incident resulted in a greater death toll than was attributed to the *Titanic*. However, its notoriety in American history seems almost a sidenote when compared to other incidents. Essentially, the incident remains practically forgotten historically. In a sense, the *Eastland* incident may be viewed in terms of a social commentary regarding the public perspectives of the period. Some speculate that the inattentiveness arose from the fact that the vessel carried primarily working class, blue-collar individuals, but its passengers were absent any well-known, famous, or aristocratic folks. Thus, national interest in the event was unattainable for any substantial period.

Most likely, the *Eastland's* passengers and crew never imagined they were merely minutes from death as they boarded the vessel and prepared for departure. Similarly, the emergency management resources of the period probably never imagined the scenario of responding to a berthed, capsized ship bearing its personnel and passengers. Given these notions, the *Eastland* incident shows the extreme types and ranges of inclement events that must be accommodated by emergency planners and responders. Essentially, emergency response resources must be capable of responding to a variety of different events, emanating from both human and natural origins, during any type of weather scenario (McElreath, et al., 2014).

However, gleaned lessons from the *Eastland* incident provide important insights regarding law, policy, and regulation within the context of emergency management. Just because a maritime vessel conforms to any safety mandates does not mean that danger is neutralized. In other words, given the ship's design and construction before the crafting of safety measures, it was not intended to conform to any specific requirements for safety. Refitting the *Eastland* to ensure alignment with safety measures intensified the dangerousness. Additionally, given its track record of observed near-accidents coupled with additional structural alterations, its chance of some catastrophe occurring was heightened.

The *Eastland* had a record of known, documented near-missed accidents, but she was generally deemed seaworthy. When viewed from hindsight, such listing incidents were symptomatic of greater problems that were concealed by numerous moments of serendipity and denial. Regardless, corrective measures were typically implemented to ensure that the *Eastland* could fulfill the requirements of its voyages. However, good fortune and luck eventually dissipated to reveal the dangerous nature of reality. Regarding the *Eastland*, its luck ran out on the morning of July 24, 1915.

Certainly, modern ship designs are far safer and more robust than those of the late 19th and early 20th centuries. Despite improved resources for crafting maritime vessels, the *Eastland* incident provides a timeless lesson that transcends industries: exercised regulatory conformance without the corresponding, appropriate infrastructure for enhanced safety is practically a meaningless scenario. From the EMC perspective, this notion embodies the concepts of prevention and preparation. Although some may have viewed the *Eastland* incident as a tragically senseless loss of life, the deaths of many became purposeful for future generations wherein vessel designs were augmented to prevent similar incidents from occurring again. From this perspective, the *Eastland* incident highlights two undeniable, immutable realities of effective EMC implementation—prevention and preparation.

REFERENCES

Bellows, A. (2007). *The fall of the Eastland*. Retrieved from https://www.damninteresting.com/the-fall-of-the-eastland

Bicknell, A. (1915). The *Eastland* disaster: An incredible tragedy in the heart of Chicago. *American Red Cross Magazine, 10*(1), 306–309.

Bonansinga, J. (2004). *The sinking of the Eastland: America's forgotten tragedy*. New York, NY: Kensington Publishing.

Chicago Sun-Times. (2016). *Editorial: 100 years later, Eastland disaster has hard lessons to offer*. Retrieved from https://chicago.suntimes.com/opinion/editorial-100-years-later-eastland-disaster-has-hard-lessons-to-offer

Dial, C. P. (2016). *Only the river remains: History and memory of the Eastland disaster in the Great Lakes region, 1915-2015*. Kalamazoo, MI: Western Michigan University.

Doss, D., Lackey, H., McElreath, D., Gokaraju, B., Tesiero, R., Jones, D., & Lusk, G. (2017a). Quantitatively assessing reported crime versus enrollment among selected higher education institutions. *Universal Journal of Educational Research, 5*(10), 1678–1683.

Doss, D.A., Lackey, H., McElreath, D., Jones, D., Goza, R., & Sheffield, J. (2017b). Revisiting violence in rural academic settings: A school killing impacting Livingston High School and considerations of policy and programs. *Journal of Studies in Social Sciences, 16*(2), 155–191.

Eastland Disaster Historical Society. (2017). *The history timeline*. Retrieved from http://www.eastlanddisaster.org/history/timeline

Hendee, S. (1915). Supreme tragedy of the lakes. *Popular Mechanics*, September, 357–361.

Hilton, G.W. (1995). *Eastland: Legacy of the Titanic*. Stanford, CA: Stanford University Press.

McElreath, D., Doss, D., Jensen, C., Wigginton, M., Nations, R., Van Slyke, J., & Nations, J. (2014). *Foundations of emergency management*. Dubuque, IA: Kendall-Hunt.

McElreath, D., Jensen, C., Wigginton, M., Doss, D., Nations, R., & Van Slyke, J. (2013). *Introduction to homeland security* (2nd ed.). Boca Raton, FL: CRC Press.

New York Times. (1916). *Eastland never tested. Builder of ill-fated ship says she was designed to carry 500.* Retrieved from http://query.nytimes.com/gst/abstract.html?res=9E04E6DF1439E233A25750C2A9679C946796D6CF&legacy=true

Newton-Matza, M. (2014). *Disasters and tragic events: Encyclopedia of catastrophes in American History.* Santa Barbara, CA: ABC-CLIO Publishing.

Powers, D.M. (2009). *Taking the sea: Perilous waters, sunken ships, and the true story of the legendary wrecker captains.* New York, NY: AMACOM.

Stranahan, S.Q. (2014). *The Eastland disaster killed more passengers than the Titanic and the Lusitania. Why has it been forgotten?* Retrieved from https://www.smithsonianmag.com/history/eastland-disaster-killed-more-passengers-titanic-and-lusitania-why-has-it-been-forgotten-180953146

Wachholz, T. (2005). *The Eastland disaster.* Charleston, SC: Arcadia Publishing.

Zavodnyik, P. (2011). *The rise of the federal Colossus: The growth of federal power from Lincoln to F.D.R.* Santa Barbara, CA: ABC-CLIO Publishing.

16

The Great Atlanta Fire (1917)

Contributed by Alexis K. Liberto, Jeffrey M. Pavlacic, and Stefan E. Schulenberg.

Figure 16.1 Great Atlanta Fire of 1917
Source: Kenneth Rogers Collection/AJC Staff

BRIEF OVERVIEW

On Monday, May 21, 1917, the Fourth Ward of Atlanta, Georgia burned to the ground. The fire raged for nearly 11 hr, only stopping after firefighters detonated houses in the fire's pathway as a means of creating a firebreak (Kahn, 2017). Approximately 300 acres and 50 city blocks were destroyed, equating to 2,000 homes and a displacement of some 10,000 people (Atlanta History Center, 2017). Because of other fires reported throughout the day, the initial response to the start of this fire was slow to form and rather limited. Contributing to this slow-moving, ineffective response was the inadequate number of firefighters available, as well as a lack of availability with respect to modern firefighting technology. Moreover, due to the combination of the large number of wooden structures in the Fourth Ward and the increasing wind, the fire spread quickly, resulting in great devastation. Total damages were estimated to be around $5 million, or more than $100 million, after adjusting for today's inflation (Burns, 2017; Kahn, 2017).

INTRODUCTION

Atlanta's experience with fires was not unusual. Previously, the city had experienced several large fires that were attributed to the tree-lined residential areas and large patches of woodlands (Waugh, 2000). The Fourth Ward (now known as the Old Fourth Ward) was founded a decade before the Civil War began when the city of Atlanta was split into 10 wards (Kuhn, 2005). At the time of the Great Fire, the Fourth Ward was, for the most part, diverse in terms of race and socioeconomic status. For example, many people in business occupations populated the area driven by its close location to Atlanta's central business district (Burns, 2017). In addition, many Black families moved into the Fourth Ward's neighborhood of Sweet Auburn at the beginning of the 1900s, exponentially increasing after the Atlanta race riots of 1906. Despite its diversity for the time, many of the Fourth Ward residents remained low-income, Black citizens living in modest, wooden-flanked houses (Burns, 2017).

The Setting

As a point of context, the United States declared war on Germany, entering World War I on April 6, 1917. Like much of the rest of the country, the city of Atlanta was fixated on the war, as opposed to being worried about "smaller" issues like fire safety. This lack of concern included failing to update the Atlanta Fire Department with new equipment such as fully motorized crews for all fire company houses. When the Great Fire occurred, only about half of the stations had motorized engines, while the other half relied on horse-drawn transportation (Campbell, 1968; Watts, 2015).

It is also noteworthy that in the summer of 1916, an ordinance was proposed to prohibit wooden shingle roofing within the city of Atlanta due to its danger posed to close-quarter living. This proposal came after conflagration spreading was attributed to wooden roofing materials in the cities of Augusta (Georgia), Nashville (Tennessee), and Paris (Texas). Unfortunately, the Atlanta City Council tabled this ordinance, termed the "Anti-Wood Shingle Law of 1916," so that lumber companies could sell the remainder of their stock (Burns, 2017; Campbell, 1968). On January 15, 1917, the Atlanta City Council suspended the resolution (Campbell, 1968).

The Disaster

The morning of May 21, 1917 was hot and sunny. With a slight breeze from the south, the day was also relatively dry, with a recorded 47% humidity (Burns, 2017; Campbell, 1968). Before the fire in the Fourth Ward broke out on Decatur Street, multiple other fires had been called in to the Atlanta Fire Department (Burns, 2017; Campbell, 1968). These conflagrations included a few houses, an apartment building, and a warehouse in the West End. At approximately 12:46 p.m., numerous phone calls began flooding operators and fire stations in the area, reporting the fire in the Fourth Ward. Lieutenant Hooks, the first to arrive on scene via horse-led apparatus, had his crew use soda and acid extinguishers to attempt to treat the blaze (Campbell, 1968). Because multiple fires were being contained in other locations, steam and motorized engines capable of attending to the Fourth Ward fire were unavailable following the initial reports of its occurrence (Burns, 2017).

Additionally, faulty communication lines worsened the chaos (Campbell, 1968). By 1:30 p.m., hose crews operated from nearby hydrants in efforts to contain the west side of the fire near Fort Street. However, these efforts were not as effective as working steam and motorized engines due to their inability to cover as large of an area with as much power (Campbell, 1968). It was not until about 1:45 p.m. that the steam and motorized engine companies were in place on both the west side and southeast side near Edgewood Avenue (Campbell, 1968). Multiple fire stations located on the north side of Edgewood Avenue also tried to contain the fire using chemical fire engines, but were unsuccessful stopping the fire's forward movement. Because many structures in the Fourth Ward were made of wood, the time it took for them to be engulfed in flames was miniscule.

Around 1:50 p.m., Fire Chief William "Bill" Cody of motor engine and motor hook and ladder truck No. 2 contacted the Mayor of Atlanta, Asa Griggs Candler, specifically requesting his help reaching out to nearby cities for assistance with fire containment. Unfortunately, the subsequent fires had resulted in a lack of available personnel and equipment (Burns, 2017; Kuhn, 2005). By 2:00 p.m., effective steam and motorized engines from surrounding Atlanta fire stations were in use. Additionally, one building, Trio Laundry on Tanner Street, was equipped with a sprinkler system that helped contain the fire (Campbell, 1968). However, despite these efforts, by about 2:45 p.m., the fire was burning churches of historical significance, as well as extensive neighborhoods of apartment buildings and free-standing houses. With the hope of trying to save important possessions from the fire, people were running frantically about and throwing their belongings from windows (Burns, 2017; Kuhn, 2005; Watts, 2015). The fire continued to spread as winds picked up to approximately 15 miles per hour.

Eventually, Mayor Candler asked Colonel Charles R. Noyes of the Seventeenth U.S. Infantry and the Officers Training Camp at Fort McPherson for military assistance (Campbell, 1968). By 5:00 p.m., martial law was declared in the greater Atlanta area. Soldiers from nearby Fort McPherson marched into the city to reinforce safety, direct traffic, and assist in the suffocation of the fire (Campbell, 1968). Chief Cody, Mayor Candler, and Colonel Noyes then made the decision to use dynamite to destroy large homes in the northern pathway of the fire with the idea of creating a firebreak (Campbell, 1968; Kuhn, 2005). By 7:00 p.m., the wind had changed direction. This aided in effectively containing the fire allowing firefighters to effectively move the flames toward areas that had already been burnt to ashes, away from unharmed areas. Several other fire departments from surrounding areas (Chattanooga and Nashville, Tennessee; Savanna, Decatur, and Macon, Georgia) had arrived to offer support. With the assistance of these various reinforcements, the Great Atlanta Fire was brought under control around 10:00 p.m. However, isolated portions continued to smolder for the rest of the week. In 11 hours, approximately 22,000,000 gallons of water had been used to put out the fire, with most hydrants running on low pressure (Burns, 2017; Watts, 2015). Economically, the Great Atlanta Fire caused around $5 million in damages (more than $100 million after adjusting for today's inflation), with Black families experiencing the greatest impact (Kahn, 2017). Miraculously, though still unfortunate, there was one death—an elderly woman who had gone into cardiac arrest after seeing her home burn (Atlanta History Center, 2017; Kahn, 2017).

AFTERMATH AND INVESTIGATION

The Great Atlanta Fire of 1917 was an unexpected tragedy in American history. As aforementioned, the Atlanta City Council had proposed and tabled an ordinance designed to prohibit the use of wooden shingles within city limits (Campbell, 1968). Close-quarter city living would allow a blaze to quickly spread. Unfortunately, however, the proposed ordinance was not enacted until after the tragedy (Campbell, 1968) because lumber companies possessed stock that they hoped to sell (Campbell, 1968). As city residents rallied to prepare for World War, no one expected something as tragic as what ensued to occur so close to home.

Fire Marshal Walthal R. Joyner attributed the Great Fire as having been caused by sparks that traveled from the Woodward Avenue fire (Campbell, 1968). Others attributed the fire to the Germans, claiming that the German Empire was attacking the city as a means of inhibiting their preparations for war (Campbell, 1968). Still, Dr. J. P. Kennedy, a health professional working in the city of Atlanta, felt that the fire began in the hospital geared specifically toward the Black population (Campbell, 1968). Much speculation exists with respect to the origins of the fire (Campbell, 1968; Kahn, 2017). After much investigation, it was determined that the fire began in a hospital storage area, and spread through an adjacent shantytown to a neighborhood of large homes (Waugh, 2000).

Beyond having problems with personnel and the most up-to-date equipment being available at the time, the blaze may have spread out of control due to other factors. Additional units were involved with fighting other fires from earlier that morning (Atlanta History Center, 2017). Thus, there was little assistance available to dispatch to the Decatur Street fire at the outset of the event when Lieutenant Hooks, the first responder in charge, asked for it (Campbell, 1968). Delays in responding, a lack of personnel, a lack of proper equipment, and faulty communication lines all may have contributed to the massive spreading of the fire (Atlanta History Center, 2017). The resulting Great Fire damaged over 1,900 buildings, nearly 1,700 of which utilized wooden shingles (Campbell, 1968).

Firefighters responding to the event were forced to wait until more sufficient equipment, such as additional hoses, arrived. Lieutenant J. A. Hooks was first to the blaze with his crew. After attempting to use hand soda and acid extinguishers to put out the fire to no avail, he attempted to notify headquarters that hoses would be necessary to treat the blaze, but received no response (Campbell, 1968). The telephone lines had collapsed, which significantly hindered communication (Campbell, 1968). Hooks was forced to send his horse-led ladder truck to the station so that it could be equipped with hoses (Campbell, 1968). The truck then had to travel approximately 5 miles before returning to the blaze (Campbell, 1968). As a result, hoses were not available until about 15 minutes after the fire had been reported (Campbell, 1968). By this time, what came to be known as the Great Atlanta Fire had spread, throughout the originating structure, as well as to other, smaller buildings by means of shrubbery and fences (Campbell, 1968).

The city contacted fire departments of neighboring cities to see whether they could spare equipment. As time went on, additional fire trucks made their way to the scene. Luckily, sprinkler systems in some buildings assisted firefighters in treating the fire (Campbell, 1968). For example, Trio Laundry on Tanner Street was the only building in the area equipped with a sprinkler system (Campbell, 1968). This building's sprinkler system helped prevent the spread of the fire to other buildings. As such, this initially helped the fire department to bring the fire under control. However, after believing that the fire had been stopped, firefighters noticed additional buildings burning to the north (Campbell, 1968). While the firefighters moved their equipment to treat the blaze, the wind helped to cause further spread. Unable to control the conflagration, Chief Cody and city officials used dynamite to destroy homes as a means of creating a firebreak. The fire raged for still more hours, although firefighters were eventually able to quell the flames (Campbell, 1968). By the time the fire was contained and extinguished, it had consumed 73 blocks of homes (Patureau, 1987).

POLICY IMPACT AND LESSONS LEARNED

The Great Atlanta Fire of 1917 galvanized city officials to expedite the passing of an ordinance to prohibit the use of wooden shingles within city limits, an ordinance that had previously been neglected and filibustered for financial reasons. What came as a result of the Great Fire was the Anti-Wood Shingle Law of 1917, which prohibited the use of wooden shingles within city limits (Campbell, 1968). In 1909 (8 years before the Great Fire), the Committee on Fire Prevention (CFP), a national organization designed to develop policy regarding the prevention of large blazes, claimed that homes utilizing wooden shingles were at high risk (Campbell, 1968). Although the Atlanta ordinance had not yet been passed, the idea had been proposed and discussed prior to the Great Fire (Campbell, 1968). Further, risks had been identified by the CFP with specific regard for the city of Atlanta. It is unfortunate that economic interests were of primary concern at the time, winning out against safety concerns.

The Anti-Wood Shingle Law of 1917 also outlined that the city fire department should be completely motorized, as the horse-led apparatus that arrived on scene inadvertently delayed the treatment of the fire, evidently contributing to its spreading (Campbell, 1968). Finally, a festival entitled "Fire in the 4th" was organized 100 years later to educate the community on fire safety, and as a method of memorializing the devastating event (Kahn, 2017). Thus, the Great Atlanta Fire of 1917 caused the city to rethink its strategies for addressing the occurrence of multiple fires developing at the same time in different locations within the confines of the city. This is an issue that continues to be discussed to this day.

In broadly summarizing policy concerns and lessons learned, the Great Atlanta Fire of 1917 spurred the city of Atlanta to fully motorize their fire department and to develop policy to prohibit the use of wooden shingles in residential areas within city limits. Safety concerns should have indeed outweighed economic concerns, though it was not until after this tragedy that this shift in values occurred. Horse-led equipment was replaced entirely with motor vehicles to expedite the arrival of firefighting crews to the scene of a fire (Atlanta History Center, 2017). Motorized vehicles were complete with an 85-ft truck ladder that would help crews to reach heights of buildings that they might otherwise not have been able to reach previously (Campbell, 1968). For these reasons, city officials felt that horse-led firefighting equipment was no longer useful (Campbell, 1968). The Great Atlanta Fire of 1917's rapid, destructive trajectory would have been mitigated had adequate equipment and personnel been readily available.

IMPLICATIONS FOR MODERN EMERGENCY MANAGEMENT

As alluded to above, given that the majority of the structures affected by the fire were built with wooden shingles, a law banning wooden shingles would have lessened, or may have even prevented, the incident altogether (Campbell, 1968). The extent of damage was in large part due to the expansive number of wooden shingles.

Thus, there was a dire need for policy restricting such construction materials, especially in major cities. Other cities already had such laws in place at the time, modeling the importance of such a policy for the city of Atlanta (Campbell, 1968).

Also, as indicated above, the Great Atlanta Fire of 1917 roused the city to provide motor vehicles for fire stations in order to expedite the arrival of fire crews on site. Firefighters were forced to divide available resources due to the number of fires reported earlier in the day, thus, contributing to their inability to respond effectively to what became the Great Atlanta Fire when it was initially reported (Campbell, 1968). Emphasis was placed on providing enough vehicles to ensure that future fire crews could effectively handle multiple fires simultaneously (Campbell, 1968). In the case of the Great Atlanta Fire, the devastation caused by the rapid spreading of the flames would have been mitigated if resources had been sufficient. By providing satisfactory, reliable firefighting equipment, the city aimed to ensure that fire crews could handle situations in the future where a "worst-case scenario" occurred. The era of horse-led firefighting technology had come to an end, as this equipment simply wasn't adequate with respect to preparing firefighters to respond to multiple fires, either presently or considering the future, as the city continued to grow in size (Campbell, 1968).

The Great Atlanta Fire also resulted in economic considerations, with further implications related to socioeconomic status and diversity. Economically, Black families experienced the greatest impact from the event (Kahn, 2017), most likely because they were less likely to receive insurance coverage than White individuals (Atlanta History Center, 2017; Burns, 2017). Many Black individuals were kept away from the rubble after the fire, threatened with violence if caught trespassing on their old properties during the period of martial law that followed. Eventually, ruins from the southern portion of the fire were turned into low-income houses, termed Buttermilk Bottoms (Atlanta History Center, 2017). Following the conflagration, Black individuals sought housing in the poor area of West End, while White residents relocated to newer neighborhoods in the northern part of the city (Atlanta History Center, 2017).

Today, scholars find economic deficiencies to be a major concern postdisaster. Beyond any physical impact, stressors such as discrimination, job loss, and being forced to move to a different geographic location enhance an individual's risk for developing a disaster-related psychological disorder (Arata et al., 2000; Fullilove, 1996; Sabucedo et al., 2009). Simply stated, disasters have the potential to negatively impact the physical and psychological well-being of survivors (Walters et al., 2013). Disaster-related economic, employment, and displacement impacts are not uncommon postdisaster, and such issues were particularly evident following the Great Atlanta Fire of 1917 (Atlanta History Center, 2017).

SUMMARY

The Great Atlanta Fire occurred on May 21, 1917, swiftly spreading across 50 blocks, between the areas of Murphy Avenue and Piedmont Park (Campbell, 1968). As the fire blazed from street to street, the city's response proved inefficient and ineffective. Implications and considerations of the event include repercussions from a lack of available modern firefighting technology, limited personnel due to the occurrence of multiple fires throughout the course of a day, communication problems, and the essential utility of policy, or the lack thereof, in this case, which resulted in the Anti-Wood Shingle Law of 1917. Finally, an annual festival entitled "Fire in the 4th" was created 100 years after the Great Fire, in order to educate the community on fire safety as well as a means to memorialize the significance of the event to the city's history.

REFERENCES

Arata, C. M., Picou, J. S., Johnson, G. D., & McNally, T. S. (2000). Coping with technological disaster: An application of the conservation of resources model to the Exxon Valdez oil spill. *Journal of Traumatic Stress, 13*, 23–29. doi:10.1023/A:1007764729337

Atlanta History Center. (2017). 100 years after the fire. [Blog post]. Retrieved from http://www.atlantahistorycenter.com/blog/new-entry

Burns, R. (2017). The second burning of Atlanta. *Atlanta Magazine.* Retrieved from http://www.atlantamagazine.com/great-reads/second-burning-atlanta

Campbell, S. B. (1968). The Great Fire of Atlanta, May 21, 1917: An official record. *The Atlanta Historical Bulletin, 13*(2). Retrieved from http://album.atlantahistorycenter.com/cdm/compoundobject/collection/AHBull/id/13818

Fullilove, M. T. (1996). Psychiatric implications of displacement: Contributions from the psychology of place. *American Journal of Psychiatry, 153*, 1516–1523.

Kahn, M. (2017). Atlanta's Old Fourth Ward burned 100 years ago on this weekend. *Curbed Atlanta.* Retrieved from https://atlanta.curbed.com/2017/5/19/15661150/great-atlanta-fire-old-fourth-ward-festival-1917-100-years

Kuhn, C. M. (2005). *Living Atlanta: An oral history of the city, 1914-1948.* Athens, GA: University of Georgia Press.

Patureau, Alan. (1987, May 17). Atlanta's Other Great Fire. *Atlanta Constitution*, pp. 1H, 5H.

Sabucedo, J. M., Arce, C., Ferraces, M. J., Merino, H., & Duran, M. (2009). Psychological impact of the Prestige catastrophe. *International Journal of Clinical Health Psychology, 9,* 105–116.

Walters, A. B., Drescher, C. F., Baczwaski, B. J., Aiena, B. J., Darden, M. C., & Johnson, L. R. (2014). Getting active in the Gulf: Environmental attitudes and action following two Mississippi coastal disasters. *Social Indicators Research, 118,* 919–936.

Watts, G. (2015). Old Fourth Ward remembers Great Atlanta Fire of 1917. *WABE.* Retrieved from https://www.wabe.org/old-fourth-ward-remembers great-atlanta-fire-1917

Waugh Jr., William L., (2000). Living with hazards dealing with disasters: An introduction to emergency management. Armonk, New York: M.E. Sharpe.

1947 Texas City Explosion

17

Figure 17.1 Oily Black Smoke Rises from the Texas City Monsanto Chemical Company Plant on the Morning of April 16, 1947

Source: Everett Historical/Shutterstock.com

BRIEF OVERVIEW

The 1947 Texas City explosion remains the nation's deadliest industrial disaster, and one of the largest nonnuclear explosions ever recorded. It began with a midmorning fire aboard the *SS Grandcamp*, a French-registered ship loaded with ammonium nitrate that was to be carried to European farmers for use as fertilizer. The fire and explosion created a chain reaction in the port, eventually leading to the explosion of neighboring ships and oil refineries. Fires continued to burn for nearly a week and it took more than a month to recover all of the bodies. At least 581 people were killed, including all but one member of the Texas City Volunteer Fire Department. More than 5,000 people were also injured, many requiring admission to area hospitals. Damage estimates at the time were estimated at about $100 million—more than 500 homes were destroyed with many more damaged-leaving 2,000 people homeless. The port was also destroyed and with it more than a thousand vehicles and a large number of railcars used to service the port.

INTRODUCTION

In 1947, Texas City was a booming deepwater port on Galveston Bay. Berthed in the port was a French ship, the *SS Grandcamp*, that contained approximately 2,200 tons of ammonium nitrate. The city experienced significant growth during World War II as its refineries and chemical plants worked to support the war effort. In addition to incendiary hazards aboard ships and within the port, all of the necessary ingredients were present for a massive explosion and subsequent destruction—nearby, there were chemical plants, grain elevators, refineries, and large oil and chemical storage tanks. On April 16, 1947, a fire aboard the *Grandcamp* served as the catalyst for igniting its deadly cargo. The combination of flammables aboard the ships and in the immediate area erupted to generate the largest explosion and deadliest industrial incident detailed in U.S. history.

The Setting

In 1947, Texas City had a population of approximately 15,000. It was industrialized heavily and boasted of the world's largest tin smelting plant. The city was located on the mainland, across Galveston Bay from Galveston, and was 11 miles from the Gulf. Petroleum products comprised much of its industry. It was a booming deepwater port on Galveston Bay along the Gulf Coast. Being a coastal city and commerce point, Texas City shared a harbor with the nearby city of Galveston. The harbor was 800 ft wide and over 5 miles long. It extended from the Gulf of Mexico between two protective breakwaters to Boliver Roads, the terminal of the Texas City, Galveston and Houston ship channels. It also was an intersection of the intracoastal canal. During 1946, regarding its commerce throughput, it reached fourth place in Texas ports, with 3,907 vessels cleared and 13,441,248 net cargo tons moved (Associated Press, 1947).

Many ships either transited or were berthed in the harbor. The *Grandcamp* was berthed in Texas City because at its previous port of call in Houston, the port authority disallowed the loading of ammonium nitrate because of the dangers associated with such a volatile cargo. In addition to ammonium nitrate, the *Grandcamp* also carried small arms ammunition, machinery, and bales of sisal twine on the deck. The *Grandcamp* was not the only ship carrying ammonium nitrate. Docked next to it in Texas City was the *SS High Flyer* which held both ammonium nitrate and sulfur. This scenario represented a deadly combination as events unfolded.

The Disaster Unfolds

Around 8 a.m. on April 16, 1947, longshoremen aboard the *SS Grandcamp* were preparing to finish loading the rest of the ammonium nitrate to the ship. When they opened the cargo hold, they noticed smoke coming from the sacks of ammonium nitrate several layers down. They attempted to extinguish the fire with regular fire extinguishers, but were unsuccessful. The Texas City Volunteer Fire Department was notified. At around 9:00 a.m., 27 of the fire department's 28 members arrived at the scene when the captain of the ship ordered the hold "steamed" in order to save the cargo. This process involved the use of steam to smother the fire. Soon thereafter, the cover blew from the hatch to the cargo hold, and reddish-orange smoke poured out followed by flames. At 9:12 a.m. the cargo exploded, killing all of the present firefighters, longshoremen, and ship's captain instantly (Fire Prevention and Engineering Bureau of Texas & National Board of Fire Underwriters, 1947). The *Grandcamp* and its crew were obliterated instantly (Associated Press, 1947).

The explosion triggered a chain reaction of equally deadly events. Cascading explosions devastated the area. The nearby Monsanto Chemical Plant lost 145 of its 150 employees who were on duty at the time of the explosion. Nearly 1,000 buildings near the port were leveled and the refineries and chemical tanks were burning out of control. The initial explosion aboard the *Grandcamp* also ignited the ammonium nitrate on board the *High Flyer* which had been torn loose from its mooring by the explosion. The *High Flyer* drifted into another ship beside it, the *SS Wilson B. Keene*, and they became intertwined in the wreckage. Firefighters could not battle the fire on the *High Flyer* once the burning sulfur began giving off gases. Some 16 hours after the first explosion, at approximately 1:10 a.m., the *High Flyer* also exploded, taking with it the *Wilson B. Keene* (Fire Prevention and Engineering Bureau of Texas & National Board of Fire Underwriters, 1947).

The magnitude of the explosions and scale of devastation were comparable to war-related incidents. The scene was similar to bomb destruction scenes that occurred in Europe. Flying toward the blaze, the smoke could be seen from Ellington Field, approximately 30 miles away, and it reached approximately 4,000 ft. A mile away from the incident location, black smoke from six roaring fires billowed nearly 5,000 ft into the air, and drifted southward over the Gulf of Mexico. A 50-acre area of devastation marked the scene where the incident occurred (Beitler, 2018).

Impact

The blasts rocked the surrounding region for approximately 150 miles. The destruction was widespread—a steel barge was swept 100 ft inland after an enormous wave was generated by the explosion. Steel shrapnel rained down in the area and bails of burning twine, also part of the *Grandcamp's* cargo, were deposited in the area causing additional fires. The *Houston Chronicle* and the *New York Times* reported that, "Bodies were tossed about like playing cards" (Associated Press, 1947). Fires burned for days after the initial explosions producing poisonous gases that impeded accessing the area. When the retrieval of bodies commenced, nearly 500 remains were removed from the debris in 1 day (Associated Press, 1947).

Ground zero was annihilated. The sound from the calamity was heard for a distance of 150 miles. Airplanes flying unluckily over the incident were destroyed by the explosion. Burning shrapnel, ejected from the blasts, flew through the air and landed among industrial areas, thereby causing fires and additional destruction (Kintz, Jones, & Carpenter, 1948). Oil tankers near the incident location, loaded with petroleum fuel, burned for days afterward. The explosion's blast wave flattened structures in all directions, and rendered approximately 2,000 individuals homeless (*Encyclopedia Britannica*, 2018).

The blasts were so severe that windows were shattered and plaster was knocked from ceilings at Galveston, about 11 miles across the bay. Many people fled Galveston fearing an earthquake (Stephens, 1997). The explosion

Figure 17.2 A Year After the Worst Peacetime Explosion in U.S. History, the Monsanto Chemical Company Plant at Texas City is Rebuilding (March 24)

Source: Everett Historical/Shutterstock.com

was the worst in Texas history, exceeding even the New London school explosion in 1937 when 294 schoolchildren were killed. It was the second worst disaster in Texas history, being exceeded only by that of the Galveston hurricane in 1900, when 5,000 to 80,000 fled (Associated Press, 1947).

During modern times, the *Grandcamp's* two anchors are testaments and memorials of the tragedy. A 2-short-ton (1.8-metric-ton) anchor was hurled about 1.62 miles and found in a 10-ft (3 m) crater, and its main 5-short-ton (4.5-metric-ton) anchor was hurled approximately one-half mile (800 m) to the entrance of the Texas City Dike. The former rests in a memorial park whereas the latter rests on a "Texas-shaped" memorial at the entrance of the Texas City Dike. Regarding the financial costs of the incident, the solitary costs of property damages were approximately $100 million. In terms of modern dollars, the equivalent amount is approximately $1.1 billion in damages (History Channel, 2018).

Containing the Disaster

The disaster was grievous and difficult to mitigate. Because of the magnitude of the explosions, destruction of response resources, resulting fires, and toxic fumes, controlling and containing the disaster were challenging endeavors. For instance, events at the Monsanto factory adjacent to the port were tedious. The gigantic Monsanto Chemical Company structure was erected during World War II at a cost of $19,000,000 for the purpose of making styrene, an ingredient of synthetic rubber. At the time of the explosion, numerous incendiary and flammable chemicals were present at the facility. Given the gutting fires and poisonous gases, the Monsanto factory became a lethal, inaccessible tinderbox. Fires raged in the Monsanto factory, and firefighters heard the screams of people trapped inside the building. Rescue was impossible because of the heat, fumes, and flames (Associated Press, 1947).

Resources necessary for quashing the fires were unavailable in the immediate area. After the cascading explosions started, flames raged unchecked because of the damaged water system. Between the locations of Galveston and Texas City, roadways, highways, and causeways became congested by frantic Texas City residents, who worked in Galveston, returning to their homes or traveling to assist with disaster response. Harbor access was impeded following the explosions, thereby increasing the difficulty of accessing the incident location. Throughout the affected area, fires raged unceasingly for days after the explosions (Associated Press, 1947).

Although the city possessed a fire department, many of its personnel and resources, which were present at the port when the explosion occurred, were destroyed. As a result, requisite people and corresponding firefighting resources were unavailable, thereby exacerbating the situation (Encyclopedia Britannica, 2018). Texas City possessed 28 volunteer firefighters; however, 27 of them perished in the initial explosion because they were attempting to quash fire aboard the *Grandcamp*. Essentially, the initial explosion eradicated the city's firefighting ability.

RELIEF AND FIRST RESPONSE EFFORTS

Texas City was immediately overwhelmed in its response capabilities; medical personnel and firefighters had to be called in from other areas to assist and had a difficult time accessing the site due to the damage and fires still burning. The U.S. Army, Navy, Coast Guard, Marine Reserves, and Texas National Guard were all called to assist. Firefighters from neighboring cities arrived from places such as Galveston and Houston. UT-Galveston provided medical personnel and nurses while emergency hospitals were set up in unconventional places such as the high school and city hall. A surviving gas station became a morgue. While 581 is often listed as the death toll, many have speculated that number to be conservative. Sixty-three of the dead were never identified, and another 113 were classified as missing. There may have been hundreds more killed whose bodies remain unlocated (Lyons, 2016, p. 11).

At the time of the explosion, the Texas City telephone services were unavailable because of a telephone operators' strike. When the operators learned of the accident, they quickly went back to work, but the strike caused an initial delay in coordinating rescue efforts. In addition to the strike, because of lines and other technological infrastructure being physically destroyed by the explosions, telephone systems were often inoperative, thereby making communication efforts largely ineffective.

Upon requesting assistance, rescue personnel and resources from all over the area responded immediately. Governor Beauford H. Jester flew to the area to direct the relief efforts. The U.S. Army, Navy, Coast Guard, Marine Reserve, and the Texas National Guard all sent personnel and medical resources, including doctors, nurses, and ambulances. The University of Texas Medical Branch at Galveston sent doctors, nurses, and medical students. Firefighters from Austin, Ellington Field, Fort Crockett, Galveston, Houston, San Antonio, and surrounding municipalities contributed to the response initiative The National Red Cross headquarters in Washington allocated the sum of $250,000 for the response effort, and dispatched 30 disaster experts. It was the second major disaster in Texas within the week. During the preceding Wednesday, a tornado traversed the Panhandle and northwestern, and it killed 132 persons in the two states. Response personnel who were stationed in the tornado area were assigned to the location of the Texas City disaster (Associated Press, 1947).

The City of Houston sent an amalgamation of doctors and nurses to provide medical assistance. It also sent both police and firefighting resources and personnel. The state of Texas sent highway patrol resources and disaster experts. The Eighth Naval District headquarters at New Orleans indicated the dispatching of a commodore, three doctors and 16 hospital corpsmen. The naval personnel were sent from the Orange Naval Station and the Naval Air Station at Hitchcock, Texas. Without waiting for an assistance call, the Fourth Army at San Antonio sent an airplane to St. Louis for blood plasma. Other aircraft were dispatched to Fort Worth, Texas to obtain 10,000 blankets. A chemical officer was sent with 500 gas masks from the San Antonio General Depot. Nurses from Ellington Field and two navy nurses assisted the injured and wounded among Galveston hospitals (Associated Press, 1947).

The high school gymnasium was converted into a temporary morgue, and McGar's garage was used as an embalming room. A number of morticians, including some students, volunteered their services as well. In the days after the explosions, 150 embalmers worked on the bodies in the garage. Students from local dental schools were aided with the identification of the dead through dental records.

AFTERMATH AND INVESTIGATION

Reconnaissance of the incident area provided a bleak report of utter destruction regarding not only the epicenter, but also the surrounding areas. A reporter for the *Houston Chronicle*, who flew over Texas City for an hour after the initial blast, reported fires along the waterfront, the Santa Fe Railroad, and within a gasoline refinery area. He reported no fires among the observed business and residential areas. Bill Nottingham, an Air Force veteran who flew air raids in World War II and who was a *Houston Post* photographer at the time of the Texas City incident, flew over the city and reported that the physical destruction resembled the bombing aftermath produced by a flight of B-29 Superfortresses. Nottingham also reported that, during the war, he had flown among raids that produce fewer damages. A reporter flying over the scene likened it to bomb destruction of European cities in the recent war. The mushrooming cloud of smoke that arose was described as resembling the aftermath of the atom bombing of Hiroshima (Associated Press, 1947).

Although disagreement existed regarding the exact number of deaths, it was estimated that between 500 and 600 people died in the explosion, and that approximately 4,000 were injured. An exact count of the dead would have been difficult to obtain because of the condition of many of the bodies, the sheer obliteration of many who were near the initial explosion, and the inaccessibility of the incident area. There were also a number of foreign seamen and noncensus laborers present who were unaccounted for and missing. The identification of bodies continued through the middle of June 1947. It was performed by the Houston Identification Bureau, the Department of Public Safety, Vic Landig (the chair of the city's Burial Committee), Judge G.P. Reddell, and citizens of Texas City (Moore Library, 2018).

Literally hundreds of lawsuits were filed on behalf of the disaster victims. In one of the most notable lawsuits, the *Dalehite Case of 1950*, several claimants compiled multiple lawsuits into one, filing jointly against the federal government. The case eventually made it to the Supreme Court, where the court ultimately ruled in favor of the U.S. government (*Dalehite v. United States*, 1953).

The event resulted in the first ever class action lawsuit against the U.S. government. During 1950, a district court found the United States responsible for negligence, involving both omission and commission, by 168 named agencies and their representatives in the "manufacture, packaging, and labeling of ammonium nitrate" as well as further errors with the "transport, storage, loading, fire prevention, and fire suppression" (Lyons, 2016, p. 11). That court victory was short-lived, however, as the Fifth Circuit Court of Appeals overturned the verdict and it was reaffirmed by the U.S. Supreme Court with a 4-3 vote. The U.S. Congress eventually stepped in after nearly a decade of litigation and awarded some compensation to those affected by the explosion and ensuing fires. By 1957, awards of nearly $17 million were given to some 1,394 recipients (Minutaglio, 2003).

At the time of the disaster, the Federal Emergency Management Agency (FEMA) did not exist, nor was there another government system in place to provide monetary aid to disaster victims who had lost their homes, possessions, or means of making a living. With this in mind, U.S. Representative Clark Thompson of Galveston introduced legislation to Congress that would provide compensation to disaster victims to help them rebuild their town and their lives. The bill was passed in 1955. It allowed about $17 million dollars to be distributed to almost 1,400 claimants. The Texas Legislature also agreed to rebate municipal and school taxes in Texas City for 3 years following the disaster, in an effort to stimulate an economic recovery in the area.

Monsanto, which sustained the worst damage, was the first refinery to pledge to rebuild. Other companies quickly followed Monsanto's lead with pledges of their own, and the people of Texas City gradually rebuilt their town. At the time of the 2010 Census, it was home to about 45,000 people, and the city was growing. During modern times, the petrochemical industry remains a major factor in the city's economy.

LESSONS LEARNED AND POLICY IMPACT

The Texas City incident was a hard lesson regarding the dangers of flammable materials. As a result of the Texas City disaster, quality control officials implemented new standards for the transportation and dispersal of ammonium nitrate. In addition to cool temperatures, new regulations required specialized containers for storage and prohibited ammonium nitrate from being stored near other reactive materials. Travel over long distances was discouraged and overseas transfer of the substance was highly restricted.

The Texas City disaster also influenced attitudes regarding disaster planning across the country. It was obvious that there was a need for a more proactive approach toward disaster planning. Many also noted that central coordination throughout the emergency response effort might have been beneficial during the earliest hours of the incident. Because of such an assessment, refineries in the Texas City area formed the Industrial Mutual Aid System (IMAS). The IMAS was a cooperative endeavor in which they agreed to cooperate during any disaster. The IMAS set a precedent for other refineries throughout Texas. Soon thereafter, they also formed various coalitions and cooperatives.

The resulting investigation showed a need for better handling procedures for nitrate materials. After months of tests and evaluation, it was concluded that conditions aboard the ship were conducive to a reaction of the ammonium nitrate, although the cause of the original fire aboard the *Grandcamp* was never identified. Because of such findings, greater attentiveness was given to transporting and handling incendiary and flammable materials.

IMPLICATIONS FOR MODERN EMERGENCY MANAGEMENT

As a result of the explosion, more serious attention was given to the packaging and transport of ammonium nitrate. Quality control officials also created new regulations requiring specialized containers for storage in order to keep the product at cooler temperatures and prohibited its storage near other reactive materials. Overseas transfer of ammonium nitrate was highly restricted and any long distance of travel was discouraged. All of these practices remain intact today.

Attitudes toward disaster planning also changed as a result of the incident. There was no centrally coordinated emergency response but had there been one, perhaps early on in the disaster a number of things might have been mitigated. As a result, refineries in Texas City formed the IMAS—a cooperative arrangement in which they agreed to assist one another in a time of disaster. This concept subsequently spread to other industrial areas across Texas (Texas City Library).

Today's emergency managers have access to annual SARA Title III reports to assist in planning and response efforts as they relate to large quantities of hazardous chemicals being stored in their communities. According to the Environmental Protection Agency, Title III of the SARA provisions is also known as the Emergency Planning and Community Right to Know Act (EPCRA) and is part of the federal Superfund Amendments and Reauthorization Act which became law in 1986. Title III requires states to promote outreach for developing local emergency preparedness programs to respond to chemical releases; receive reports from the regulated community; and organize, analyze, and disseminate the resulting information on hazardous chemicals to local governments and the public. These reports include facility chemical inventories in the form of Tier I and Tier II reports.

SUMMARY

The event showed the unpredictability of calamities. Any human-made incident may occur unexpectedly with speed, surprise, and violence. In some cases, depending on the severity of the incident and its characteristics, varying levels of controllability exist that may constrain the ability of responders to effectively manage the situation. The incident also provided another lesson regarding emergency resources. In some events, the resources necessary for addressing the situation may be destroyed by the event itself. After all, the harbor explosion killed 27 of the 28 available firemen necessary for responding to incidents as well as destroying their equipment.

This event was significant for reasons that extended beyond the fact that it remains the nation's deadliest industrial disaster and one of the largest nonnuclear explosions ever recorded. Reports written following the disaster contained significant changes to the handling, storage, and transportation of material as flammable and explosive as ammonium nitrate, yet history has demonstrated that there remains a difference between lessons "observed" and lessons "learned." Had lessons truly been learned, the 2013 West Fertilizer Plant Explosion would never have occurred. Despite the passage of the EPCRA and SARA Title III reporting requirements, enforcement and compliance remain issues faced by emergency managers every day who must plan for response to incidents involving these often deadly chemicals.

Despite its destructiveness to the locality and its impact forever on the lives of residents, the incident also affected national perspectives of emergency management. Because of the tragedy, national attitudes changed toward preparing for potential calamities. Nationally, organizations began cooperating to devise emergency processes and resources that were envisioned as necessary should another catastrophe occur. Thus, it incited a more proactive attitude toward emergency management across the country.

REFERENCES

Associated Press. (1947). *Blasts and fires wreck Texas City of 15,000; 300 to 1,200 dead; thousands hurt, homeless; wide coast area rocked, damage in millions.* Retrieved from https://archive.nytimes.com/www.nytimes.com/learning/general/onthisday/big/0416.html#article

Beitler, S. (2018). *Texas City, TX, Disaster—Explosion and fire, APR 1947.* Retrieved from http://www.gendisasters.com/texas/1949/texas-city-tx-disaster-explosion-fire-apr-1947?page=0,3

Encyclopedia Britannica. (2018). *Texas City explosion of 1947.* Retrieved from https://www.britannica.com/event/Texas-City-explosion-of-1947

Fire Prevention and Engineering Bureau of Texas & National Board of Fire Underwriters. (1947, April 29). *The Texas City Disaster.* Retrieved November 1, 2017 from http://www.local1259iaff.org/report.htm

History Channel. (2018). *Fertilizer explosion kills 581 in Texas.* Retrieved from https://www.history.com/this-day-in-history/fertilizer-explosion-kills-581-in-texas

Kintz, G., Jones, G., & Carpenter, C. (1948). Carpenter. *Explosions of Ammonium Nitrate Fertilizer on Board the S.S. Grandcamp and S.S. High Flyer at Texas City, Tex., April 16,17, 1947*. Report of Investigations. U.S. Dept. of the Interior. February 1948.

Lyons, C. (2016, June/July). The 1947 Texas City Explosion. *History Magazine*, pp. 8–11.

Minutaglio, B. (2003). *City on Fire: The explosion that devastated a Texas Town and ignited a historic legal battle.* New York, NY: Harper Collins.

Moore Library. (2018). *Fire on the Grandcamp*. Retrieved from http://www.texascity-library.org/disaster/fire.php

Stephens, H. (1997). *The Texas City Disaster, 1947*. Austin, TX: University of Texas Press.

Texas City Library. *1947 Texas City Disaster, April 16 & 17, 1947*. Retrieved from http://www.texascity-library.org/disaster/index.php

U.S. Environmental Protection Agency. *What Is ECPRA?* Retrieved from https://www.epa.gov/epcra/what-epcra

The Centralia Mine Fire (1962)

18

Figure 18.1 Black and White Image of the Cracked and Buckled Highway of a Ghost Town in Centralia, Pennsylvania. The Result of an Underground Mine Fire that has been Burning Since 1962.

Source: dmvphotos/Shutterstock.com

BRIEF OVERVIEW

Coal proved to be gold for Pennsylvania for over a century. Though there is some debate as to how it exactly started, what is not debated is that since 1962 a coal vein has been burning under the Pennsylvania mining town of Centralia. It is estimated that there is enough fuel to keep it burning for another 250 years. The fire is burning at depths of up to 300 ft and has spread more than 8 miles, infringing on the neighboring town of Byrnesville which is also now abandoned (Krajick, 2005). Throughout the years, attempts at remediation were unsuccessful and as a result, in 1984 Congress allocated more than $42 million to relocate the town's residents.

In 1992, Governor Bob Casey evoked eminent domain on all property within the town, condemning all properties. A few residents refused to leave and after continued litigation efforts, they were allowed to live out the remainder of their lives and upon death the property will be claimed under eminent domain.

INTRODUCTION

From its humble beginnings in 1866 as an anthracite coal mining town, Centralia was no different from many other small coal mining towns scattered throughout Pennsylvania. By the 1950s, Centralia's population was comprised of some 1,000-plus residents who were working class poor, most employed by coal mining companies. As that industry dried up, so did the jobs and many residents relocated to other areas of the state. Left behind, according to a 1983 survey was an aged population—40% of the residents left behind were over age 65, 52% had lived in Centralia for more than 25 years, and 42% of them were earning less than $10,000 a year (Kroll-Smith and Couch, 1987).

The Setting

Centralia was initially inhabited by Native American tribes and the land was sold to colonial agents in 1749. Early settlers would soon discover that there was an abundance of anthracite coal in the region. In 1842, the Locust Mountain Coal and Iron Company purchased Centralia's land. During the construction of the Mine Run Railroad in 1854, the rich deposits of coal were discovered. In 1856, the first two mines opened—the Coal Ridge Mine and the Locust Run Mine. Others would follow in quick succession—the Hazeldell Colliery Mine (1860), the Centralia Mine (1862), and the Continental Mine (1863). The railroads expanded lines into Centralia as a result of the mining boom. Centralia was officially incorporated as a borough in 1866 (DeKok, 2009).

According to census records, Centralia reached its peak population in 1890 when it reached 2,761 residents. The town, at its peak, has five hotels, 27 saloons, several churches, two theaters, a bank, post office, 14 general and grocery stores along with schools. Coal production in Pennsylvania would reach its peak nearly four decades later. With the onset of World War I, many young miners from Centralia joined the military. Once the 1929 crash of the stock market occurred, many of the Centralia mines were closed and bootlegging became commonplace. Eventually in the 1960s, the remaining coal companies shut down and rail service ended not soon after in 1966 (Krajick, 2005).

The Disaster Unfolds

Scientists and scholars can't seem to agree exactly when the fire started; however, most agree that it began when the town decided to clean up an unsightly landfill (O'Carroll, 2010). It is believed that the fire department attempted a controlled burn of the dump as it had done in the past at a dump in another location. However, this time they failed to extinguish the fire and it spread into an unsealed opening that fed into the abandoned coal mines. Another theory posed was that the fire started after hot ash or coal was dumped into the dump site. Borough council minutes dated June 4, 1962 make reference to two separate fires burning at the dump, including one started by the firefighters who submitted bills for fighting the fire at the landfill (Quigley, 2007). Residents soon came to realize that the fire was not going to be extinguished, or ever burn itself out—at least not until all the interconnected coal veins in eastern Pennsylvania were no longer able to provide fuel for the fire. More than $5 million was spent in the 1970s and 1980s by the state and federal governments in an attempt to gain control of the fire, but to no avail.

Impact

The fire has since caused significant environmental impacts on Centralia and surrounding communities (McElreath, Jensen, et al., 2013). The release of deadly carbon monoxide gas from the fire was detected first in 1969 when three families were evacuated from their homes due to extremely high levels of gas present. In 1976, just outside of another Centralia home, emanating from an open bore hole, carbon monoxide flowing at a rate 20 times greater than that needed to kill a human being was detected. Then, in 1979, a gas station owner realized that the underground gas tanks at his service station were being heated above 170-degrees and the station was ordered closed. Residents began complaining of physical ailments associated with the fire and the federal government began purchasing properties deemed unsafe. Finally, on Valentine's Day in 1981, a young boy fell into a sinkhole caused by the fire that opened in his own backyard and nearly lost his life. He managed to grab onto a tree root and his cousin was able to pull him out. It was later determined that a lethal amount of carbon monoxide was flowing from the sinkhole (Morton, 2014).

The fire in Centralia is not an isolated incident. Thousands of coal fires are burning across the globe. Once they get started, they are nearly impossible to extinguish. These underground fires threaten homes and roads, poison the air, soil, and water sources, and many believe impact global warming as well. The fires are ignited in a number of different ways, including spontaneous combustion, lightning strikes, or human-caused means (McElreath, Doss, et al., 2014). The United States has the world's largest coal reserves, with hundreds of fires burning, including 38 alone in Pennsylvania. However, this pales in comparison to China and India which have thousands of fires burning (Krajick, 2005).

RELIEF AND FIRST RESPONSE EFFORTS

For 20 years after the fire got burning, firefighters attempted to extinguish it eight separate times. Like forest fires, they tried to dig out trenches but the fire outran them. They next attempted "flushing," which rarely ever works. It is the process by which holes are drilled into or ahead of the fire which is followed by the pouring of wet sand and gravel into the holes to try and cut off oxygen fueling the fires. However, it is difficult to fill all gaps which is what would be needed for success. State and federal geologists then drilled hundreds of exploratory boreholes to determine the exact location of the fire. They followed with a huge trench across what they thought was the supposed path of the fire. But by the time they got the trench dug, the fire had already spread beyond it. Many experts believe that the digging actually helped ventilate the fire and contributed to additional spread (Krajick, 2005). The $42 million dollar government relocation program begun in 1984 finally ended in 2005, and in 2013 the seven remaining residents reached an agreement with the state allowing them to remain in Centralia until they died, at which point their properties would be condemned (Rubinkam, 2013).

AFTERMATH AND INVESTIGATION

By 2000, the fire had moved into the cemetery, with smoke wafting up from around the gravestones. In 2002, the town had its zip code revoked (Krajick, 2005). The only answer to the problem was determined to be the use of eminent domain to relocate residents and allow the fire to simply burn itself out, since all previous attempts had failed and realistically the location cannot even be determined any longer. According to a geologist with the state's Office of Surface Mining, if you can't put the fire out you have no choice but to relocate the people (Krajick, 2005). That is exactly what the state and federal governments decided to do with Centralia.

LESSONS LEARNED AND POLICY IMPACT

Lessons learned from this disaster are limited. The harsh reality is that once these types of fires start, there is little that can be done to stop them. So far, the government has opted to simply apply eminent domain and relocate residents impacted by the deadly consequences of these fires. What we have learned is a number of significant environmental impacts are a result of these fires, including air, water, and soil contamination as well as possible contributions toward global warming (Ratchford, 2013). Additionally, after generations of engineers and geologists have studied these massive fires, it has been determined that total excavation is usually the only tactic to work in extinguishing these types of fires, but they must be aggressively challenged early before they reach the size of Centralia's fire (Krajick, 2005). As a result, the federal Bureau of Land Management has been employing this strategy as part of its firefighting policies.

Figure 18.2 Centralia Pennsylvania Mine Fire Dump Site
Source: SimcoePix/Shutterstock.com

IMPLICATIONS FOR MODERN EMERGENCY MANAGEMENT

Modern emergency management is challenged by the size and scope of these fires. They are not contained to short events, but generally long-term events that may span multiple jurisdictions over a period of time. Emergency managers who serve jurisdictions impacted by coal mines should maintain situational awareness and request appropriate resources at the start of an event, as it has been determined that the most successful suppression techniques need to be executed early in the event or it may not be able to be contained. There should be close cooperation with state emergency management agencies as well as those state agencies dealing with environmental issues such as environment control. Additional support may be needed from federal agencies such as the Bureau of Land Management and Environmental Protection Agency among others. All-hazards planning for such events should also be included in the jurisdictions emergency operations and mitigation plans.

SUMMARY

While Pennsylvania and many mining companies thrived and profited from coal production, Centralia provided an example of how an abandoned industry could haunt its residents for eternity. Today, Centralia has become a morbid tourist attraction, with visitors snapping photos and stealing souvenirs. Empty plots of brown grass dot the area and steaming sinkholes are scattered about, releasing toxic gases. Some roads continue to buckle and fissures continue to open releasing fumes. Scientists also come to Centralia to study the volcano-like minerals forming around cracks in the soil and to probe for bacteria that thrive in the heat. The media still occasionally has a presence in town with TV and newspaper reporters looking for stories.

REFERENCES

DeKok, D. (2009). *Fire underground: The ongoing tragedy of the Centralia Mine Fire*. Guilford, CT: Globe Pequot Press.

Krajick, K. (2005). Fire in the Hole. *Smithsonian Magazine*. Retrieved March 17, 2018 from https://www.smithsonianmag.com/science-nature/fire-in-the-hole-77895126/#myW0qJWqEHsKGid5.03

Kroll-Smith, J. and Couch, S. (1987). A chronic technical disaster and the irrelevance of religious meaning: The case of Centralia, Pennsylvania. *Journal for the Scientific Study of Religion, 26*(1): 25–37.

McElreath, D., Doss, D., Jensen, C., Wigginton, M., Nations, R., Van Slyke, J., & Nations, J. (2014). *Foundations of emergency management*. Dubuque, IA: Kendall-Hunt.

McElreath, D., Jensen, C., Wigginton, M., Doss, D., Nations, R., & Van Slyke, J. (2013). *Introduction to homeland security* (2nd ed.). Boca Raton, FL: CRC Press.

Morton, E. (2014). "How an Underground Fire Destroyed an Entire Town". *Slate Magazine*. Retrieved March 17, 2018 from http://www.slate.com/blogs/atlas_obscura/2014/06/04/centralia_a_town_in_pennsylvania_destroyed_by_a_mine_fire.html

O'Carroll, E. (2010, February 5). Centralia, Pa.: How an underground coal fire erased a town. *The Christian Science Monitor*. Retrieved March 17, 2018 from https://www.csmonitor.com/Environment/Bright-Green/2010/0205/Centralia-Pa.-How-an-underground-coal-fire-erased-a-town

Quigley, J. (2007). *The day the earth caved in: An American mining tragedy*. New York: Random House.

Ratchford, D. (2013, October 30). Agreement reached with remaining Centralia residents. *WNEP 16 News*. Retrieved March 17, 2018 from http://wnep.com/2013/10/30/agreement-reached-with-remaing-centralia-residents

Rubinkam, M. (2013, October 31). Pa. Residents Living Above Mine Fire Cleared to Stay. *Associated Press*. Retrieved March 17, 2018 from https://finance.yahoo.com/news/pa-residents-living-above-mine-200344503.html

The West Fertilizer Company Explosion (2013)

19

Figure 19.1 The West Fertilizer Company Explosion
Source: U.S. Chemical Safety and Hazard Investigation Board

BRIEF OVERVIEW

On April 17, 2013, a fire and explosion occurred at the West Fertilizer Company in West, Texas. The explosion killed 12 first responders and three civilians while injuring more than 250 others. The blast destroyed the company and caused widespread damage to more than 150 buildings throughout the town, including homes, schools, and other structures. The West Fertilizer Company began operations in West in 1962 and stored and distributed fertilizers, chemicals, grains, and other farming supplies to the largely rural community located 18 miles north of Waco, Texas. It was a relatively small facility with just nine employees.

INTRODUCTION

The world saw firsthand what ammonium nitrate could do when Timothy McVeigh loaded up a Ryder rental truck and detonated it in front of the Alfred P. Murrah Building in Oklahoma City on April 19, 1995. So when a fire erupted at the West Fertilizer Company on Aril 17, 2013, the end result should not be surprising—the stockpile of between 40 and 60 tons of ammonium nitrate detonated, killing the 12 emergency responders, three civilians, injuring more than 250 others while triggering a 2.1 magnitude earthquake in the area. Nearly half of the city's buildings received damage or were destroyed.

The Setting

West is located in McLennan County, Texas, and has a population just under 3,000. The rural farming town is located 18 miles north of Waco, 70 miles south of the Dallas-Fort Worth area and 120 north of the state capitol Austin. The West Fertilizer Company opened in 1962 and according to the U.S. Chemical Safety and Hazard Investigation Board's (CSB) Final Investigation Report on the West Fertilizer Company Fire and Explosion (Report 2013-02-I-TX), the company was originally built with no other buildings encroaching the facility. The city of West literally grew around the company over the years and because there were so few reported accidents, there was very little concern to regulate the growth around the company. Eventually, apartment complexes, parks, nursing homes, and schools would come to be built within a 600-ft radius of the company. The community's only concern from the company had been the risk of an accidental release of anhydrous ammonia gas that was being stored in tanks on the company's grounds—it was unaware of the storage of mass quantities of ammonium nitrate and the consequences a fire would cause.

The Disaster

At approximately 7:29 p.m. on April 17, 2013, the 911 dispatch center in West began receiving reports of smoke and fire at the West Fertilizer Company. A West Police Department patrol officer was the first to respond to the incident. While on his routine patrol he reported smelling smoke while driving through the city park which was not far from the company; however, he could not confirm a fire at that time. A citizen then advised that smoke was venting from the top of the West Fertilizer Company and he contacted dispatch and requested that the West Volunteer Fire Department be dispatched to the facility. Upon arrival at the company, the officer observed a working structure fire and relayed that information back to dispatch.

While en route, the West Volunteer Fire Department (WVFD) requested the West Police Department (WPD) officers to assist with traffic so that hoses would not get run over. However, the officer opted to begin evacuations of the park and nearby area which included the schools. During this time, onlookers started gathering in the area in their vehicles as firefighters began arriving on the scene. Upon their arrival, other members of the police department who observed the fire growing began further evacuations fearing an anhydrous ammonia gas leak.

The WVFD responded with two fire engines, two brush trucks, a water tanker and a number of volunteers arrived in personally owned vehicles (POVs). An EMT-Basic class had been going on at the WVFD Emergency Management Support building where a number of members from mutual aid departments were in attendance. When the sirens were activated for the fire, they all made their way to the site. In all, 10 volunteer firefighters including a career member from the Dallas Fire Department plus two civilians arrived on scene to help fight the blaze. In just 20 min, at 7:51 p.m., the company exploded, immediately killing all of the first responders on the scene.

AFTERMATH AND INVESTIGATION

Due to the nature of the event, multiple local, state, and federal agencies were involved in the response and subsequent investigations were conducted. The Bureau of Alcohol, Tobacco, Firearms and Explosives (ATF), the U.S. Chemical Safety and Hazard Investigation Board (CSB), and the National Institute for Occupational Safety and Health (NIOSH) all conducted separate investigations into the explosion and released their findings. The most comprehensive of the reports was produced by the CSB, while the NIOSH report focused on

Figure 19.2 West Fertilizer Explosion
Source: U.S. Chemical Safety and hazard Investigation Board

the firefighter line of duty deaths. The ATF was charged with determining if the explosion was criminal in nature, accidental, or perhaps even an act of terrorism. The ATF was assisted by the Texas State Fire Marshal's Office (SFMO) in a joint investigation into the incident. Additional agencies responding included the U.S. Occupational Safety and Health Administration (OSHA), Texas Commission on Environmental Equality, the U.S. Federal Emergency Management Agency (FEMA), the U.S. Environmental Protection Agency (EPA), and numerous other state and local law enforcement and emergency response agencies.

At the request of the SFMO, the ATF deployed to the scene to conduct a joint investigation into the explosion. The agencies sought to determine the origin and cause of the fire and explosion and determine whether the fire was intentionally set. The two agencies secured the scene for 4 weeks. Together the agencies interviewed witnesses, excavated the company site, reconstructed the electrical systems, searched for evidence of any criminal activity, and looked closely at the three possible causes according to inspectors: faulty electrical wiring, a short circuit in a golf cart stored in the company, or arson. Finally, in May 2016, the ATF announced that all reasonable accidental and natural causes had been eliminated and that the fire that led to the explosion was a criminal act of arson. To date, no one has been arrested and charged for the crime (Warren 2016).

LESSONS LEARNED AND POLICY IMPACT

Two distinct areas of lessons learned emerged from this disaster—one involving emergency planning, hazard awareness, storage practices, and land use planning and zoning that are primarily the responsibility of the municipality and local government along with other agencies that govern the way things are managed at that particular level. The second area in which significant lessons were learned involved the emergency response by first responders including police, fire, and emergency management. As a result of the West Fertilizer Company explosion, policy recommendations were made at all levels of government to include local, state, and federal agencies (National Institute for Occupational Safety and Health, 2014). The two most enlightening and comprehensive reports included the CSB's final report and the NIOSH report that detailed the events leading up to the deaths of the first responders. To a large degree, the CSB report supported the findings of the NIOSH report.

The CSB's report focused on a number of key areas which included regulatory oversight, hazard awareness, emergency planning and response, fertilizer grade ammonium nitrate storage practices, and land use planning and zoning. Based on the Board's findings, the following key recommendations were made (U.S. Chemical Safety and Hazard Investigation Board, 2016):

- The EPA should develop a guidance document on Emergency Planning and Community Right-to-Know Act (EPCRA) requirements that is issued annually to State Emergency Response Commissions and Local Emergency Planning Committees (LEPCs) to ensure everyone understands which chemicals are exempt and which must be reported—this was a major failure in the West Fertilizer Company's reporting procedure leading up to the explosion. This document would also detail planning requirements, placing an emphasis on annual training exercises and drills for local emergency response agencies—none of which were ever conducted in West, Texas.

- OSHA should require automatic fire sprinkler systems and detection systems for indoor fertilizer grade ammonium nitrate storage areas. They should also define adequate ventilation for these same indoor storage areas. These same storage areas should be isolated from the storage of other combustible, flammable, and contaminating materials. Finally, separation distances should be established between fertilizer grade ammonium nitrate storage areas and other hazardous chemicals, processes, and facility boundaries—all of these factors drastically contributed to the explosive event. The activation of a suppression system may very well have prevented the ammonium nitrate from ever igniting in the first place.
- FEMA, through a new or existing program in conjunction with other training partners, should create and implement a mechanism to help fund training at the regional, state, and local career and volunteer fire departments on how to respond to fire and explosions at facilities that store fertilizer grade ammonium nitrate. This training should also incorporate preincident planning which was clearly absent in West, Texas.
- The Texas Commission on Fire Prevention (TCFP), which is the agency responsible for training and certifying firefighters in the state of Texas and the State Firefighters' and Fire Marshals' Association of Texas (SFFMA), should develop minimum standards in course curricula which includes hazard awareness of fertilizer grade ammonium nitrate for those departments that have facilities in their jurisdictions or respond to jurisdictions with such facilities. The curricula should include lessons learned, pre-incident planning, hazards posed by other materials generally stored in conjunction with fertilizer grade ammonium nitrate, on-scene emergency response duties such as risk assessment, scene size-up and situational awareness, and the National Incident Management System (NIMs) and Incident Command System (ICS)—all of these elements were missing at the time of the West incident. The curricula could also lead to a training certification program recognized by the state.

The NIOSH report listed five key recommendations followed by two additional recommendations that governing agencies at the federal, state, regional, and local levels should consider. They included the following:

- Fire departments should have a written risk management plan and use risk management principles at all structure fires, especially ones involving high-risk hazards such as those present at the West Fertilizer Company—none were applied at this incident.
- Fire departments should develop, implement, and enforce a written Incident Management System (as required by federal guidelines) to be followed at all emergency operations—at no time did anyone call for or establish command at this incident nor did all responders have proper ICS training according to their training records.
- Fire departments should ensure that firefighters wear full personal protective equipment appropriate for the assigned tasks—a number of first responders who arrived on scene did not don the appropriate gear as they had come in POVs from class.
- Fire departments should ensure that firefighters are trained to standards that meet or exceed NFPA 1001 Standard for Fire Fighter Professional Qualifications—not all firefighters to arrive on scene were certified firefighters, therefore lacked experience in handling hazmat situations and even basic firefighting strategies and tactics.
- Consider requiring automatic sprinkler systems, performing regular fire inspections, preplanning facilities, and other types of active fire prevention methods in commercial facilities, especially those posing a high risk—the West Fertilizer Company was not sprinklered and had not been preplanned by the WVFD.
- Consider following the most current safe handling procedures for ammonium nitrate fertilizer storage and handling—because no inspections were being performed by any agency, the plant was not properly handling, storing, and disposing of ammonium nitrate properly over time which aggravated the situation on the night of the explosion.

IMPLICATIONS FOR MODERN EMERGENCY MANAGEMENT

Modern emergency management stands to learn a great deal from the West Fertilizer Company explosion. First and foremost is knowing what resides within your jurisdiction. Never assume that your commercial enterprises adhere to all local, state, and federal regulations—if they did we would not require regulatory agencies and governing bodies. In the case of the West Fertilizer Company, over the years there were minor safety and other violations—nothing significant was ever found during an inspection of the plant. OSHA last inspected the plant in 1985 (Hananel 2013) and it was fined $30 for improper storage of anhydrous ammonia. The plant was also cited for violations of respiratory protection standards but was never fined. Additional fines were levied against the company in subsequent years by the EPA and Texas Commission on Environmental Quality for various violations, but never amounted to anything serious.

It wasn't until after the explosion that the then Homeland Security Secretary Janet Napolitano told Senate investigators that the company never reported its ammonium nitrate stock to her department (McElreath, Doss, et al., 2014). Following the Oklahoma City bombing, federal law requires that the Department of Homeland Security be notified whenever anyone has more than one ton of ammonium nitrate on hand or 400 lb when combined with other combustible material. In this case the West Fertilizer Company had 100 more times the quantity on hand necessary to be reported and failed to do so (Gebrekidan & Schneyer 2013).

The other implication for modern emergency management this disaster brought to light was the significance of the LEPC and the role it plays in keeping our communities safe. The CSB report noted that the West Fertilizer Company had not attended an LEPC meeting in 21 years, something officials should have been more cognizant of and proactively engaged the company in participation considering the concerns over the anhydrous ammonia leaks of the past. However, the biggest error as noted in the CSB report was the LEPC's belief that the company was exempt from reporting requirements because it was inadvertently left off of Annex Q: Hazardous Material and Oil Spill Response and not being regulated by the EPCRA of 1986. However, it should have been listed and had adequate planning been done and the company more involved as part of the LEPC, the outcome on that fateful day might have turned out differently. Therefore, engaging all stakeholders is paramount for successful emergency management and ensuring accurate interpretation of regulations is also imperative.

SUMMARY

The West Fertilizer Plant Explosion demonstrated gaps in the enforcement of SARA Title III of the EPCRA. The explosion mirrored many of the aspects seen in the 1947 Texas City Explosion (see Chapter 17). One major theme uncovered through the various postincident investigations was the failure of the company to report the type and quantities of material being stored at the plant. A lack of professional emergency management within the town allowed the company to skirt the SARA Title III reporting requirement, as there was no official taking accountability to ensure the company was in compliance. Finally, the first responders were not equipped nor properly trained to suppress a fire of this type and magnitude. When all of these factors came together, West, Texas became another casualty of ineffective practices—many of which were repeated despite the learning that should have occurred following the Texas City Explosion.

REFERENCES

Gebrekidan, S. and Schneyer, J. (2013, May 3). *At Texas fertilizer plant, a history of theft, tampering.* Reuters. Retrieved from https://www.reuters.com/article/us-usa-explosion-texas-idUSBRE94206020130503

Hananel, S. (2013, April 19). *Texas fertilizer plant had last OSHA inspection in 1985.* The Huffington Post. Retrieved from https://web.archive.org/web/20130422074548/http://www.huffingtonpost.com/2013/04/18/texas-fertilizer-plant-ha_n_3113117.html?

McElreath, D., Doss, D., Jensen, C., Wigginton, M., Nations, R., Van Slyke, J., & Nations, J. (2014). *Foundations of emergency management.* Dubuque, IA: Kendall-Hunt.

National Institute for Occupational Safety and Health. (2014). *9 volunteer fire fighters and 1 off-duty career fire captain killed by an ammonium nitrate explosion at a fertilizer plant fire—Texas (Report #F2013-11).* Retrieved from https://www.cdc.gov/niosh/fire/pdfs/face201311.pdf

U.S. Chemical Safety and Hazard Investigation Board. (2016). *Investigation report west fertilizer company fire and explosion final report* (Report 2013-02-I-TX). Retrieved from http://www.csb.gov/west-fertilizer-explosion-and-fire-

Warren, D. (2016, May 11). *Authorities say a fire that caused the 2013 explosion at a Texas fertilizer plant that killed 15 people was intentionally set.* U.S. News & World Report. Retrieved from https://www.usnews.com/news/us/articles/2016-05-11/official-fire-in-texas-fertilizer-Company-blast-was-criminal

The Johnstown Flood (1889)

20

Figure 20.1 The 1889 Johnstown Flood Destruction
Source: Library of Congress

BRIEF OVERVIEW

The most infamous dam failure in U.S. history occurred on Friday, May 31, 1889, when the South Fork Reservoir Dam collapsed (Wall, 2000). The poorly designed and neglected maintenance of this earthen dam located on a resort lake resulted in its failure during heavy rains, which resulted in a massive wave of approximately 20 million tons of water that swept Johnstown, Pennsylvania (Frank, 2004). Some observers reported that Lake Conemaugh drained in as little as 37 min to as much as 75 min (McCullough, 1968; Kaktins, Davis, Wojno & Coleman, 2013). Contributing to this disaster, there was excessive snow and rain precipitation during the months prior to the flood, particularly heavy rain fell days before the tragedy, from May 28 through May 31 (Frank, 2004). Several residents at the time claimed that water levels rose about 2 ft overnight (Frank, 2004). As a result, the dam's spillways couldn't withstand the overflow, which further influenced the dam's failure (Frank, 2004). The American Red Cross and several countries sent aid to this disaster. The town was destroyed and over 2,200 people lost their lives (Frank, 2004).

Johnstown had seen its share of downpours over the years leading to this specific event in 1889 (Frank, 1988). Floods occurred almost annually in the 1880s, which caused erosion, riverbank damage, and increased reservoir levels (Frank, 1988). Nothing was done to alleviate the damage done by these deluges, and the constant stress on the integrity of the South Fork Dam was taking its toll (Frank, 1988). Over-cutting timber and channel obstructions created by landowners upstream also added to the disaster (Frank, 1988).

The legendary flood occurred on Friday, May 31, 1889, after severe rains had engorged the South Fork Reservoir (McGough, 2002). The dam gave way, discharging 20 million tons of water generating an enormous wave that struck the Conemaugh Valley which smashed Johnstown and its neighboring towns with such force that over 2,200 lives were lost during its onslaught and aftermath (McGough, 2002). All manner of vegetation, vehicles, structures, boulders, railroad cars and tracks, and any other obstacles were rushed away as the undulation hurtled down the valley toward Johnstown (McGough, 2002). The wave also destroyed a barbed wire manufacturing company and picked up miles of barbed wire which intensified the devastating power of the wreckage (McGough, 2002). A small number of structures in Johnstown survived the 1889 flood; those that did were typically masonry constructed (McGough, 2002).

The surge crashed into Yonder Hill causing a backwash into town and the debris accumulated under the stone arch railroad bridge, holding water, creating a temporary lake in downtown Johnstown (Johnson, 1889). Water eventually seeped through the bridge and drained from the temporary lake into the lower part of the town, leaving a complete catastrophe of standing water, mud, and distorted wreckage (Johnson, 1889). People were drowned by the flood, crushed by the debris, and some burned to death by fires that erupted due to the devastating wreckage (Johnson, 1889).

INTRODUCTION

The Pennsylvania Main Line was a rail and canal system that transported commodities from the east to the west with Johnstown as a crucial port along the way. By the end of the Civil War, Johnstown flourished from trade and industry and was the country's largest manufacturer of iron at the time. The canal connecting Johnstown to Pittsburgh would run dry in the summer if there wasn't sufficient precipitation. It took 18 years to build a large reservoir in the mountains above Johnstown to store water during dry periods. Shortly after the reservoir's completion, the railroad put the Pennsylvania Main Line rail and canal system out of business in 1854, thus abandoning the newly constructed reservoir and dam (Wellington & Burt, 1889, Wharton-Michael, 2012).

Johnstown canal was a retreat for some of Pittsburgh's most affluent. As such, an entrepreneur decided to develop the site and bought the dam, establishing the South Fork Fishing and Hunting Club, which resulted in the private ownership of the earthen dam. Without considering the critical infrastructure and safety factors for developing and maintaining the construction of the dam, the club lowered the dam to create an extensive road with adequate width to handle two carriages (Frank, 1988). The reservoir was stocked with fish and a fish screen was installed across the spillways to keep the fish from escaping (Frank, 1988). The recreational club failed to maintain the integrity of the dam's structure; the changes made by the club weakened the South Fork Dam's stability (Frank, 1988).

The industrial city of Johnstown was built on a floodplain that had three converging rivers at the base of the Laurel Highlands and Allegheny Mountain region. In 1889, during a particularly wet month of May, an intense rainstorm hit Johnstown on May 30. By noon the next day, the streets of Jamestown were inundated with a couple feet of water. On May 31, 1889, the South Fork Dam in Cambria County failed, sending a wall of water 20–30 ft high that smashed downtown Johnstown, pushing ahead of it automobiles, trees, boulders, houses, and even an 80-ton locomotive all of which killed 2,209 people. Johnstown, South Fork, Mineral Point, and East Conemaugh were reduced to rubble. Boilers exploded at Woodvale's Gautier Wire Works, and miles of barbed wire were rushed into the surge adding to its destructive force. The flood swept onward to the Conemaugh like a tidal wave, over 20 ft in height, to Johnstown, 6 or 8 miles below, gathering force as it tore along through the wider channel, and quickly swept everything before it (The New York Times, June 1, 1889a). Houses, factories, and bridges were overwhelmed in the twinkling of an eye and with them human occupants were carried in a vast chaos down the raging torrent (The New York Times, June 1, 1889a). Firsthand accounts are that the dam looked as though it melted away as water overflowed and raced down the mountains.

The Setting

The industrial town of Johnstown was founded by Joseph Johns in 1791, which is near the base of the Allegheny Mountains of Pennsylvania at the conjunction of the Conemaugh and the Stone Creek River. Johnstown is 60 miles east of Pittsburgh and roughly 30,000 people resided there, many working in the iron mills which thrived due to the persistent demand of iron for war. Johnstown was considered the great industrial center of Pennsylvania by the end of the First World War (Johnson, 1889).

Johnstown is located on a floodplain and standing water in the streets was nothing new because of the region's proneness to floods; a 918-ft-long by 72-ft-high by 10-ft-wide at the crest and 220-ft-wide at the base earth dam made of dirt and rock was built in 1840, on the Little Conemaugh River, which is 14 miles upstream from Johnstown (Frank, 2004). Lake Conemaugh was the largest man-made lake which was created by the largest earth dam in the United States.

The dam was eventually named the South Fork Dam which was reconstructed in 1852, and created a water supply for the western division of the Pennsylvania Canal from Allegheny Portage Railroad at Johnstown to Pittsburgh (Frank, 1988). A culvert made of stone and a control tower with five relief valves were incorporated into the dam's structure to discharge water to the canal, as needed (Frank, 1988). On the eastern side of the dam, an 85-ft-wide spillway was included (Frank, 1988). The original earth dam was part of an extensive rail and canal system that was repaired and rebuilt several times over the years and became obsolete in 1852, as new technology, the private Pennsylvania Railroad (PRR), which was faster and less expensive and replaced the canal system as the major method of transporting freight (Frank, 1988). By 1854, the rail and canal system was closed, the abandoned canal system fell into dereliction, and preventative and predictive maintenance on the dam was disregarded (Frank, 1988).

Figure 20.2 Johnstown Flood, 1889: General View of Debris
Source: Library of Congress

In 1862, a segment of the culvert collapsed triggering a section of the dam to be swept away. Substandard fill materials were used to repair the damaged sections and the dam's future repairs were not done to proper engineering standards (Frank, 1988). Subsequent washouts occurred furthering the erosion of the embankments (Frank, 1988). The fill materials used to repair the collapsed culvert eventually settled creating a low point along the dam crest near the center (Frank, 1988). The outlet tunnel eventually collapsed and was not repaired, so no relief ports were available to drain the lake during torrential downpours (Frank, 1988). The top portion of the dam was widened to permit two carriages to pass by which reduced the height of the dam having a negative effect on spillway capacity (Frank, 1988).

The Disaster Unfolds

In 1889, Johnstown had a population of 30,000 people, with many working in the growing steel industry. Leading up to the disaster, Johnstown had a rainy month in 1889, and the steady heavy rain over the course of the previous day, May 30, had caused major capacity issues with the dam. The spillways were clogged with debris and the South Fork Fishing and Hunting Club fish screens that were installed across the spillways were rapidly clogging up preventing the ability to lower the lake levels (Frank, 1988). The reduction of the dam crest height for the installed two-carriage roadway, and the low point near the center of the dam due to settlement of debris from previous embankment erosion, made the conditions perfect for a disaster (Frank, 1988).

The unrelenting rains saturated the ground and the runoff from surrounding hillsides filled Lake Conemaugh beyond its capacity so that eventually the dam overflowed, and the man-made earth dam washed away, releasing 20 million tons of water from the lake into the Little Conemaugh River to descend on the towns of Johnstown, South Fork, Mineral Point, and East Conemaugh, Pennsylvania (Rose, 2013). The South Fork Dam collapsed at 3:10 p.m. on May 31, 1889, causing a massive wall of water (Frank, 1988). People, homes, structures, trees, boulders, locomotives, and other vehicles in the path of the rushing flood waters were crushed and swept away (The New York Times, (1889a, b, c)"). Tall buildings were destroyed by the hazards created by the raging waters (Frank, 1988). Over 30 train engines and miles of barbed wire from the Woodvale's Gautier Wire Works were wrenched into the onslaught, and multiple whirlpools added to its destructive force (Frank, 1988).

A stone bridge downstream from the town snared heaps of debris and then proceeded to catch fire. Some of the survivors who were on top of the floating rubble were burned to death in the fire. The bridge debris bottleneck also created a temporary lake in the town. All 20 million tons of water from the man-made Lake Conemaugh rushed forward at speeds up to 40 miles per hour, sweeping away everything in its path. About 6 o'clock, Superintendent Robert Pitcairn of the PRR telegraphed from Sang Hollow, at the gap in the Laurel Ridge mountains west of Johnstown, that he had seen about 200 persons afloat on gondola cars, shanties, etc. and that the disaster was appalling (The New York Times, June 1, 1889a). One telegraph operator says he counted 63 bodies in 20 min floating past his office (The New York Times, June 1, 1889a).

A temporary martial government was established over the ruined city of Johnstown, under the Adjutant General of Pennsylvania, assisted by military companies from Pittsburg and by volunteer officers (The New York Times, June 3, 1889b). Attempts at disorder and violence by small gangs of tramps were vigorously suppressed, and several marauders were lynched and shot to death, for the people in the solemn earnestness of their work of succor and rescue have not the patience to wait the tedious process of law (The New York Times, June 3, 1889b).

Impact

The river flowed over several hundreds of putrefying bodies, besides acting as a sewer for the entire district (The New York Times, June 7, 1889c). There was no effort made to dig below the surface of the city as it was left by the deluge, and from the fact that, in the immediate vicinity, bodies had been found in the debris at the rate of five an hour, the logical inference is that in all likelihood there were scores of bodies mixed with the heavy timbers and other wreckage which were driven down upon the town by the flood and then covered with a thick layer of mud from the river bottom (The New York Times, June 7, 1889c). Thus, at a distance not more than two feet below the surface there were corpses which were a continual menace to the health of any person who remained in the vicinity (The New York Times, June 7, 1889c). It took 5 years to rebuild

Johnstown: approximately 2,209 people died, more than 1,600 homes were destroyed, $17 million dollars in damage was reported in 1889, and bodies were being discovered 22 years later as far away as Cincinnati, Ohio (Haseleu, 2012).

RELIEF AND FIRST RESPONSE EFFORTS

The Johnstown Flood was a major catastrophic event in American history (McElreath, Jensen, et al., 2014). Technologic advancements were still taking place in the late 1800s, and safety and engineering codes weren't developed to improve the quality and stability of the structure (Frank, 1988). In 1862, the abandoned dam had another washout and flood; it left the dam untouched and dry for an additional 13 years, until a new owner decided to create an upscale resort near the dam called the South Fork Fishing and Hunting Club (Frank, 1988).

The repair plan was to have it returned to the original specifications, but not taking into account dam construction and safety, the club lowered the dam by 2 ft to create an extensive road with adequate width to handle two carriages (Frank, 1988). The reservoir was stocked with fish and a fish screen was installed across the spillways to keep the fish from escaping (Frank, 1988). The recreational club failed to maintain the integrity of the dam's structure, the changes made by the club weakened the South Fork Dam's stability; this was one of the major causes of the collapse in 1889. The cause of the calamity, admitted by the President of the South Fork Fishing Club, the proprietor of the artificial Conemaugh Lake, was the weakness of the dam alone (The New York Times, June 3, 1889b). No cloudburst or waterspout occurred to compel it—the frailty of the dam and the tremendous pressure of water behind it was the only cause of the catastrophe (The New York Times, June 3, 1889b).

AFTERMATH AND INVESTIGATION

There were several causes leading up to the dam failure. Years of negligence, improper design and repair, constant erosion and a series of imposing rain falls in 1889 led to the washout of the earth and rock dam. Additional factors contributing to the Johnstown's flood included the following:

- The Conemaugh River valley's ridgeline captures and slows down prevailing storms from the west causing considerable amounts of precipitation.
- The Little Conemaugh and Stonycreek Rivers converge in Johnstown to form the Conemaugh River. Historically, sections of these river channels were too shallow and narrow to handle peak flows.
- Deforestation, erosion and encroachment by the town's population expansion were deemed to be dangerous to the floodplain.

The demise of Johnstown was due to the massive man-made earth dam which required an expensive upgrade with new stronger fill materials, which never took place, and continuous maintenance. Several minor and major floods after 1889 destabilized the South Fork Dam and very little was done to prevent it (Frank, 1988). It became too much for the residents to bear and the once great industrial town became more of a ghost town.

LESSONS LEARNED AND POLICY IMPACT

The Johnstown Dam failure encouraged engineers to focus more on stability and durability through catastrophic events, like the 100-year storm (JAHA, 2017). Could the collapse have been prevented if the South Fork Fishing and Hunting Club had maintained the integrity of the dam (McGough, 2005)? The 1889 South Fork Dam failure was proclaimed an "act of God" and no one was held accountable (JAHA, 2017). At the time of the collapse, Pennsylvania had no dam inspection laws (JAHA, 2017).

In 1911, another fatal dam collapse in Austin, PA occurred, which finally prompted state officials to consider implementing some regulations. Individuals made wealthy by America's new industrialization after the Civil War had a strong influence in the political arena and were able to prevent the passage of laws regulating industry (Cook 1972, Rose, 2013). This was partly due to ownership interests and control of newspapers that portrayed industrial growth in a positive light whereas some of it was due to corruption in local government and politicians who governed while working for the very industries needing regulation (Cook, 1972, Rose, 2013). This conflict of interest made it difficult for those elected to protect the interests of the people, to place restrictions on the industries that gave their communities economic life (Cook, 1972, Rose, 2013). As the turn of the century approached, the interest in the needs of the people and how the wealthy industrialists were achieving their wealth at the expense of the public and exploited workers began to be exposed more openly in magazines and newspapers (Cook, 1972, Rose, 2013).

The Flood Control Act was initiated in 1936, and its purpose was to focus efforts on dam construction and safety, preventing the likelihood of flooding. The development of Pennsylvania's dam safety laws was directly related to three significant dam failures:

1. The 1889 failure of South Fork Dam—2,209 deaths—no laws were enacted in the 1880s; however, this started the evolution of the future laws.
2. The 1911 failure of Austin Dam—78 deaths—led to the passing of the nation's first dam safety law in 1913.
3. The 1977 failure of Laurel Run Dam—40 deaths—led to the passing of Pennsylvania's Dam Safety and Encroachments Act in 1978.

The original dam design, with a crest ~0.9 m higher and the added capacity of an auxiliary spillway and five discharge pipes, had a discharge capacity at overtopping more than twice that of the reconstructed dam (Coleman, Kaktins, & Wojno, 2016). A properly rebuilt dam would not have overtopped and would likely have survived the runoff event, thereby saving thousands of lives (Coleman, Kaktins, & Wojno, 2016). The 1889 Johnstown Flood tragedy did not result in the establishment of any laws regarding dam safety in the Commonwealth (Rose, 2013). Twenty-two years later in 1911, the failure of Austin Dam in Potter County claimed 78 lives, which resulted in Pennsylvania becoming the first state to enact dam safety legislation in 1913 (Rose, 2013). Pennsylvania's dam safety laws remained unchanged until 1978 when Pennsylvania's Dam Safety and Encroachments Act (Act 325 of 1978) was enacted (Rose, 2013). The legislative action of 1978 came soon after the 1977 Johnstown Flood in which Laurel Run Dam failed claiming 40 lives (Rose, 2013). Why the 1889 South Fork Dam failure did not directly result in legislation, but the 1911 Austin Dam failure, and 1977 Laurel Run Dam failure did is of interest from an historical perspective (Rose, 2013).

IMPLICATIONS FOR MODERN EMERGENCY MANAGEMENT

The May 31, 1889 South Fork Dam collapse in Johnstown creating the fatal flood that killed 2,209 people was the first major peacetime relief effort for the American Red Cross. On June 5, 1889, Clara Barton, the founder of the American Red Cross, and five Red Cross workers arrived from Washington, D.C., only 5 days after the Johnstown 1889 tragedy. Along with the Philadelphia Red Cross and a team of 50 doctors, nurses, and relief workers who specialized in medical relief, the American Red Cross provided furniture, food, shelter, and supplies for the over 25,000 flood survivors. The first Red Cross Hotel, where the homeless were the primary residents, was built in Johnstown on the old site of the flood-destroyed St. Mark's Episcopal Church. Clara Barton stayed in Johnstown until October 24, 1889, 5 months after the disaster, assisting those in need and the organization has been there ever since. The intensive news coverage of the 1889 Johnstown Flood and heroic relief efforts of those early members of the Red Cross established the organization as the major disaster relief agency in the United States.

Since the 1889 flood, the Red Cross in Johnstown has been a guiding light for those in need. The American Red Cross' mission is to prevent and alleviate human suffering in the face of emergencies, including disaster relief, preparedness education, cardiopulmonary resuscitation (CPR) training, blood collections, and to prevent, prepare, and respond to emergencies.

The development and evolution of Pennsylvania's dam safety laws is directly related to three significant dam failures that resulted in a substantial loss of life (United States Bureau of Reclamation, 1998). The historical context of each of these failures influenced the reaction of the public and legislature (Rose, 2013).

Although each of the dam failures could have likely been prevented if the dams in question had been properly designed, constructed, and maintained, the lack of regulations or the shortcomings of state oversight resulted in dams that were destined for failure (McConnell, 1973). Each failure provides technical lessons for the dam engineering profession, as well as renewed appreciation of why dam safety laws are needed and must be properly enforced (Lewis, Bink, & Muller, 2009).

Devastating and costly floods during the early 20th century led the U.S. Congress to pass the Flood Control Act of 1917 (FEMA, 2016). After a long period of unregulated and poorly constructed levees, the Flood Control Act of 1928 brought into focus public awareness and advances in flood control theory and production by expanding the policy to include flood control, spillways, and channel improvements (FEMA, 2016). The Flood Control Act of 1936 continued to recognize flood control as a national priority, which also authorized the United States Army Corps of Engineers (USACE), and other agencies, to construct flood control structures such as levees and floodwalls, and to perform channel improvements. The 1936 Act also committed the Federal government to protecting people and property (FEMA, 2016). The Flood Control Act of 1936 established an enormous commitment by the federal government to protect people and property on approximately 100 million acres (Arnold, 1988; (McElreath, Doss, et al., 2014). The only limitations on federal flood control projects were that the economic benefits had to exceed the costs (Arnold, 1988). Since 1936, Congress has authorized the Corps of Engineers to construct hundreds of miles of levees, floodwalls, channel improvements, and approximately 375 major reservoirs (Arnold, 1988). These remarkable engineering projects today comprise one of the largest single additions to the nation's physical infrastructure rivaled only by the highway system (Arnold, 1988). They have saved billions of dollars in property damage and protected hundreds of thousands of people from anxiety, injury, and death (Arnold, 1988). They stand today as one of the more significant marks of our technical skill and humane spirit (Arnold, 1988).

SUMMARY

The incontrovertible evidence of the infamous Johnstown Flood of 1898 is indicative of the disaster being a man-made catastrophe (Wall, 2000). Primarily, due to the poorly designed earthen dam and the negligence of the private dam owners that catered to the affluent lake resort area, it was not a matter of "if," but "when" such an unfortunate incident would occur. The legendary flood occurred on Friday, May 31, 1889 after severe rains had engorged the South Fork Reservoir (Rose, 2013). The dam gave way discharging 20 million tons of water generating an enormous wave that struck the Conemaugh Valley which smashed Johnstown and its neighboring towns with such force that over 2,200 lives were lost during its onslaught and aftermath (Rose, 2013). Pennsylvania's history of fatal dam failures had a direct influence on the implementation of dam safety laws in the state (Rose, 2013). The legislators were torn between the promise of industrial development and economic growth vs. the need for laws and regulations that protect the public (Rose, 2013). At the time, laws that would cause hardship on industrial enterprises and would hinder industrial growth were difficult to accept (Rose, 2013).

Pennsylvania's first dam safety law, the *Water Obstructions Act*, was enacted in 1913, largely due to the failures of both the South Fork Dam and the Austin Dam (Rose, 2013). The first dam safety legislation was introduced in the Pennsylvania Legislature shortly after the Austin Dam failure calling for state oversight of dams (US Army Corps of Engineers, n.d). The legislation was stalled in the legislature as lawmakers were reluctant to force regulations on businesses and members of industry who helped them get elected (Rose, 2013).

REFERENCES

Arnold, J. L. (1988). *The evolution of the 1936 Flood Control Act*. Washington, D.C.: Library of Congress Cataloging in Publication Data. KF5588.A316Al62

Coleman, N. M., Kaktins, U., & Wojno, S. (2016). *Dam-Breach hydrology of the Johnstown flood of 1889—challenging the findings of the189 1investigation report*. Heliyon 2 (2016) e00120. Elsevier.

Cook, F. J. (1972). *The muckrakers: Crusading journalists who changed America*. Garden City, NY: Doubleday & Co.

FEMA. (2016). *History of levees*. Retrieved from https://www.fema.gov/media-library/assets/documents/22863

Frank, W. (1988). *The cause of the Johnstown Flood.* Civil Engineering, May 1988, pp. 63–66.

Frank, W. S. (2004). *The Cause of the Johnstown Flood: A new look at the Historic Johnstown Flood.* Retrieved from http://smoter.com/flooddam/johnstow.htm

Haseleu, S., J. (2012). *The Johnstown Flood: Understanding why the South Fork Dam failed.* Chatham University. June 25, 2012. Retrieved from https://rhetoricallyurs.files.wordpress.com/2013/02/the-johnstown-flood-understanding-why-the-south-fork-dam-failed2.pdf

Johnson, W. F. (1889). *History of the Johnstown Flood.* Edgewood Publishing Co. 1889. Retrieved from https://archive.org/details/historyofjohnsto01john

Johnstown Area Heritage Association (JAHA). (2017). Retrieved from http://www.jaha.org/FloodMuseum

Kaktins, U., Davis, T. C., Wojno, S., & Coleman, N. M. (2013). *Revisiting the timing and events leading to and causing the Johnstown Flood of 1889. Pennsylvania History: A Journal of Mid-Atlantic Studies, 80*(3), 335–363.

Lewis, K. H., Bink, D. L., & Muller, L. M. (2009). *The inspection, maintenance, and operation of dams in Pennsylvania.* Harrisburg, PA: Pennsylvania Department of Environmental Protection, Division of Dam Safety.

McConnell, C. H. (1973). Dam safety program in Pennsylvania. *Proceedings, Inspection, Maintenance and Rehabilitation of Old Dams, ASCE,* 139–156.

McCullough, D. (1968). *The Johnstown Flood.* New York, NY: Simon and Schuster.

McElreath, D. H., Doss, D. A., Jensen, C. J., Wigginton, M. P., Nations, R., Van Slyke, J., & Nations, J. (2014). *Foundations of emergency management.* Dubuque: Kendall Hunt.

McElreath, D. H., Jensen, C. J., Wigginton, M. P., Doss, D. A., Nations, R., & Van Slyke, J. (2014). *Introduction to homeland security.* Boca Raton: CRC Press.

McGough, M. R. (2002). *The 1889 flood in Johnstown, Pennsylvania.* Gettysburg, PA: Thomas Publications, p.180.

McGough, M. R. (2005). *The Club and the 1889 flood in Johnstown, Pennsylvania.* Johnstown, PA: Friends of the Johnstown Flood National Memorial, p.124.

Rose, A. T. (2013, January). *The influence of dam failures on dam safety laws in Pennsylvania.* Paper presented at the ASDSO Annual Conference, September 8–12, Providence, RI.

The New York Times. (1889a, June 1). *Hundreds of lives lost: A waterspout's dreadful work in Pennsylvania. The New York Times.*

The New York Times. (1889b, June 3). *The Desolated Valley. The New York Times.*

The New York Times. (1889c, June 7). *Peril after the Flood. The New York Times.*

United States Bureau of Reclamation (USBR). (1998). *Concrete dams: Case histories of failures with back calculations.* Report DSO-98-005, USBR, Denver, CO, 94 p.

US Army Corps of Engineers. (n.d.). *Johnstown local flood protection project major rehabilitation.* Philadelphia, PA: Historic American Engineering Record.. HAER No. PA-413. Compiled by Carlisle, R.C., Brown, E.S.

Wall Jr., William L., (2000). *Living with hazards dealing with disasters: An introduction to emergency management.* Armonk, NY: M.E. Sharpe.

Wellington, A. M., & Burt, F. B. (1889). *The South Fork Dam and Johnstown Disaster. Engineering News and American Railway Journal, 21,* 540–545.

Wharton-Michael, P. (2012). The Johnstown Flood of 1889: The Johnstown Tribune's commonsense coverage vs. common-practice sensationalism. *Journalism History, 38*(1), 23–33.

21

The Iroquois Theater Fire (1903)

Figure 21.1 In front of the Theater at the Time of the Fire, December 30, 1903, 4 p.m.

Source: Publishers Union of America/1904

BRIEF OVERVIEW

For any city in the nation, the Iroquois Theatre was a grand marvel. Opening on November 23, 1903, it reflected in many ways the promise of the new century. Its construction seemed to incorporate the latest in technological advances, and it was even proclaimed by its owners as the safest theater in the nation, and the term "fireproof" became analogous with the theater. Unfortunately, as grand as the theater was, the unthinkable occurred on December 30, 1903, when during a matinee performance, the Iroquois Theatre was engulfed in flames, claiming more lives than any other fire in Chicago history.

Figure 21.2 Aftermath of the Fire
Source: Library of Congress

The investigation into the disaster would reveal mistakes that would border on criminal negligence on the part of the owners and contractors as they rushed to meet the opening deadlines, which resulted in compromising building codes. Theater management failure to complete the instillation of the critical fire precaution measures in a greedy rush to open for the upcoming holiday seasons would prove fatal. The "Absolutely Fireproof" theater would fall well short of that claim of safety.

There is little doubt that those attending the performance on December 30, 1903, in Chicago's Iroquois Theatre believed they were in for an experience of a lifetime. Waiting for the theatergoers were 400 performers and stagehands prepared to host an audience of about 2,000 who were enjoying a lighthearted musical comedy entitled Mr. Bluebird (Flynn, 2012). Soon the performance would turn deadly serious.

INTRODUCTION

The advertised, state-of-the-art "fireproof"-constructed Iroquois Theatre in Chicago was fashioned after the Opéra Comique in Paris, which was originally founded back in 1714. Mimicking the Opéra Comique, the Iroquois Theatre building was designed to give the audience a sense of awe prior to the performance, decorated with frescos, gilded artworks, and ornate wooden trim. Similar to Paris, management strived to introduce children to the world of theater holding matinee performances designed for a younger audience and for families.

The Iroquois Theatre claimed its 1,724 seat audience could empty the theater in less than 5 min utilizing its abundant 27 exits. The owners wanted to open prior to the Thanksgiving, Christmas, and New Years' holiday seasons, but construction was grossly behind schedule. Ignoring safety, the impatient owners made the catastrophic decision to open the theater under construction on November 23, 1903. Following the various building construction trade sequence, the advertised "absolutely fireproof" protection features had not been installed, prior to the grand opening. On Wednesday afternoon on December 30, 1903, Eddie Foy was appearing as the leading comedy character in "Mr. Blue Beard." The theater was packed. Estimates of the crowd admitted for the performance ranged between 2,000 to 2,300 theatergoers, far beyond the planned capacity of the 1,724 seats in the theater.

Shortly after 3 p.m., during the beginning of the second act, William McMullen, a stagehand corroborated that a stage spotlight ignited a red velvet curtain. Subsequently, the fire quickly spread to the stage scenery made of oil-painted canvas (Groves, 2006). Utilizing a Kilfyre, which is a patent powder fire extinguisher, McMullen attempted to extinguish the flames, but the fire rapidly ignited the stage backcloth above. The disaster occurred during the second act, as Eddie Foy entered the stage during a double octette called "In the Pale Moonlight." Eddie Foy attempted to calm the audience while fragments of the burning backdrop collapsed during the performance. The fire-resistant asbestos curtain failed to drop all the way to the floor, which should have saved lives by separating the audience from the stage fire. The 400–500 performers and stagehands began fleeing the theater through the backstage exits, allowing outside air to fuel the fire engulfing the audience and further igniting all combustibles. A known total of 602 people died during the 1903 Iroquois Theatre fire in Chicago, 575 burned to death in the fire, of which 212 were children, and another 27 succumbed later to their injuries and burns sustained during the blaze. It still remains the most lethal theater infernos in U.S. history, with about one in every three audience members perishing in the fire.

The Setting

In Chicago on November 23, 1903, the new theater, the Iroquois, was opened with Klaw & Erlanger's Drury Lane spectacular production, "Mr. Bluebeard" (The Indianapolis Journal, 1903). It is owned by Will J. Davis and Harry J. Powers, of Chicago; Nixon & Zimmerman, of Philadelphia; and Klaw & Erlanger, of New York; and represents, including the site and an office building, an investment of $1,000,000, the theater costing $500,000 (The Indianapolis Journal, 1903). The Iroquois Theatre was one of the most completely equipped playhouses in the United States (The Indianapolis Journal, 1903).

The seating capacity is 1,744 and the stage is 110 ft deep by 60 ft wide (The Indianapolis Journal, 1903). The foyer, 53 ft high and 90 ft in length, is a striking feature, being patterned after the Opéra Comique in Paris and the Congressional library (The Indianapolis Journal, 1903). The staircases are supported by 10 columns of Pavonazzo marble, and over 2,000 electric lights are required for illuminating purposes (The Indianapolis Journal, 1903). The spectacle, a most elaborate one, was witnessed by a packed house, particular interest being taken by local theatergoers from the fact that three of the principal actors, Eddie Foy, Bonny McGinn, and Harry Gilfoil, are natives of Chicago (The Indianapolis Journal, 1903).

The Iroquois Theatre was located on the north side of Randolph Street near the corner of Dearborn (Norbury, 2003). The building was L-shaped, with the entrance foyer, ticket windows, offices, and facilities for the audience (checkrooms, restrooms, etc.) off of Randolph extending to the north (Norbury, 2003). The stage and seating were in the other arm of the L, extending east and west (Norbury, 2003). On the far north side, a loading ramp for props and two huge doors opened onto an alley called "Couch Place" (Norbury, 2003). In the middle of these huge doors was an ordinary-sized door for the cast to enter the theater (Norbury, 2003). Fire escapes were located on the north wall, which led to Couch Place (Norbury, 2003). Another door off the west end of the stage led to a vacant lot (Norbury, 2003).

The Iroquois Theatre was 2 months old when it burned (Flynn, 2012). The owners promoted the theater as being "Absolutely Fireproof" (Flynn, 2012). During the construction, the theater was plagued with delays and labor unrest (Flynn, 2012; Hatch, 2003). The Iroquois had three levels of seating: The main floor, called the parquet (what we call the orchestra today); the first balcony or dress circle, with two boxes on each side of the regular balcony seats; and the second balcony, or gallery, above (Norbury, 2003). The theater program for the grand opening described the Iroquois Theatre as absolutely fireproof. In addition, it was mentioned that the exits were "far more numerous, the entire north frontage being available for such service in case of emergency," which were the exits leading to Couch Place (Norbury, 2003).

On Wednesday afternoon December 30, 1903, the Iroquois Theatre in Chicago was packed due to a bargain-priced matinee (NFPA Journal, 1995). Eddie Foy was appearing in "Mr. Blue Beard" playing the leading comedy character Sister Anne (NFPA Journal, 1995). The theater was one of the finest in the country; it was a plush palace of marble, plate glass, mahogany, and gilding (NFPA Journal, 1995). It had a magnificent promenade foyer, like an old-world palace hall, with a ceiling 60 ft from the floor and grand staircases ascending on either side (NFPA Journal, 1995). The backstage was commodious, and able to hold hundreds of performers (NFPA Journal, 1995).

The owners boasted that the theater was extremely efficient, convenient, and, most important of all, safe (NFPA Journal, 1995). While the building itself was marketed as being as fireproof as possible, due to an interest in budgetary savings, certain standard fire code requirements were omitted during construction, which invariably resulted in the creation of a fool's paradise (NFPA Journal, 1995).

The Iroquois Theatre was one of the prodigious entertainment sites on the performing circuit, drawing the celebrated performers of its day in Chicago's flourishing industrial downtown. Much of the scenery used was of a very flimsy character (NFPA Journal, 1995). Hanging suspended by a forest of ropes above the stage and touching each other were about 280 drops, several of which were necessary to each set, all painted with oil colors, the great majority of them cut into delicate lacery, and some of them sheer gauze (NFPA Journal, 1995). The play went merrily through the first act (NFPA Journal, 1995). At the beginning of the second act, a double octette, which involves eight men and eight women, had a cute number called "In the Pale Moonlight" (NFPA Journal, 1995). The stage was flooded with bluish light while the performers sang and danced. It was then that the trouble began (NFPA Journal, 1995).

The Disaster Unfolds

A standing-room-only audience with an attendance estimated at 2,000, crowded the Iroquois Theatre to see the comedian Eddie Foy who was starring in the matinee of the musical comedy, "Mr. Bluebeard" (Everett, 1904). The auditorium and stage was darkened during the second act for the popular song "The Pale Moonlight" (Everett, 1904). Eight men and women in showy costumes strolled through the measures of the piece, bathed in a flood of dazzling light (Everett, 1904). Overhead, a spotlight was shooting forth its brilliant ray of concentrated light while a tiny flame crept over the inside of the proscenium drape, where it ignited a portion of paint-saturated muslin (Everett, 1904). Members of the octette cast their eyes above, and saw the spark growing into a whirling maelstrom of fire; however, they didn't consider that it wouldn't easily be put out (Everett, 1904). The actors and stagehands were familiar with minor, occasional fires that broke out during the course of a season, and they tried to put out the fire, but the flames raced up the drapery and spread across the curtains above the stage, and then began raining scraps of burning fabric down on the actors below (Everett, 1904).

In a vain attempt to calm the audience, Eddie Foy had the band continue playing while fragments of the burning backdrop collapsed during the performance (Everett, 1904). Unfortunately, Foy was unsuccessful in his attempt to calm the crowd as smoke began filling the auditorium due to the ever-growing calamity and the terrified audience struggled to escape the burning stage (Everett, 1904). The flame grew into an inferno as blazing embers began to shower upon the merry dancers on stage (Everett, 1904). The manila lines by which the carloads of scenery in the loft above was suspended gave way before the fire, paper, and canvas backdrop, and great wooden batons fell like thunderbolts upon the now deserted stage (Everett, 1904). The fire quickly spread to the oil-painted set pieces made of wood and canvas that hung in the catwalks above the stage (Everett, 1904). The asbestos fire curtain failed to completely drop snagging on stage equipment and lodged partially down creating a gap of several feet between the bottom of the fireproof curtain and the stage floor (Everett, 1904).

As the fire continued to spread through freshly painted suspended stage backdrops, the fleeing performers panicked and threw open the double equipment rear exit doors behind the stage in order to escape (Everett, 1904). The sudden invasion of air ignited a ball of fire that curled under the asbestos curtain, detonated the fire up into the balconies, and the now fiery inferno swept over the auditorium in a crushing explosion of death (Everett, 1904). Understandably, the terrified audience immediately stampeded for the exits (Everett, 1904). Doors, windows, hallways, and fire escapes were all jammed in a moment with people struggling and fighting for their lives. Several of the doors were jammed almost instantly so that no human power could make egress possible (Everett, 1904). The ventilation system was incomplete and the smoke hatches on the roof were nailed shut, inhibiting the venting of smoke and hazardous fumes to the outside (Everett, 1904).

Immediately following the rush of flames, an explosion lifted the entire roof of the theater from its walls, shattering the great skylight into fragments (New-York Tribune, 1903). As soon as the flames first appeared beyond the curtain, a man in the rear of the hall shouted "Fire! Fire!," and the entire audience rose as one person and made for the doors (New-York Tribune, 1903). It is believed that the explosion was caused by the flames coming in contact with the gas reservoirs of the theater, causing them to burst (New-York Tribune, 1903). It was the opinion of the firemen that found numerous people sitting in their seats, and whose faces were directed

Figure 21.3 Door to the Fire Escape That Could not Be Opened; many Died Here
Source: Publishers Union of America/1904

toward the stage as if the performance was still going on, that these people had been suffocated at once by the flow of gas, which came from behind the asbestos curtain (New-York Tribune, 1903).

Unfortunately, the fleeing audience members were trapped (Grove, 2006). Various exits were locked or hidden behind heavy, decorative curtains and drapes that were not marked (Grove, 2006). Certain doors were unlocked, but they only opened inward, trapping the panicked victims in the jam-packed hallways, whereas other doors were locked to thwart gate-crashers (Grove, 2006). Several doors were latched with Bascule locks, which were not popular in America at the time, requiring the unapprised, frightened audience to use a small, unremarkable lever to open (Grove, 2006). A number of spectators were even trapped by illegal, accordion-style metal gates that the theater management locked during the shows to keep audience members in the upper balconies from sneaking downstairs for a better view of the show (Grove, 2006). As the flames spread into the auditorium, hundreds of theatergoers were burned to death, while hundreds more trapped in the hallways suffocated from the smoke (Grove, 2006).

Charles B. Taylor, a janitor, was washing windows in the University building, directly across the areaway where the theater fire escape was located (The Plymouth Tribune, 1904). He describes a terrible scene (The Plymouth Tribune, 1904). Immediately, after the cry of fire was raised, he said, a score more men, women, and children clambered onto the platform of the fire escape, each crushing the other in the mad flight to climb to a place of safety (The Plymouth Tribune, 1904). The terrible crush prevented anyone from descending (The Plymouth Tribune, 1904). While the heartrending appeals for help rose high above the din inside the theater, and while the awful fight for mastery was going on the small iron platform, two huge iron doors, which up to that time had held the flames inside, burst open and a wave of fire swept upon the mass of humanity (The Plymouth Tribune, 1904). With shrieks, mingled with curses and pleadings, the 20 or more were swept from the platform (The Plymouth Tribune, 1904). One woman nearest the outside fell headlong to the pavement, alighting upon her head. In a moment, a pile of human beings was heaped upon her. According to Taylor, he believes that all in that terrible crush were killed (The Plymouth Tribune, 1904).

Figure 21.4 Front Rows of Seats and Front of Stage of the Lroquois Theatre after the Fire
Source: Publishers Union of America/1904

Impact

Subsequent investigation of the fire determined that the major cause of death was from smoke inhalation and burning to death, while the remainder was trampled in the rush to exit the theater. A stagehand had to run and notify the nearest fire department about the fire at the theater because there was no fire alarm installed. Smoke was not visible from the outside of the Iroquois Theatre for almost 15 min, which was all the time needed for the blaze to claim the lives of 602 people. Firemen were able to extinguish the fire in less than 30 min because the highly combustible material had already been consumed.

Containing the Disaster

The fire spread rapidly, fueled not only by the theater but also by the victims themselves. Dressed in heavy clothing that is to be expected in a December in Chicago, the flames intensified as it swept across the panicked crowd. With many doors locked or blocked, panic and confusion even further inhibiting escape, the disaster would run its deadly course. With no fire alarm in the theater to automatically notify nearby firehouses of a incident, when the disaster unfolded, a stagehand had to run to the nearest firehouse to report that the theater was in flames. When the first firefighters arrived at the theater, they found it difficult to enter the auditorium because of the number of bodies stacked up at the doors, inhibiting entrance into the theater itself (Groves, 2016).

RELIEF AND FIRST RESPONSE EFFORTS

During the second act of the show, a stage overhead floodlight burst and sparks set fire to a velvet curtain, which quickly spread killing 602 people (Flynn, 2012). The fire started around 3:15 p.m. when an arc lamp shorted out and caught a muslin curtain on fire (Flynn, 2012). Attempts were made to extinguish the fire with extinguishers, but to no avail (Flynn, 2012). The stage manager attempted to lower the asbestos fire curtain, but it got stuck half way down on a projecting light reflector (Flynn, 2012; Hatch, 2003). Attempts were made by actor Eddie Foy to calm the audience, but to no avail (Flynn, 2012). Full panic was on as flames and smoke began to spread from under the half-lowered fire curtain (Flynn, 2012). That curtain was later discovered to be made of wood pulp and asbestos and determined to be of no help had it lowered properly (Flynn, 2012; Hatch, 2003).

Firemen attempted to extinguish the fire by entering the theater but were met by a wall of burned bodies (Grove, 2006). The doors were squeezed shut by the throng of bodies and the firemen toiled to wrench the corpses that amassed in piles at the doorways to gain access to the inside of the theater (Grove, 2006). Firefighters claimed that at one exit they counted almost 200 scorched bodies heaped 10 high and twenty deep (Grove, 2006). In an attempt to escape the disaster, the frantic audience members rushing to the exits trampled and crushed by each other as they tore at each other trying to escape the inferno (Brandt, 2006). Firemen found scorched bodies on the outside of the piles, but bodies in the middle were unscathed by the blaze but had suffocated under the overwhelming weight of the dead and of smoke inhalation (Grove, 2006). The exterior fire escape ladders were not installed on the upper floor's fire escapes, so as some of the frantic audience opened those doors and windows they had to jump into the alley beside the theater, obviously dying on impact with pavement. Over a 100 dead bodies piled in the alley below, eventually cushioning the landing of others and saving lives.

Northwestern University's dental school employees in the building adjacent to the theater quickly assembled an improvised footbridge utilizing a ladder and several planks, spanning the buildings' rooftops saving close to a dozen people who bravely escaped the deadly fire, but several continued to fall to their deaths (Highland Recorder, 1904). Ladders, planks, ropes, poles, everything that could possibly serve to assist these poor creatures in their battle for life were rigged and turned into bridges, but few got across alive (Highland Recorder, 1904). Rapidly, one blackened corpse after another was passed along until every building on the north and west side was filled with them (Highland Recorder, 1904).

The greatest loss of life, however, was due to the panic and the wild rush for the exits, and the stairways leading to them (The Minneapolis Journal, 1903). Many of the fire escapes had not yet been put on (The Minneapolis Journal, 1903). The struggle at these choked channels of egress must have been frightful, echoing Darwin's survival of the fittest (The Minneapolis Journal, 1903). It was a battle for life with everyone for himself (The Minneapolis Journal, 1903).

AFTERMATH AND INVESTIGATION

The responsibility of the fire is ascribed to careless electric light operators who placed their apparatus too near the scenery. Then there was the fireproof curtain of asbestos that failed to work at the critical moment (The Minneapolis Journal, 1903). A Chicago Fire Department captain who made an unofficial tour of the theater days before the official opening noted that there were no extinguishers, sprinklers, alarms, telephones, or water connections; the only firefighting equipment available were six canisters of a dry chemical called "Kilfyre," which was normally used to douse residential chimney fires (Flynn, 2012; Hatch, 2003). An editor of Fireproof Magazine had toured the building during construction and had noted "the absence of an intake, or stage draft shaft; the exposed reinforcement of the (proscenium) arch; the presence of wood trim on everything and the inadequate provision of exits" (Flynn, 2012; Hatch, 2003).

In a rush to get the theater opened quickly, the theater management skipped many basic fire precautions (Groves, 2006). Most notably, the theater had no fire alarm or sprinklers and the emergency smoke vents above the stage were nailed shut (Groves, 2006). Six weeks later, the Iroquois Theatre was home to the deadliest fire in Chicago history (Groves, 2006). Scores of exits were locked or hidden behind heavy, decorative curtains (Groves, 2006). Numerous doors were unlocked, but they only opened inward, trapping the victims in the jam-packed hallways (Groves, 2006). The subsequent investigations into the fire uncovered numerous troubling facts, including the lack of sprinklers and fire alarms and the locked exits (Groves, 2006).

From the Verdict of the Coroner's jury and the testimony presented after the Iroquois Theatre disaster noted in Marshall Everett's 1904 detailed recount of events, several causes of the fire were identified:

- Grand drapery coming in contact with electric flood or arc light, situated on iron platform on the right hand of stage, facing the auditorium (Everett, 1904).
- City laws were not complied with relating to building ordinances regulating fire-alarm boxes, fire apparatus, damper or flues on and over the stage and fly galleries (Everett, 1904).

- A distinct violation of ordinance governing fireproofing of scenery and all woodwork on or about the stage (Everett, 1904).
- Asbestos curtain totally destroyed; wholly inadequate, considering the highly inflammable nature of all stage fittings, and owing to the fact that the same was hung on wooden bottoms (Everett, 1904).
- Building ordinances violated inclosing aisles on each side of lower boxes and not having any fire apparatus, dampers or signs designating exits on balcony (Everett, 1904).
- Building ordinances violated regulating fire apparatus and signs designating exits on dress circle (Everett, 1904).
- Building ordinances violated regulating fire apparatus and signs designating exits on balcony (Everett, 1904).
- Generally, the building is constructed of the best material and well planned, with the exception of the top balcony, which was built too steep and therefore difficult for people to get out of especially in case of an emergency (Everett, 1904).
- We also noted a serious defect in the wide stairs in extreme top east entrance leading to ladies' lavatory and gallery promenade, same being misleading, as many people mistook this for a regular exit, and going as far as they could, were confronted with a locked door which led to a private stairway, preventing many from escape and causing the loss of fifty to sixty lives (Everett, 1904).
- The employees were not trained sufficiently for all emergencies (Everett, 1904).
- The Owners, Building Commissioner, Inspector, Fire Marshall and firemen were all cited with gross neglect of duty and glaring incompetency in reporting the Iroquois Theater [sic] was OKAY on a most superficial inspection and not enforcing the city ordinances. (Everett, 1904)

Benjamin H. Marshall was the architect of the Iroquois Theatre in Chicago who allegedly researched previous structural fire disasters, so he could apply proper safety precautions into his blueprints. In his final design plans, he forfeited safety for appearance. During the subsequent investigations, Marshall's plans were never questioned. The theater owners failed to finish construction of all of the building elements in Marshall's designs, in their greedy rush to open the facility for the holiday seasons. It was later discovered that city inspectors had been given free tickets to ignore obvious fire code violations.

The coroner's jury, which for 3 weeks had been investigating the responsibility for the fire in the Iroquois Theatre, returned a verdict by which the following persons were held to await the action of the grand jury (Evening Star, 1904):

- Carter H. Harrison, mayor of Chicago. (Evening Star, 1904)
- Will J. Davis, proprietor in part and manager of the theater. (Evening Star, 1904)
- George Williams, building commissioner of Chicago. (Evening Star, 1904)
- Edward Laughlin, building inspector under Williams. (Evening Star, 1904)
- William Sailers, fireman in the theater. (Evening Star, 1904)
- James E. Cummings, stage carpenter. (Evening Star, 1904)
- William H. Musham, chief Chicago fire department. (Evening Star, 1904)
- William McMullen, who had charge of the light that caused the fire (Evening Star, 1904)

Civil lawsuits were issued, but unfortunately there were no convictions of wrongdoing by any involved (Norbury, 2003). The coroner's jury called many witnesses and several people involved in ownership and management of the theater were later indicted by a grand jury (Norbury, 2003). None was convicted of any offense (Norbury, 2003). Several civil damage suits were filed but only one led to compensation (Norbury, 2003). This suit was against the George F. Fuller Company, the contractors who built the theater (Norbury, 2003). The suit was finally settled out of court in January 1909, 5 years after the fire (Norbury, 2003). Thirty-nine claimants received $750 each (Norbury, 2003). The ensuing investigations into the Iroquois Theatre fire exposed several alarming facts that could have prevented such a deadly event, including:

- no sprinkler system in the theater and no automatic sprinklers installed above the stage area, which was a municipal requirement
- fire alarm box was never installed and no fire alarm system was present

- fancy drapery was installed obscuring working exit signs
- excessive use of wood trim, which was highly combustible
- unfinished ventilators not in operation above the stage were nailed shut resulting in the back draft of the fire; these ventilators were intended to vent out smoke and noxious fumes during a fire but were nailed shut to keep out precipitation
- unfinished fire escapes without ladders of egress
- several exits not properly marked and missing
- no backstage phone for emergencies
- obstructed and inadequate asbestos fire curtain—the asbestos curtain lodged into stage equipment during its deployment and failed to perform its intended task; the asbestos curtain was insufficiently made of cotton and other combustible fabrics
- flammable wood and glass doors
- no fire protection devices such as fire extinguishers and standpipes and useless firefighting equipment (Kilfyre instead of fire extinguishers)
- entrances/exits to the upper level balcony and gallery were locked by iron gates during performances to prevent audience members from sneaking into other levels
- exit doors opened inward into the theater resulting in the inability of the audience members to open them as the frantic horde lodged against the doors trying to escape
- management bolted the exterior doors shut to keep nonpaying customers out
- several exit signs were turned off during the show
- code violations were overlooked during construction by fire inspectors who took bribes

The cause of the "jamming" of the asbestos curtain, considered the cause of the disaster, was explained by an inspector for the underwriters who made a tour of the house and found that the wire on which the queen of the aerial ballet flew out over the audience in the second act held the asbestos curtain in place and prevented it from being lowered (The Minneapolis Journal, 1903). It was shown by the inspector that the wire on which the most spectacular feature of the show was made practically caused the holocaust (The Minneapolis Journal, 1903).

LESSONS LEARNED AND POLICY IMPACT

The Iroquois disaster had an immediate impact on theater construction and operation nationwide, and it prompted many cities to enforce their theater ordinances with greater diligence (Hayes, 1999). In the coming months and years, many laws were enacted in response to the Iroquois Theatre fire, including laws requiring mandatory theater upgrades including outward-opening doors that remain unlocked, exit lights, automatic sprinklers, fire alarm systems, and flame-resistant scenery, props, and curtains (Groves, 2006).

The devastating fire at the Iroquois Theatre helped facilitate several fire precaution regulations and federal and state code standards for exit doors, exit signs, clearly marked exiting pathways, seating capacity maximums, and the use of the panic bar. The investigation that followed led to a variety of fire safety improvements requiring mandatory theater upgrades including:

- push to open exit doors, not pull to open
- unlocked outward-opening doors
- exit signs with lights must always be ON
- clearly marked exits
- automatic sprinklers
- fire alarm systems
- flame-resistant scenery, props, and curtains
- panic bars on emergency exits
- maximum seating capacities
- ventilation standards
- numerous regulations involving clear pathways

- The building department should have classified printed lists, to be filled out by an inspector, then signed by the head of the department, before any public building can secure amusement license, and record kept thereof in a duplicate carbon book (Everett, 1904)
- All fire escapes should have separate passageways to the ground, without passing any openings in the walls (Everett, 1904)
- All scenery and paraphernalia of any kind kept on the stage should be absolutely fireproof (Everett, 1904)
- Asbestos curtains should be reinforced by steel curtains and held by steel cables (Everett, 1904)
- There should be two electric mains entering all places of amusement, one from the front, with the switchboard in the box office, controlling entire auditorium and exits, and one on stage, to be used for theatrical purposes (Everett, 1904)
- All city officials and employees should familiarize themselves with city ordinances as they relate to their respective departments, and pass a rigid and signed examination on the same before they are given positions. This same rule should apply to those holding office (Everett, 1904)

Carl Prinzler was the Manager of the Vonnegut Hardware Company Builders Hardware Department, and was scheduled to be at that ill-fated matinee, but unforeseen circumstances led him to miss the show avoiding the disastrous 1903 Iroquois Theatre fire (Ingersoll Rand Company, 2005). Prinzler spent the next 5 years working with Henry Dupont, an architectural engineer, developing hardware to preclude a similar horror (Ingersoll Rand Company, 2005). Together they fashioned a lever that kept doors locked on the outside, yet unlocked from the inside (Ingersoll Rand Company, 2005). The pivotal element was the crossbar which opened the door outward once any part of the body touched it (Ingersoll Rand Company, 2005). In 1908, the team (Vonnegut Hardware Company, DuPont, and Prinzler) finalized the design of door hardware known as the panic release bar, anti-panic device, or the panic bar and combined their names to call it the Von Duprin model 88 Panic Release Bar (Ingersoll Rand Company, 2005). Many other safety devices have been developed over the years to keep occupants safe (Ingersoll-Rand Company, 2005).

IMPLICATIONS FOR MODERN EMERGENCY MANAGEMENT

The deadliest single building fire in the United States on the records for the National Fire Protection Association (NFPA) is the Iroquois Theatre (Ingersoll Rand Company, 2005). The devastating fire at the Iroquois Theatre helped facilitate several fire precaution regulations and federal and state code standards (Ingersoll Rand Company, 2005). Emergency management is the restraint dealing with risk and risk avoidance, is essential to the security of our daily lives, and is a vital role of government (Ingersoll Rand Company, 2005). Countless innovations and building code changes have been effective in reducing meaningless deaths (Ingersoll Rand Company, 2005). Inventions and building code enhancements including balanced construction, providing life safety through the application of fire protection features, and by improving the safety and well-being of others have reduced the number of fire deaths (Brannigan & Carter, 2017). It is now the responsibility of all of us and not just an architect, code official, or owner to provide adequate building security, while meeting ever-changing fire and life safety codes (Ingersoll Rand Company, 2005).

The NFPA 101 Life Safety Code & NFPA 500 Building Construction & Safety Code include many important regulations that if in place at the time of the Iroquois disaster, fewer lives would have been lost (Ingersoll Rand Company, 2005). Many codes have been revised over the last century and address several of the issues that caused the mass casualties of the Iroquois Theatre fire (Ingersoll Rand Company, 2005). These codes address classifications of (Ingersoll Rand Company, 2005):

- occupancy
- building construction
- safety code
- providing a secure building with safe egress
- Hazard of Contents classification: (Low, Ordinary, High)

- fire protection system—approved sprinkler system, fire alarm system
- locks, latches, and alarm
- door in a stair enclosure
- stair to roof
- stair enclosures
- door-releasing devices
- panic hardware
- inspection of door openings
- doors required to swing in the direction of egress
- door testing and inspection no less than once per year, etc.

SUMMARY

A known total of 602 people died during the 1903 Iroquois Theatre fire in Chicago, 575 burned to death in the fire, of which 212 were children, and another 27 succumbed later to their injuries and burns sustained during the blaze. It still remains one of the most lethal infernos in U.S. history, with about one in every three audience members perishing in the fire. The allegedly "fireproof" Iroquois Theatre opened in Chicago on November 23, 1903. Disaster struck several weeks later on December 30, 1903; the Iroquois Theatre was engulfed in flames, claiming more lives than any other fire in Chicago history.

The devastating fire at the Iroquois Theatre eventually led to the creation and implementation of several new federal and state fire code regulations for exit doors, exit signs, and required clearly marked exiting pathways, seating capacity maximums, and the use of the panic bar. The investigation that followed led to a variety of fire safety improvements requiring mandatory theater upgrades including: exit doors must open outward; exit signs with lights must always be ON with exits clearly marked; fire notification and fire suppression systems in place and functional to include, but not limited to automatic sprinklers, fire alarm systems, panic bars on emergency exit doors; flame-resistant scenery, props, and curtains must be in public venues; maximum seating capacities established and posted; ventilation standards; fire escapes have separate passageways to the ground, without passing any openings in the walls; in the theater setting, asbestos curtains should be reinforced by steel curtains and held by steel cables.

Bottom line, those responsible for constructing any type of public venue must closely adhere to, and follow, all preconstruction guidelines, regulations, and codes necessary to ensure the safety and welfare of the public, which includes providing the necessary and due diligent options needed for evacuation of the venue, and for facilitating first responder operational capability.

REFERENCES

Brandt, N. (2006). *Chicago death trap: The Iroquois Theater Fire of 1903*. Carbondale, IL: Southern Illinois University Press. ISBN: 080932721X

Brannigan, F. L., & Carter, H. R. (2017). *Fire Disasters: What Have We Learned?* Retrieved from https://www.emergencydispatch.org/articles/firedisasters.html

Chicago Eagle. (1904a, January 9). Harrison to blame. *Chicago Eagle*, p. 1, Image 1, Col. 1–7.

Evening Star. (1904, January 26). Harrison arrested: Chicago Mayor and other officials held. *Evening Star*, p. 1, Image 1, Col. 6.

Everett, M. (1904). *The Great Chicago Theater Disaster: The complete story told by the survivors . . . profusely illustrated with views of the scene of death before, during and after the fire*. Washington, D.C.: Publishers Union of America.

Flynn, J. (2012). *Theater fires—Political and legal foundations of fire protection*. University of Cincinnati. Retrieved from http://ceas.uc.edu/content/dam/aero/docs/fire/Theater%20Fires.pdf

Groves, A. (2016). *Iroquois Theater Fire, Chicago: December 30, 1903*. Retrieved from http://guides.library.illinois.edu/c.php?g=416856

Hatch, A. P. (2003). *Tinder box: The Iroquois Theatre Disaster, 1903*. Chicago, IL: Academy Chicago Publishers. ISBN 0897335147

Hayes, A. M. (1999). *The Iroquois Theatre Fire: Chicago's other great fire.* ETD collection for University of Nebraska. Retrieved from http://digitalcommons.unl.edu/dissertations/AAI9951294

Highland Recorder. (1904, January 8). Over 500 fire victims. *Highland Recorder*, p. 1, Image 1, Col. 2–4.

Ingersoll Rand Company. (2005). *Von Duprin auxiliary hardware & general information form.* Dublin, Republic of Ireland: Ingersoll-Rand Company. VD-GN-1000 Rev 3/05

New-York Tribune. (1903, December 31). Over five hundred die in fire. *New-York Tribune*, p. 1, Image 1, Col. 2–6.

NFPA Journal. (1995). A tragedy remembered. *NFPA Journal*, July/August 1995, 75–79.

Norbury, F. (2003). *A matinee to remember—The Iroquois Theater Fire.* Illinois Periodicals Online. Originally Published in Illinois Heritage magazine. Retrieved from http://www.lib.niu.edu/2004/ih031004.html

The Indianapolis Journal. (1903, November 24). Opening of the Iroquois. *The Indianapolis Journal*, p. 3, Image 3, Col. 3.

The Minneapolis Journal. (1903, December 31). Horrors of holiday holocaust in Chicago's finest theater. *The Minneapolis Journal*. p. 1, Image 1, Col. 1–5.

The Plymouth Tribune. (1904, January 7). Greatest horror of a century. *The Plymouth Tribune,* p. 6, Image 6, Col. 1–7.

Deepwater Horizon Oil Spill (2010)

22

Contributed by Brandy J. Baczwaski, Bethany J. Aiena, and Stefan E. Schulenberg.

Figure 22.1 Deepwater Horizon Well Site, Louisiana
Source: Breck P. Kent/Shutterstock.com0

BRIEF OVERVIEW

On April 20, 2010, the Deepwater Horizon oil rig exploded in the Gulf of Mexico. More than 200 million gallons of crude oil spilled into the waters over the span of 87 days. The explosion killed 11 workers and seriously injured an additional 17 individuals (Robertson & Krauss, 2010). The spill affected over 16,000 miles of Texas, Louisiana, Mississippi, Alabama, and Florida coastlines (Boxall, 2011; Polson, 2011). Within the first 6 months, over 8,000 animals were found dead (including many from the endangered species list). More than 30,000 people aided in the cleanup, but these efforts were only moderately effective as oil continues to be found years later (Beaumont, 2012; National Wildlife Federation, 2015). Gulf Coast industries, such as tourism and fishing, declined, with the total economic impact estimated to be in excess of $22 billion (Dermansky, 2013; Oxford Economics, 2010). British Petroleum, that was leasing the Deepwater Horizon rig, is expected to pay a total of approximately $61.6 billion in fines, settlements, and cleanup costs by 2019 (Bomey, 2016).

INTRODUCTION

Disaster events are often categorized as natural, man-made, and technological. Hurricanes such as Katrina, Irma, and Harvey; the Great Midwestern Floods; the Tohoku earthquake; as well as tornadoes, tsunamis, blizzards, and ice storms are classified as natural disasters (Rubin, Amlot, Wessely, & Greenberg, 2012; Sumer, Karanci, Berument, & Gunes, 2005). Man-made disasters are commonly known as international or domestic terrorist attacks, with the World Trade Center terrorist attacks, the Sandy Hook elementary school shooting, the Oklahoma City bombing, the Boston Marathon bombing, and the Orlando nightclub shooting serving as examples (Ahern & Galea, 2006; Meewisse, Ollf, Kleber, Kitchiner, & Gersons, 2011). Lastly, technological disasters consist of events such as explosions (e.g., Three Mile Island), plane crashes, and oil spills (Gallacher, Bronstering, Palmer, Fone, & Lyons, 2007; Palinkas, Downs, Petterson, & Russell, 1993). The effect of technological disasters can be delayed and long-lasting, and efforts to repair are often unfeasible or impossible (Baum, Fleming, & Davidson, 1983; Couch & Coles, 2011; Gallacher et al., 2007). While these types of disasters have different causes, they all have the potential to result in traumatic experiences among those affected (Meewisse et al., 2011).

The Setting

The Deepwater Horizon was a semisubmersible drilling rig built in 2001 for deepwater oil drilling (Transocean, 2010). It was owned and operated by Transocean and leased by British Petroleum (Meigs, 2016). Located offshore of Louisiana and Mississippi in the Macondo Prospect of the Gulf of Mexico, the oil rig was reviewed and determined unlikely to result in accident or oil spill, and therefore was exempt from undergoing an environmental impact analysis in 2009 (Eilperin, 2010; Meigs, 2016). On the evening of April 20, 2010, as the crew was entering final phases of drilling the well, water, mud, and methane gas erupted from the drill and ignited, resulting in a series of explosions that could be seen from miles away (Schwartz & Weber, 2010). This event resulted in 11 people killed and 17 people injured, causing what would become the largest oil spill recorded in U.S. history (Robertson & Krauss, 2010).

The Disaster

After the explosions, the Macondo well continued to leak over the course of 5 months, until September 19, 2010, despite being capped on July 15 (Weber, 2010). Throughout this time, approximately 4.9 million barrels worth of oil spilled into the Gulf of Mexico, affecting 68,000 square miles (Norse & Amos, 2010; Robertson & Krauss, 2010). A variety of cleanup efforts were undertaken in attempts to remove the oil from the Gulf of Mexico. These methods included skimmer ships, floating booms, controlled burns, and almost 2 million gallons of chemical dispersants containing toxins harmful to marine life (Blair, 2011; Coastal Response Research Center, 2010; Grissett, 2010; National Commission on the BP Deepwater Horizon Oil Spill and Offshore Drilling, 2010). It was estimated that up to one-third of the oil settled into the ocean sediment (Beech, 2010; Schrope, 2013).

History has seen many oil spills, such as the *Exxon Valdez*, the *Sea Empress*, the *Atlantic Empress I*, and the *Castillo de Bellver*. However, the nature of "typical" oceanic spills occurs when a vessel that is transporting oil is damaged and a limited amount of oil is spilled onto the surface of the water. These types of spills tend to be easier to contain. Oil spills such as the Deepwater Horizon oil spill and Ixtoc I were deepwater spills that continuously leaked oil from the ocean floor (Fountain, 2013; Koch, 2015). This type of spill is difficult to manage due to the large amount of oil, as well as the presence of the oil in all areas of the water, including the surface and sediment (Wilson et al., 2017).

AFTERMATH AND INVESTIGATION

The damage caused by the Deepwater Horizon oil spill was widespread, affecting 16,000 miles of Louisiana, Mississippi, Alabama, and Florida shorelines (Boxall, 2011; Polson, 2011). As a result of the oil itself and the cleanup efforts, which included toxic dispersants, over 8,000 animals were found dead over the first 6 months (including many from the endangered species list). Significant harm, mutation, or death occurred among fish, spiders, corals, crabs, sharks, pelicans, and dolphins (Beaumont, 2012; Gunderson, 2012; White et al., 2011;

Whitehead et al., 2011). The spill also resulted in the widespread blight of plant life in coastal marshlands (Boxall, 2011; National Wildlife Federation, 2015).

The Deepwater Horizon oil spill directly affected the economy of the Gulf Coast. For example, immediately after the spill, there was a decrease in tourism due to dirty beaches (Dermansky, 2013; Oxford Economics, 2010). The seafood industry was directly impacted, with fewer crab, oysters, and shrimp being caught and businesses losing up to an estimated $2.5 billion in revenue (Gill, Picou, & Ritchie, 2012; Walsh, 2010). The coastal fishing industry continues to be negatively impacted years after the event, with marine life deteriorating at an abnormally high rate (Elliot, 2015; Gallucci, 2015; National Wildlife Federation, 2015).

In addition to the spill adversely affecting the health of the wildlife, the environment, and the economy, many Gulf Coast residents experienced health-related physical issues, such as rashes, respiratory problems, memory loss, lesions, liver and kidney damage, hypertension, and miscarriage (Blair, 2011; Juhasz, 2012). Research into the aftermath of the *Exxon Valdez* oil spill suggests that these health effects may be long-lasting (Wang, 2010). Moreover, while it was expected that the Deepwater Horizon oil spill would increase the risk of physical health conditions in coastal residents, research has also examined the event with regard to mental health effects. For example, Fan and colleagues (2015) conducted the Gulf States Population Survey in which they surveyed residents of Louisiana, Mississippi, Alabama, and Florida via telephone. In their sample of 36,669 individuals, 12.7% of men and 17.7% of women reported symptoms of depression. They also found depressive symptoms among 13.0% of those identified as White and 59.8% of those identified as non-White (African American, Latino, or other). Varner, Langhinrichsen-Rohling, and Bell (2016) found in a study of Alabama coastal residents 18 months after the spill higher prevalence rates of generalized anxiety disorder in the area as compared to the general public, and those who had direct contact with the oil (e.g., individuals involved in cleanup efforts) reported significantly higher anxiety levels than those who did not have contact. A study by Osofsky and colleagues (2014) found that among the 355 adults they were treating at the time of the spill, there was an increase in depressive symptoms. Research has also shown that individuals living below the poverty line in Mississippi reported a higher level of distress than those living above the poverty line in terms of depression, anxiety, stress, and posttraumatic stress (Drescher, Schulenberg, & Smith, 2014). Among those reporting the presence of symptoms of posttraumatic stress, higher levels of resilience and perceived meaning in life were shown to predict lower levels of symptomatology (Aiena, Buchanan, Smith, & Schulenberg, 2016).

In addition to psychological distress, research into protective factors found that Mississippi coastal residents affected by the disaster reported self-efficacy to be predictive of life satisfaction following the spill (Drescher et al., 2012). This research group also found that female adolescents had higher rates of perceived meaning in life than male adolescents affected by the spill (Schulenberg, Smith, Drescher, & Buchanan, 2016). Further, it was found that perceived meaning in life and perceived resilience were significant, negative predictors of posttraumatic stress symptoms and that meaning may be an essential aspect of what makes an individual resilient to posttraumatic stress symptoms following a technological disaster (Aiena et al., 2016).

The Deepwater Horizon oil spill was unique from other oil spills in a number of ways (IOM, 2010). For instance, job loss brought on by the spill often affected multiple generations in one family (e.g., families of fisherfolk). Further, the population diversity in the Gulf Coast area could have led to more untreated mental health issues in some cultural groups (e.g., Vietnamese fisherfolk) due to the stigma of mental illness being more prevalent in some populations in comparison to others. Furthermore, many of the people living on the Gulf Coast were still in the process of recovering from previous disasters (Hurricanes Katrina, Rita, and Gustav), which raises the risk of psychological distress due to the cumulative effects of multiple traumas (i.e., retraumatization; IOM, 2010).

LESSONS LEARNED AND POLICY IMPACT

Shortly after the oil began spilling into the Gulf of Mexico, a 6-month moratorium on drilling was issued by the Obama administration (Department of Interior, 2010). This moratorium was issued in May 2010 by the U.S. Secretary of the Interior, Ken Salazar, in order to develop new safety requirements. This moratorium was opposed by some coastal residents, state officials, and political leaders due to concern that it would negatively impact the economy of the Gulf Coast due to the loss of jobs. Alternatively, several environmental groups supported the moratorium, citing that deepwater drilling was not well regulated. When the moratorium was lifted in October 2010, new regulations were implemented that required drilling operations to provide documentation on third-party inspections of equipment, as well as how plans to explore and develop new wells would safeguard against blowouts in the future.

In June 2014, the U.S. Chemical Safety Board (CSB) released information regarding their investigation of the failure of safeguards that contributed to the Deepwater Horizon oil spill (CSB, 2014). Deepwater oil rigs are equipped with a blowout preventer that is designed to seal the oil well in cases of emergency. In the case of the Deepwater Horizon oil rig, the blowout preventer failed due to a combination of factors. As a drill bores into subsea rock and sediment, it may pass trapped water, crude oil, and natural gas known as well fluids. If these well fluids unexpectedly enter the borer, they can cause a dangerous "kick" that can lead to a blowout, an uncontrolled release of oil and gas from the well. Once the oil and gas reach the rig on the surface, they have the potential of igniting and causing an explosion. To prevent a blowout, drill mud is pumped into the riser (i.e., drill pipe) to create a barrier. If this barrier fails, the blowout preventer is a multifaceted device that seals the riser to prevent continued uncontrolled release of oil and gas from the well. The first method of sealing the riser is to seal the annular space around the drill pipe with metal pipe rams that are manually closed by the crew. Next, rubber annular preventers can also be manually closed by the crew. If the first two methods are ineffective, the blind shear ram is employed which cuts and seals the pipe and can be activated manually by the crew or through automated emergency systems.

About an hour before the explosion of the Deepwater Horizon oil rig, a kick occurred and, unbeknownst to the crew, oil and gas flowed up the riser and past the blowout preventer (CSB, 2014). When the drill mud shot out of the riser into the rig, crew members manually shut the upper annular preventer. However, this did not stop the flow of oil and gas. They then manually shut one of the pipe rams, which temporarily sealed the well. The oil and gas that was already in the riser continued to shoot out of the rig on the surface and while the well was sealed, oil and gas continued to force itself into the lower part of the riser below the seal. This created a pressure differential between the inside of the pipe and the annular space around the pipe, causing it to bend inside the blowout preventer. Once the rig exploded and power was lost, the automatic system was activated to cut the drill pipe and seal the well with the blind shear ram. This system is operated by two identical battery-operated control systems (the yellow and blue pod, each controlled by two 9-V batteries and one 27-V battery) that were created to increase reliability in ensuring the well would be sealed in an emergency. The CSB's investigation found that the blue pod had been wired incorrectly prior to installation, causing its 27-V battery to drain. As a result, it did not activate the signals necessary to close the blind shear ram. Additionally, the yellow pod had also been wired incorrectly. If the yellow pod had activated correctly, the faulty wiring would have caused the system to fail entirely (not severing the pipe or stopping the flow of oil). However, one of the 9-V batteries had also drained due to faulty wiring, which resulted in the system being only partially activated, but which was enough to close the blind shear ram. Unfortunately, because the drill pipe had bent due to the pressure differential and was off center, the blind shear ram only partially cut the pipe. It did not seal it.

The CSB (2014) found in its investigation that the drill pipe likely bent because of effective compression. Effective compression occurs due to unseen irregularities in the pipe that can cause it to buckle when the pressure inside and outside are unequal. Effective compression is commonly known among other types of drilling but has not been considered with respect to drill pipe. As a result, the CSB warned other, comparable rigs of this issue as a similar accident could occur during oil drilling operations in the future, even if the crew or automated systems successfully seal the well. These findings suggest that blowout preventers are not as effective as previously thought (CSB, 2014).

Findings from the CSB have led to numerous new regulations in oil drilling proposed by the Obama administration in 2015 and the Bureau of Safety and Environmental Enforcement (BSEE) in 2016 (BSEE, 2016; Muskal, 2015). These regulations were to be phased in over time, and include the following emphases:

1. drilling operations must use the most up-to-date industry standards in developing and maintaining blowout preventers
2. greater control must be exercised as to blowout preventer maintenance and repair

3. a second blind shear ram must be added in case the first fails
4. technology must ensure that the drill pipe is centered so that the blind shear ram will be effective
5. specific criteria must be developed for use by independent third parties with respect to verifying the effectiveness of blind shear rams
6. the operational functionality of the blowout preventer in sealing the well needs to be increased
7. the monitoring and data storage of deepwater, high-risk drilling operations must occur in real time
8. specific criteria for testing and inspecting well equipment must be developed and implemented
9. drill operations must report blowout preventer failure data to the BSEE
10. specific criteria for safe drilling margins must be developed and implemented
11. drill operations must use accepted engineering principles and equipment that is intended for expected conditions
12. remotely operated vehicles to activate the components of the blowout preventer must have a standardized protocol implemented
13. centralized cementing of the well must be required
14. the testing frequency of blowout preventers should be increased throughout all operations

IMPLICATIONS FOR MODERN EMERGENCY MANAGEMENT

Following the explosion of the Deepwater Horizon oil rig, oil spilled into the Gulf of Mexico continuously for 87 days due to failed efforts to cap the well. As a result, well containment companies were developed with examples being the Marine Well Containment Company and the Helix Well Containment Group (Dulisse, 2014). These containment companies are comprised of several oil companies, including BP, Chevron, and ExxonMobil, with the goal of developing methods to place structures over an uncontrolled spill in order to efficiently and effectively contain the oil well.

Once an oil spill occurs, cleanup efforts have largely proved to be futile (Nikiforuk, 2016). In the case of the Deepwater Horizon oil spill, it was estimated that BP only recovered 25 percent of the oil lost, including 5% from controlled burning, 3% through skimming, and 17% from siphoning. Additionally, a large amount of

Figure 22.2 Grand Isle, LA—June 5: A sign on the Front Lawn of a Home on June 5, 2010 in Grand Isle, LA. The BP Oil Spill That Began April 20, 2010, Has Resulted in Oil Washing up on the Shores of Louisiana
Source: Katherine Welles/Shutterstock.com

Corexit, a chemical dispersant, was used to break down the oil. However, the dispersant has been found to be toxic to wildlife, with little effectiveness in eliminating oil. As a result, Sweden and the United Kingdom have banned its use.

In the report written by the National Commission on the BP Deepwater Horizon Oil Spill and Offshore Drilling, organized by President Obama following the event, industrial and governmental recommendations were made for addressing the causes and consequences of the disaster in an effort to decrease the chances and potential harm of future spills (Graham et al., 2011). With regard to industry-targeted recommendations, a complete reinvention of the safety culture in the petrochemical industry was recommended. One way that this could be done is by creating self-policing safety organizations similar to that done after the Three Mile Island nuclear disaster. More specifically, the Institute of Nuclear Power Operations (INPO) was created with the mission of promoting high levels of safety and reliability in the industry. This organization establishes performance objectives, criteria, and guidelines for the nuclear power industry; conducts regular, detailed evaluations of nuclear power plants; and assists these plants in continually improving their safety strategies, producing successful industry-wide improvements in safety regulations and implementation (Institute of Nuclear Power Operations, 2017). A self-policing organization for the oil and gas industry could have a similar, positive impact with respect to promoting a culture of safety, connecting individuals involved in the day-to-day implementation of safety procedures with senior management, enabling a flow of communication leading to an improvement in operations.

With regard to the role of the government, multiple recommendations were made to improve the safety of offshore operations; safeguard the environment; strengthen oil spill response, plan, and capacity; advance well containment capabilities; overcome the impacts of the event (the restoration of the Gulf of Mexico); ensure financial responsibility of the companies involved; and promote congressional engagement (Graham et al., 2011). More specifically, one recommendation was that the Department of the Interior develop an individualized, proactive, and risk-based performance approach for each drilling operation by developing an independent agency housed within the Department of the Interior with the authority to oversee all aspects of offshore drilling safety, both operational and occupational. The newly created organization would have much stronger oversight of onshore and offshore petrochemical companies in order to enforce safety regulations and plan for the effective development and implementation of oil spill cleanup and response efforts. The commission also recommended increased state and local involvement in oil spill contingency planning to better represent citizens' interests and make more personalized management plans for coastal regions. Furthermore, the commission recommended that all offshore operators would have to submit reports for approval of current and future wells that demonstrate adequate safety and contingency planning that mitigates risk during potential catastrophic failures in equipment (as seen on the Deepwater Horizon oil rig). Additional information on these recommendations can be found in the National Commission on the BP Deepwater Horizon Oil Spill and Offshore Drilling report titled "Deep Water: The Gulf Oil Disaster and the Future of Offshore Drilling" (Graham et al., 2011).

SUMMARY

On April 20, 2010, the Deepwater Horizon oil platform exploded, killing 11 rig workers and injuring another 17 people. It became the largest oil spill in U.S. history. The event resulted in the death of land and marine wildlife and the widespread blight of plant life in coastal marshlands. Industries along the Gulf Coast, such as fishing and tourism, were significantly, negatively impacted, and have yet to fully recover (Elliot, 2015; Gallucci, 2015; Gill, Picou, & Ritchie, 2012). Following the spill, both physical (e.g., rashes, respiratory problems, memory loss, hypertension) and psychological (e.g., increases in depression, anxiety, stress) impacts were seen in Gulf Coast residents. While the full impact of the Deepwater Horizon oil spill remains unknown, the event inspired many changes in onshore and offshore drilling procedures and safety regulations. Well containment companies such as the Marine Well Containment Company and Helix Well Containment Group were developed to increase the industry's ability to cap wells in the future. Many organizations have developed more efficient oil cleanup methods. However, new research and developmental efforts need to be undertaken to validate solutions as being efficacious. Lastly, in an effort to decrease the chances of future spills and improve recovery, the National Commission on the BP Deepwater Horizon Oil Spill and Offshore Drilling was created by the Obama administration. After studying the event, the commission made both industrial and governmental-level recommendations to correct for flaws in current methods of safety implementation, as well as provide improved oversight for petrochemical companies.

REFERENCES

Ahern, J., & Galea, S. (2006). Social context and depression after a disaster: The role of income inequality. *Journal of Epidemiology and Community Health, 60*, 766–770. doi:10.1136/jech.2006.042069

Aiena, B. J., Buchanan, E., Smith, C. V., & Schulenberg, S. E. (2016). Meaning, resilience, and traumatic stress following the Gulf oil spill: A study of Mississippi coastal residents seeking mental health services. *Journal of Clinical Psychology, 72*, 1264–1278. doi:10.1002/jclp.22232

Baum, A., Fleming, R., & Davidson, L. M. (1983). Natural disaster and technological catastrophe. *Environment and Behavior, 15*, 333–354.

Beaumont, P. (2012, March 13). Gulf's dolphins pay heavy price for Deepwater oil spill. *The Observer.* Retrieved from http://www.guardian.co.uk/environment/2012/mar/31/dolphins-sick-deepwater-oil-spill?mobile-redirect=false

Beech, C. (2010). Gulf Coast oil spill. In K. B. Penuel & M. Statler (Eds.), *Encyclopedia of disaster relief* (pp. 252–255). Los Angeles, CA: Sage Publications.

Blair, K. (2011, September 26). Dispersants: Chemicals BP used may cause cancer. *Pensacola News Journal*, p. A1. Retrieved from http://pqasb.pqarchiver.com/pnj/access/2468189491.html?FMT=ABS&date=Sep+26%2C+2011

Bomey, N. (2016). BP's Deepwater Horizon costs total $62B. *USA Today.* Retrieved from https://www.usatoday.com/story/money/2016/07/14/bp-deepwater-horizon-costs/87087056

Boxall, B. (2011, April 20). Post-spill Gulf environment appears to have escaped catastrophe, for now. *Los Angeles Times*, p. 1A.

Bureau of Safety and Environmental Enforcement. (2016). Well control rule. Retrieved from https://www.bsee.gov/guidance-and-regulations/regulations/well-control-rule

Chemical Safety Board. (2014). CSB releases new computer animation of 2010 Deepwater Horizon blowout. Retrieved from http://www.csb.gov/csb-releases-new-computer-animation-of-2010-deepwater-horizon-blowout

Coastal Response Research Center. (2010). Deepwater Horizon dispersant use meeting report. Retrieved from http://www.crrc.unh.edu/dwg/dwh_dispersants_use_meeting_report.pdf

Couch, S. R., & Coles, C. J. (2011). Community stress, psychosocial hazards, and EPA decision-making in communities impacted by chronic technological disasters. *American Journal of Public Health, 101*(S1), S140–S148. doi:10.2105/AJPH.2010.300039

Department of Interior. (2010). Salazar calls for new safety measures for offshore oil and gas operations; orders six month moratorium on deepwater drilling. Retrieved from https://www.doi.gov/news/pressreleases/Salazar-Calls-for-New-Safety-Measures-for-Offshore-Oil-and-Gas-Operations-Orders-Six-Month-Moratorium-on-Deepwater-Drilling

Dermansky, J. (2013, April 20). Three years after the BP Spill, tar balls and oil sheen blight Gulf Coast. *The Atlantic.* Retrieved from http://www.theatlantic.com/national/archive/2013/04/three-years-after-the-bp-spill-tar-balls-and-oil-sheen-blight-gulf-coast/275139

Drescher, C. F., Baczwaski, B. J., Walters, A. B., Aiena, B. J., Schulenberg, S. E., & Johnson, L. R. (2012). Coping with an ecological disaster: The role of perceived meaning in life and self-efficacy following the Gulf oil spill. *Ecopsychology, 4*, 56–63. doi:10.1089/ec0.2012.0009

Drescher, C. F., Schulenberg, S. E., & Smith, C. V. (2014). The Deepwater Horizon oil spill and the Mississippi Gulf coast: Mental health in the context of a technological disaster. *American Journal of Orthopsychiatry, 84*, 142–151.

Dulisse, C. (2014). Marine well containment company: An industry collaboration advancing deepwater well containment technology and readiness efforts. *International Oil Spill Conference Proceedings, 2014*, 300281. doi:10.7901/2169-3358-2014-1-300281.1

Eilperin, J. (2010). U.S. exempted BP's Gulf of Mexico drilling from environmental impact study. *Washington Post.* Retrieved from http://www.washingtonpost.com/wp-dyn/content/article/2010/05/04/AR2010050404118.html

Elliot, D. (2015, April 20). *5 years after BP oil spill, effects linger and recovery is slow.* Retrieved from http://www.npr.org/2015/04/20/400374744/5-years-after-bp-oil-spill-effects-linger-and-recovery-is-slow

Fan, A. Z., Prescott, M. R., Zhao, G., Gotway, C. A., & Galea, S. (2015). Individual and community-level determinants of mental and physical health after the Deepwater Horizon oil spill: Findings from the Gulf States Population Survey. *Journal of Behavioral Health Services & Research, 42*, 23-42. doi:10.1007/s11414-014-9418-7

Fountain, H. (2013). Lessons from the *Exxon Valdez* oil spill. *The New York Times.* Retrieved from http://www.nytimes.com/2013/12/09/booming/lessons-from-the-exxon-valdez-oil-spill.html

Gallacher, J., Bronstering, K., Palmer, S., Fone, D., & Lyons, R. (2007). Symptomatology attributable to psychological exposure to a chemical incident: A natural experiment. *Journal of Epidemiology and Community Health, 61*, 506–512. doi:10.1136/jech.2006.046987

Gallucci, M. (2015, April 16). *BP oil spill has lasting economic toll five years after Deepwater Horizon explosion.* Retrieved from http://www.ibtimes.com/bp-oil-spill-has-lasting-economic-toll-five-years-after-deepwater-horizon-explosion-1883832

Gill, D. A., Picou, J. S., & Ritchie, L. A. (2012). The *Exxon Valdez* and BP oil spills: A comparison of initial social and psychological impacts. *American Behavioral Scientist, 56*, 3–23. doi:10.1177/0002764211408585

Graham, B., Reilly, W. K., Beinecke, F., Boesch, D. F., Garcia, T. D., Murray, C. A., & Ulmer, F. (2011). *Deep water: The Gulf oil disaster and the future of offshore drilling. Report to the President: National Commission on the BP Deepwater Horizon Oil Spill and Offshore Drilling.* Washington, DC: Author.

Gunderson, D. (2012). BP oil spill residue found on pelicans in Minnesota. *Minnesota Public Radio.* Retrieved from http://minnesota.publicradio.org/display/web/2012/05/16/environment/oil-residue-found-on-pelicans

Institute of Nuclear Power Operations (2017). About us. Retrieved from http://www.inpo.info/AboutUs.htm

IOM (Institute of Medicine), 2010. *Assessing the effects of the Gulf of Mexico oil spill on human health: A summary of the June 2010 workshop.* Washington, DC: The National Academies Press.

Juhasz, A. (2012, April 18). Investigation two years after the BP spill, a hidden health crisis festers. *The Nation.* Retrieved from http://www.thenation.com/article/167461/investigation-two-years-after-bp-spill-hidden-health-crisis-festers#axzz2WX6nVyzY

Koch, W. (2015). Is deepwater drilling safer, 5 years after worst oil spill? *National Geographic.* Retrieved from https://news.nationalgeographic.com/2015/04/150420-bp-gulf-oil-spill-safety-five-years-later

Meewisse, M. L., Ollf, M., Kleber, R., Kitchiner, N. J., & Gersons, B. P. (2011). The course of mental health disorders after a disaster: Predictors and comorbidity. *Journal of Traumatic Stress, 24*, 405–413. doi:10.1002/jts.20663

Meigs, J. B. (2016). Blame BP for Deepwater Horizon but direct your outrage to the actual mistake. *Slate.* Retrieved from http://www.slate.com/articles/health_and_science/science/2016/09/bp_is_to_blame_for_deepwater_horizon_but_its_mistake_was_actually_years.html

Muskal, M. (2015). Five years after BP spill, new rules for offshore drilling aim to boost safety. Retrieved from http://www.latimes.com/nation/la-na-gulf-anniversary-20150418-story.html

National Commission on the BP Deepwater Horizon Oil Spill and Offshore Drilling. (2010). The use of surface and subsea dispersants during the BP Deepwater Horizon Oil Spill. Retrieved from http://media.mcclatchydc.com/smedia/2010/10/06/18/_Staff_Report_No._4.source.prod_affiliate.91.pdf

National Wildlife Federation. (2015). *Five years and counting: Gulf wildlife in the aftermath of the Deepwater Horizon disaster.* Retrieved from http://www.nwf.org/~/media/PDFs/water/2015/ Gulf-Wildlife-In-the-Aftermath-of-the-Deepwater-Horizon-Disaster_Five-Years-and-Counting.pdf

Nikiforuk, A. (2016). Why we pretend to clean up oil spills. Retrieved from https://www.smithsonianmag.com/science-nature/oil-spill-cleanup-illusion-180959783

Norse, E. A., & Amos, J. (2010). Impacts, perception, and policy implications of the Deepwater Horizon oil and gas disaster. *Environmental Law Reporter, 40*, 11058–11073.

Osofsky, H. J., Osofsky, J. D., Wells, J. H., & Weems, C. (2014). Integrated care: Meeting mental health needs after the Gulf oil spill. *Psychiatric Services, 65*, 280–283. doi:10.1176/appi.ps.201300470

Oxford Economics. (2010). Potential impact of the Gulf oil spill on tourism. Retrieved from http://www.ustravel.org/sites/default/files/page/2009/11/Gulf_Oil_Spill_Analysis_Oxford_Economics_710.pdf

Palinkas, L. A., Downs, M. A., Petterson, J. S., & Russell, J. (1993). Social, cultural, and psychological impacts of the *Exxon Valdez* oil spill. *Society for Applied Anthropology, 52*, 1–13.

Polson, J. (2011, July 5). BP oil spill still ashore one year after end of Gulf spill. *Bloomberg.* Retrieved from http://www.bloomberg.com/news/2011-07-15/bp-oil-still-washing-ashore-one-year-after-end-of-gulf-spill.html

Robertson, C., & Krauss, C. (2010, August 2). Gulf spill is the largest of its kind, scientists say. *The New York Times*, p. A14.

Rubin, G. J., Amlot, R., Wessely, S., & Greenberg, N. (2012). Anxiety, distress and anger among British nationals in Japan following the Fukushima nuclear accident. *The British Journal of Psychiatry, 201*, 400–407. doi:10.1192/bjp.bp.112.111575

Schrope, M. (2013, January 26). Dirty blizzard buried Deepwater Horizon oil. *Nature: International Weekly Journal of Science.* Retrieved from http://www.nature.com/news/dirty-blizzard-buried-deepwater-horizon-oil-1.12304 doi:10.1038/nature.2013.12304

Schulenberg, S. E., Smith, C. V., Drescher, C. F., & Buchanan, E. M. (2016). Assessment of meaning in adolescents receiving clinical services in Mississippi following the Deepwater Horizon oil spill: An application of the Purpose in Life Test-Short Form (PIL-SF). *Journal of Clinical Psychology, 72*, 1279–1286. doi:10.1002/jclp.22240

Schwartz, N., & Weber, H. (2010). Bubble of methane triggered rig blast. *The Associated Press.* Retrieved from http://www.scpr.org/news/2010/05/08/14902/bubble-methane-triggered-rig-blast

Sumer, N., Karanci, A. N., Berument, S. K., & Gunes, H. (2005). Personal resources, coping self-efficacy, and quake exposure as predictors of psychological distress following the 1999 earthquake in Turkey. *Journal of Traumatic Stress, 18*, 331–342. doi:10.1002/jts.20032

Transocean. (2010). Deepwater Horizon. Retrieved from https://www.scribd.com/document/62876589/Deepwater-Horizon-Spec-Sheet

Varner, S., Langhinrichsen-Rohling, J., & Bell, T. R. (2016). Persisting generalized anxiety disorder and physical health symptoms 18 months after the Deepwater Horizon oil spill: A community sample. *Journal of Community Psychology, 44*, 1011–1026. doi:10.1002/jcop.21824

Walsh, B. (2010). With oil spill (and blame) spreading, Obama will visit Gulf. *Time.* Retrieved from http://www.time.com/time/health/article/0,8599,1986323,00.html

Wang, M. (2010). Health effects after *Exxon Valdez* went unstudied. *ProPublica.* Retrieved from http://www.propublica.org/blog/item/health-effects-after-exxon-valdez-went-unstudied

Weber, H. R. (2010, September 9). Blown-out BP well finally killed at bottom of Gulf. *The Associated Press.* Retrieved from http://www.boston.com/news/nation/articles/2010/09/19/blown_out_bp_well_finally_killed_at_bottom_of_gulf

White, H. K., Hsing, P. Y., Cho, W., Shank, T. M., Cordes, E. E., Quattrini, A. M., . . . Fisher, C. R. (2011). Impact of the Deepwater Horizon oil spill on a deep-water coral community in the Gulf of Mexico. *Proceedings of the National Academy of Sciences of the United States of America, 109*, 20303–20308. doi:10.1073/pnas.1118029109/-/DCSupplemental

Whitehead, A., Dubansky, B., Boinier, C., Garcia, T. I., Miles, S., Pilley, C., . . . Galvez, F. (2011). Genomic and physiological footprint of the Deepwater Horizon oil spill on resident marsh fishes. *Proceedings of the National Academy of Sciences of the United States of America, 109*, 1–5. doi:10.1073/pnas.1109545108/-/DCSupplemental

Wilson, M., Graham, L., Hale, C., Moung-Douglass, E., Sempier, S., Skelton, T., & Swann, L. (2017). Oil spill science: Deepwater Horizon—Where did the oil go. GOMSC-G17-006.

The Saint Francis Dam Failure (1928)

23

Figure 23.1 St. Francis Dam Under construction Near Los Angeles
Source: USGS

BRIEF OVERVIEW

At approximately 11:56 p.m., on March 12, 1928, the St. Francis Dam failed, which resulted in sending a torrent of water (12 billion gallons), from an elevation of 1,834 ft above sea level through San Francisquito Canyon (SFC), and the Santa Clara Valley in Southern California (Jackson & Hundley, 2004). The deluge destroyed dozens of farms and communities before it reached the Pacific Ocean near Ventura, 55 miles downriver (Jackson & Hundley, 2004). It is believed that approximately 400–600 people were killed, and debris from the floodwaters scarred the Santa Clara Valley for years (Hoffman, 2016).

The collapse released a massive wave of floodwaters downstream that reached a height of 140 ft with average velocities estimated at 26 ft per second, sweeping away everything it encountered. By 5:30 a.m. on March 13 floodwaters reached the Pacific Ocean at Montalvo (near Ventura), California (Guyer, 2011). Property damage from the St. Francis Dam disaster was in the millions of dollars and is considered the greatest civil engineering catastrophe in modern U.S. history (Jackson & Hundley, 2004).

Figure 23.2 Flood Aftermath After Dam Collapse
Source: USGS

This was the second deadliest dam failure in the nation, which was preceded by the collapse of South Fork Reservoir Dam that caused the 1889 Johnstown Flood in Pennsylvania, resulting in the deaths of an estimated 2,200 individuals (Jackson & Hundley, 2004).

INTRODUCTION

The St. Francis Dam was a curved concrete gravity dam built by the City of Los Angeles Bureau of Water Works and Supply in the mid-1920s (Rogers & McMahon, 1993). Soil conditions and engineering miscalculations contributed to the disaster. Postdisaster investigations identified, among other things, the failure to properly identify the threat posed by a fault which extended beneath the abutment of the dam and the geology of the SFC which consists of sandstones, clay, coarse and finer-grained, fossiliferous sedimentary rock. This material can soften and swell under climatic conditions, further destabilizing the dam's integrity (Rogers & McMahon, 1993).

The study of geology and soil mechanics was rudimentary in the 1900s, insufficient exploration and testing was limited, and there was little consultation with geologists during the project (Guyer, 2011). The record indicates that William Mulholland, who was a self-educated engineer, who in the early 1900s envisioned and was responsible for the construction of the of the infrastructure that provides water to the Los Angeles area. In the early 1920s, building upon his success as an engineer, he directed the construction of the St. Francis Dam (Guyer, 2011).

Mulholland arrived in Los Angeles as a 22-year-old poor Irish immigrant (Jackson & Hundley, 2004). After a failed attempt prospecting for gold in Arizona, he returned to Los Angeles the following year and worked as a laborer digging ditches for the Los Angeles City Water Company (Jackson & Hundley, 2004). The president of the company, Frederick Eaton, favored his work ethic and promoted him to foreman and eventually superintendent and chief engineer (Jackson & Hundley, 2004). Mulholland lacked university training and was basically self-taught, obtaining his hydraulic-engineering knowledge from on-the-job experience and knowledge gained from technical books and articles covering engineering and geology (Jackson & Hundley, 2004). Mulholland had very little training and education; he dropped out of a Dublin high school at age 15 (Mulholland, 2002).

Despite the limitations of Mulholland's technical background, he resisted scrutiny by his engineering peers or consultation with them during his career (Alvi, 2013). There is no evidence that there was any outside peer review of the design or plans for the dam (Alvi, 2013). During the design and construction of the dam,

Mulholland's time was largely spent on other projects, and thus he delegated most of his responsibilities related to St. Francis Dam to his unqualified subordinates, typically following his general directions without question, and who were themselves inexperienced in the design of concrete gravity dams (Alvi, 2013).

The Setting

The St. Francis Dam was a significant component of the Los Angeles (California) Aqueduct system. Designed and constructed by the City of Los Angeles' Department of Water and Power (LADWP), it created a reservoir about 40 miles northwest of Los Angeles, on San Francisquito Creek near the present-day city of Santa Clarita, California. The dam stored water from the Owens Valley Aqueduct in east central California serving the growing Los Angeles area (Guyer, 2011). The reservoir was the largest the south of the Owens Valley, storing 38,168 acre-ft of water (Rogers, 1992).

The St. Francis Dam was built in SFC, about 7 miles upstream of its mouth, between 1924 and 1926, by the City of Los Angeles Bureau of Waterworks and Supply (BWWS) (Rogers, 1992). The city's Department of Water & Power (LADWP) was created by combining BWWS with the Bureau of Power & Light (BPL), in January 1931 (Rogers, 1992). The Los Angeles Aqueduct descends more than 2,500 ft from the intake dam near Independence to its entry into the San Fernando Valley at Sylmar (Jackson & Hundley, 2004). The biggest part of that drop comes when the aqueduct leaves Fairmount Reservoir and pierces the Sierra Madre escarpment separating the western Mojave Desert, or Antelope Valley, from coastal southern California (Jackson & Hundley, 2004).

The 5-mile-long Elizabeth Tunnel draws water from Fairmount Reservoir, and feeds into San Francisquito Power Plant No. 1. From there, water flows south through a 6-mile-long tunnel in the east canyon wall before dropping down to Power Plant No. 2 along the banks of San Francisquito Creek (Jackson & Hundley, 2004). About one and a half miles upstream from Power Plant No. 2 lies a broad, open area confined by a narrow gorge at the southern end, which became the site of the St. Francis Dam and Reservoir (Jackson & Hundley, 2004).

The city of Los Angeles' population was increasing at a rapid pace in the 1920s (Jackson & Hundley, 2004). The need for additional water during drought years was evident. Thus, Mulholland developed plans for a reservoir and dam in SFC (Jackson & Hundley, 2004). However, the rock along the eastern side of SFC was exceedingly rough, the difficult geological character of the east canyon ridge threatened slips and is where the location of the left or east abutment of the dam rested (Jackson & Hundley, 2004).

California's 1917 dam safety law exempted municipalities from supervision by the State Engineer when building dams allowing Mulholland to choose the site and heighten the dam twice, from its original design, without substantive review by anyone outside his immediate control (Jackson & Hundley, 2004).

Construction of the St. Francis Dam began in 1924; the dam's height was raised twice during construction with minimal consideration for the width and thickness of the base (Guyer, 2011). Ten ft was first added to the height to increase the original design's storage capacity of the reservoir from 30,000 to 32,000 acre-ft (Guyer, 2011). In 1925, when construction was about 50% complete, a decision was made to increase the height of the dam an additional 10 ft to 195 ft, thereby increasing the reservoir's storage capacity to more than 38,000 acre-ft (Guyer, 2011). The dam's new height increase required construction of a dike along the top of the ridge of the western abutment to prevent water from spilling over the ridge (Guyer, 2011). Again, there was no increase in the width or thickness of the dam base (Guyer, 2011). The strength of gravity dams depends on a height/width ratio; increasing the height without also increasing the base width created a "dangerous situation" (Rogers, 2006).

External water pressure on the upstream face of the dam is obviously increased as the vertical height of the water increases and acts horizontally on the upstream face (the pressure diagram is triangular). Water pressure below the base of the dam, called uplift pressure, also exists. The uplift pressure is created when water enters the porous foundation at the base of the dam and then seeps under and through trying to emerge out on the downstream end. The seeping water under the dam's base creates hydraulic gradient between the upstream and downstream side of the dam, causing vertical upward pressure or uplift. Many other forces exist like earthquake forces, earth and silt pressure, and wave pressure. When designing a dam, all forces must be carefully calculated and considered. Modifying the height and reservoir capacity changes the forces exerted on the original dam design and building a dam on questionable geology are serious risks that were apparently overlooked during the construction of the St. Francis Dam.

The Disaster Unfolds

Late design changes during construction of the dam added height without increasing the thickness of its base compromised its integrity (Rogers & McMahon, 1993). Thermal expansion and contraction cracks further stressed the dam. Soon leaks became apparent as the reservoir filled. The cracks were inspected and believed to be insignificant for a dam the size of the St. Francis Dam (Guyer, 2011).

The cracks and leaks were filled with hemp, oakum, and cement grout to seal off seepage, but the postdisaster investigation revealed the repairs were done incorrectly. In March 1928, the reservoir reached full capacity. Cracks and sagging of the roadbed along the east shore were reported near the dam's east abutment (Guyer, 2011).

On Monday, March 12, 1928, a substantive leak was discovered and reported to LADWP management, and Mulholland was called to the site and he spent about 2 hours inspecting conditions, cracks, leaks, and water levels (Guyer, 2011). Again, it was determined that the leaks were relatively minor and normal for a concrete dam. The dam was again pronounced safe by LADWP management (Guyer, 2011).

Shortly before the St. Francis Dam failure, employees witnessed a 12-in. soil failure crack bisecting the road along the eastern side of the canyon, and heard crashing sounds half a mile along the length of the reservoir. The east abutment was about to fail with deadly results (Guyer, 2011).

Impact

The failure of the dam occurred about midnight on March 12, 1928, just days before the water level had reached near its crest (Rogers, 1992). As the dam ruptured, water cascaded through the narrow SFC at a rate estimated to reach 70 miles per hour (Dyrud, 2013). Much of the land in the path of the water was farmland that was swept clean.

Lights flickered in Los Angeles, a sharp drop in voltage was observed by BPL, and a 60,000 volt line shorted out to Lancaster, California. There were no surviving eyewitnesses to the dam's collapse. A wall of water 125 ft high was instantly released and 12 billion gallons of water surged down SFC destroying everything in its path, including Power Plant No. 2. The flood surge traveled south down SFC and west into the Santa Clara River inflicting a path of destruction washing with the flow homes and other buildings, equipment, cattle, and hundreds of people (Dyrud, 2013). With a surge about 2 miles wide and traveling at speeds up to 70 miles per hour; bodies of victims were recovered from the Pacific Ocean as far south as the Mexican border (Guyer, 2011). Portions of Valencia, Newhall, Bardsdale, Fillmore, Castaic Junction, and Santa Paula were flooded before the water flowed into the Pacific Ocean (Guyer, 2011).

Containing the Disaster

As the dam failed, more than 500,000 cubic yards of the left abutment washed away, leaving a void where the left abutment had been shortly before. Nothing could be done to stop or divert the flow of the water. The power was enormous. Modem analyses of buoyancy effects and rapid unsteady channel flow within the canyon accounted for portions of the dam being carried miles downstream (Rogers & McMahon, 1993). The area in the path of the wave was devastated. Hundreds were injured and killed. Millions of dollars in damages had been inflicted. Relief operations began immediately.

RELIEF AND FIRST RESPONSE EFFORTS

The resulting flood swept down SFC, with an initial depth of 140 ft above the streambed (Rogers, 1992). About 5 min later, the flood destroyed BPL Powerhouse 2, approximately 7,300 ft downstream, killing 126 of the 129 people living there (Rogers, 1992).

The flood wave sped down the canyon, widening substantially joining with the Santa Clara River coming out of Soledad Canyon (Rogers, 1992). It continued through Castaic Junction down to the Santa Clara River

Valley, demolishing a Southern California Edison construction camp at Kemp siding, which killed 84 of the 140 workers encamped there. Thereafter, the flood engulfed the migrant farm community at Camulos, before skimming the lower elevations of Fillmore, Santa Paula, Saticoy, and Montalvo (Rogers, 1992).

By the time the surge reached the most distant towns and farms, warnings had been given to the residents in its path by phone and police motorcyclists (despite being at risk themselves). These actions undoubtedly saved many lives. The force of the water was so strong that it tore the clothes off of victims and survivors, surprising rescuers who arrived later to pull people out of trees and off of floating debris (Outland, 1977). Huge pieces of the dam, some weighing 10,000 tons, were washed more than half a mile down the canyon (Guyer, 2011).

Relief work began immediately. Hundreds rushed to the scenes of devastation. Search and rescue operations began immediately and quickly became more organized. Food and shelter for the survivors were provided and mortuary services provided for the dead (Scott, 2012). Groups providing relief support included the American Legion, the Women's Auxiliary of the American Legion, the Boy Scouts, and the County Federation of Women's Clubs, the American Red Cross, and the Los Angeles Times set up a Times Flood Relief Fund (Scott, 2012).

AFTERMATH AND INVESTIGATION

The aftermath of St. Francis Dam disaster immediately focused attention on the official in charge of the failed dam's design and construction William Mulholland, the longtime head of the city's Bureau of Water Works and Supply (Jackson & Hundley, 2004). The catastrophic failure of the St. Francis Dam gave birth to the practice of engineering geology in California, professional registration for engineers, and the creation of a state agency for assessment of dam and reservoir safety, among other codes and regulations (Rogers, 1992).

According to Rogers (1992), the St. Francis Dam failure can be attributed to several factors, which included the presence of the San Francisquito fault beneath the dam's right (east) abutment, and the geology of the surrounding area including sandstone, slake, gypsum, and ancient landslide. In addition to the unsatisfactory soil conditions at the site, a significant design deficiency involved failure to compensate for an additional 20 ft added to the height of the dam (Guyer, 2011).

Safety was repeatedly compromised in order to reduce cost, expedite construction, meet political expectations, and serve the interests of those who had a financial stake in the project (Alvi, 2013). The plan for the St. Francis Dam was largely adapted from the Hollywood Dam designed a few years prior, but those plans were not modified to address the specific conditions of the location. Even as construction began, the rapid growth of Los Angeles led to the decision to enlarge the dam, increasing it in height, but without increasing the width or thickness of the dam. These changes compromised the stability of the dam (Alvi, 2013).

The investigation also revealed that the dam should have been substantially trenched into the canyon walls, but only minimal trenching was performed, possibly to reduce cost or expedite construction. To control cracking, grouted contraction joints were commonly used in concrete dams during that era, but no such joints were used in St. Francis Dam (Alvi, 2013). Mulholland conducted his own testing on the dam during construction and believed it safe.

LESSONS LEARNED AND POLICY IMPACT

The consequences of a disaster eventually lead to the development of required regulatory procedures and guidelines that serve as preventative measures to reduce the propensity of repeating mistakes. Although the practice of civil engineering had progressed accordingly, the process of professional licensing for engineering was not in place in California at the time of the St. Francis Dam collapse (Guyer, 2011). Mulholland and members of his design team would not have qualified as professional engineers under today's requirements. The failure of the St. Francis Dam in 1928 triggered legislation for professional engineering licensing in California the following year, and eventually in other states as well (Alvi, 2013). The civil engineering profession gained much from the public outcry and notoriety accompanying the St. Francis Dam disaster. Sound engineering geologic input on dams became commonplace in the 1930s (Rogers, 1992).

Some of the changes resulting from the disaster included a review of federal dams, which included an order by the U.S. Bureau of Reclamation that all federal dams be promptly examined by a competent engineer, and that there was appropriate documentation of safety compliance, or lack thereof. This led to enactment of the Dam Safety Act of August 14, 1929. Also, a movement gained momentum to license and register professional engineers.

Within 5 years of the St. Francis dam catastrophe, construction began on the Boulder Canyon (Hoover Dam). In order to relieve fears about this construction project, Congress created a Colorado River Board to review the plans of the proposed Boulder Canyon Project in late May 1928, with two engineers and two geologists charged with reviewing existing documents and making a recommendation on where to site the great dam along the Colorado River. These precautions were a direct result of the St. Francis Dam failure (Rogers, 1992).

IMPLICATIONS FOR MODERN EMERGENCY MANAGEMENT

Flooding remains our nation's most frequent and costly natural disaster. According to the Federal Emergency Management Agency, 90% of all natural disasters in the United States involve flooding (McElreath, Jensen, et al., 2013). While dams such as the St. Francis and Hoover are frequently used to generate power, dams are more commonly used to simply protect against a certain level of flooding (McMullen, 2004). As seen with many dam failures throughout history, these structures can erode and weaken over time, and they can also be overtopped, or even fail during large-scale flood events. In 2008, heavy rains and rapid snowmelt in the Midwest caused dam and levee breaches across the region. The resulting floods caused significant devastation to homes and business.

Emergency managers must understand floodplains and include hazard-specific planning for dams as part of their communities' all-hazards planning approach to emergency management. Geographer Gilbert F. White, often referred to as the "Father of Floodplain Management" uttered these famous words while serving at the University of Chicago in 1945: "Floods are 'acts of God,' but flood losses are largely acts of man." His body of work focused on the importance of sound water management practices in modern society. Through his research, both structural (dams, levees, etc.) and nonstructural (codes, land-use planning and zoning, etc.) mechanisms were paramount to planning for and mitigating flood-related disasters. White served on committees under President Lyndon B. Johnson that eventually led to the establishment of the National Flood Insurance Program (NFIP).

Today's emergency managers do not need to be engineers or experts on dam construction. FEMA has taken the federal lead in dam safety and oversees the nation's National Dam Safety Program (NDSP). The NDSP provides assistance to states in order to better their programs through increased inspections, emergency action planning, and the purchase of needed equipment. The NDSP also supports dam safety training, provides grant assistance supporting the improvement of dam safety programs that regulate most of the 87,000 dams in the United States, and also funds dam safety research (McElreath, Doss, et al., 2014).

To assist emergency managers, FEMA provides a guiding document entitled Federal Guidelines for Dam Safety: Emergency Action Planning for Dams (McElreath, Doss, et al., 2014). It walks emergency managers step-by-step through a planning process to address potential dam failures. Every emergency manager with a dam in his or her jurisdiction must plan appropriately by drafting an emergency action plan that specifically addresses dams as stipulated in FEMA's guidelines.

SUMMARY

The catastrophic aftermath of the St. Francis Dam collapse remains one of the worst civil engineering disasters in America, and the second greatest loss of life in California, after the 1906 San Francisco earthquake and fire. The study of geology and soil mechanics was rudimentary in the 1900s, and insufficient exploration and testing was limited, and there was little consultation with geologists during the process of the St. Francis Dam building project (Guyer, 2011). The record indicates that Mulholland was in complete control of the project, and there was little involvement of professional engineers in the decision-making (Guyer, 2011). Evidence shows several deficiencies causing the dam's collapse to include the following:

- no external reviewers and consultants during design and construction
- too much confidence placed on Mulholland's expertise
- lack of formal civil engineering education by Mullholland and his team
- inexperience with concrete gravity dams by design and construction team
- poor site selection
- inadequate measures to control uplift and other forces
- lack of contraction joints
- insufficient width of the dam relative to its final height (Alvi, 2013).

As with all disasters, procedures and requirements evolve and improve to prevent or reduce the chance of repeating errors. Accordingly, civil engineering has progressed because of a lack of professional licensing for engineering, which was not in place in California at the time of the St. Francis Dam collapse (Guyer, 2011). As such, the failure of the St. Francis Dam in 1928 triggered legislation for professional engineering licensing in California the following year, and eventually in other states as well (Alvi, 2013). Thus, the civil engineering profession gained much from the public outcry and notoriety that accompanied the St. Francis Dam disaster, and sound engineering geologic input on dams became commonplace in the 1930s (Rogers, 1992).

REFERENCES

Alvi, I. A. (2013). Human factors in dam failures. *Association of State Dam Safety Officials Annual Conference 2013, Dam Safety 2013*, 1. pp. 778–788.

Dyrud, M. A. (2013). The limits of professional autonomy: William Mulholland and the St. Francis dam. *Proceedings of the 2013 Conference for Industry and Education Collaboration Copyright © 2013 American Society for Engineering Education*. Session ETD 455.

Guyer, J. P. (2011). *Ethical issues from the St. Francis Dam failure. CED Continuing Education and Development, Inc.* El Macero, CA: Guyer Partners.

Hoffman, A. (2016). The St. Francis Dam Disaster and its Historiography. Los Angeles City Historical Society. July 2016.

Jackson, D. C., & Hundley, N. (2004). Privilege and responsibility: William Mulholland and the St. Francis Dam disaster. *California History, 82*(3), 8–47. doi: 10.2307/25161743

McElreath, D., Doss, D., Jensen, C., Wigginton, M., Nations, R., Van Slyke, J., & Nations, J. (2014). *Foundations of emergency management*. Dubuque, IA: Kendall-Hunt.

McElreath, D., Jensen, C., Wigginton, M., Doss, D., Nations, R., & Van Slyke, J. (2013). *Introduction to homeland security* (2nd ed.). Boca Raton, FL: CRC Press.

McMullen, T. (2004) "The St. Francis Dam Collapse and its Impact on the Construction of the Hoover Dam," MS Thesis, University of Maryland. Retrieved from https://docuri.com/download/umi-umd-1629_59c1ce0af581710b28632760_pdf

Mulholland, C. (2002). *William Mulholland and the Rise of Los Angeles*. Berkeley, CA: University of California Press.

Outland, C. F. (1977). *Man-made disaster: The story of St Francis dam*. Norman, OK: A.H. Clark Company.

Rogers, J. D. (1992). The 1928 St. Francis Dam Failure and its impact on American Civil Engineering. In R. Proctor and B. Pipkin (Eds.), *Engineering Geology Practice in Southern California*, (Special Publication No. 4). San Francisco, CA: Association of Engineering Geologists

Rogers, J. D. (2006). Lessons learned from the St. Francis Dam failure. *Geo-Strata Magazine, 6*(2), 14–17

Rogers, J. D., & McMahon, D. J. (1993). *Reassessment of the St. Francis Dam Failure. ASDSO Annual Conference*. Kansas City: Association of State Dam Safety Officials.

Three Mile Island (1979)

24

Figure 24.1 Three Mile Island: By United States Department of Energy
Source: U.S. Department of Energy

BRIEF OVERVIEW

On March 28, 1979, the Three Mile Island Nuclear Generating Station Unit 2 (TMI-2) reactor in Dauphin County, Pennsylvania partially melted down, becoming the most serious nuclear accident in the nation's commercial nuclear power plant operating history. Initially, it was determined that the source of the problem was caused by the polisher, which removes resins so that the water is pure enough not to damage the turbines. However, a small leak from the polisher caused the turbine to shut down automatically. This would not have been a problem if the emergency feed water pumps had come on. The secondary cooling system was supposed to compensate for the shutdown of the primary system. Unfortunately, valves in each pipe had been left closed during maintenance 2 days earlier, and the operator did not notice that the water was being fed into a closed pipe, rather than into the cooling system (Perrow, 1984).

Although its small radioactive releases had no detectable health effects on plant workers or the public (U.S.NRC Backgrounder [USNRC], 2013), the disaster led to a series of comprehensive changes nuclear power plant operations followed (USNRC, 2013). The aftermath of the disaster prompted the Nuclear Regulatory Commission (NRC) to intensify its regulatory oversight, significantly improving reactor safety in the nation (USNRC, 2013).

The disaster at Three Mile Island Unit-2 (TMI-2) was caused by a sequence of equipment failures, human errors, and the inability of plant control room operators to acclimatize to the reactor's condition at specific times during the incident (Nuclear Energy Institute [NEI], 2014). A series of events caused the loss of cooling water to the reactor's heat-generating, radioactive, core leading to partial melting of the fuel rod casing (cladding) and the uranium fuel, releasing a small amount of radioactive material to the atmosphere (Ireland, Scott, & Stratton, 1981). Below is a brief summary of what occurred (Ireland, Scott, & Stratton, 1981; Lochbaum, 2016; NEI, 2009, 2014; USNRC, 2013):

1. On March 28, 1979, at 4:00 a.m., a condensate pump stopped working, causing feedwater pumps to automatically shutoff,
2. water levels dropped in the steam generators, which automatically engaged emergency protective devices to trip the main turbine because it could not remove heat from the steam generator,
3. the control rods automatically dropped into the reactor stopping the fission process and started shutting down the reactor.
4. the loss of the cooling water to the reactor's heat-producing core and the radioactive fission products still producing heat caused the temperature and pressure to continue to rise, causing a pilot-operated relief valve (PORV) to open,
5. steam and water started to escape from the PORV to a drain tank in the basement of the reactor building
6. the PORV should have closed when the pressure stabilized, but it stuck in the open position, discharging about 700 gpm of coolant; also, the control panel indicator showed the PORV was closed, but it wasn't.
7. An emergency feedwater system turned on; however, the pump had been tested 42 hr prior and maintenance personnel forgot to open the valves after the test, therefore the pump was isolated from the steam generator, so emergency water did not flow. The emergency feedwater system started but the closed valves prevented the supply of water.
8. The reactor was losing water and getting hotter as the pressure dropped, some water turned to steam in the reactor forcing water into the pressurizer. Water filled the pressurizer and steam surrounded some of the reactor fuel pellets.
9. Steam bubbles appeared in the coolant pumps causing vibration, so the operators turned those off.
10. Operators turned off emergency cooling water pumps because the pressurizer was full of water.
11. An operator concluded that the PORV was open; they closed an upstream block valve and introduced water into the reactor.
12. The fuel pellets started to melt resulting in partial meltdown.
13. Combustible hydrogen gas was produced and escaped from the reactor into the containment building. The hydrogen ignited and burned for a few seconds.
14. The 700,000 gallons of radioactive cooling water that escaped through the PORV had filled the drain tank and overflowed onto the floor in the auxiliary building. Due to the melting of the uranium fuel pellets, radioactivity had escaped to the coolant and some of that water was now in the auxiliary building. Some of the radioactivity was converted into noble gases that could not be contained so they soon leaked into the atmosphere, thus exposing the public to low doses of radiation.
15. Operators activated a primary pump and achieved forced circulation, ending the immediate emergency.
16. The scientific community estimated that about 17 million Curies escaped the reactor to the auxiliary building which acted like a holding tank allowing some of the radioactivity to decay before entering the atmosphere. As a result of these noble gas releases, about 9 million Curies made it to the environment, the public received some radiation dose (25–50 mrem range), according to testimony in the TMI Litigation.

The nuclear reactor had previously shut down 12 times since April 1978 due to feedwater system problems (Lochbaum, 2016). The TMI-2 accident caused no injuries or deaths. Experts concluded that the amount of radiation released into the atmosphere was too small to result in discernible direct health effects to the population in the vicinity of the plant (NEI, 2014).

There was mass hysteria and misinformation reported and published about the Three Mile Island accident, and potential danger was considerably overstated by the media (USNRC, 2013). During and after the accident, the reactor containment systems at Three Mile Island remained stable; even though the fuel was severely damaged, the reactor vessel and secondary containment systems remained intact (USNRC, 2013). There wasn't serious danger of catastrophic failure at TMI; however, the immediate effect of the accident raised questions and slowed down the development of new nuclear power plants in the United States (USNRC, 2013). Shortly after the accident, the industry established the Institute of Nuclear Power Operations (INPO) to promote excellence in operator training, and plant management and operation (USNRC, 2013).

INTRODUCTION

In September 1978, the TMI-2 plant generated power for distribution for the first time and commenced commercial operation a few months later on December 30, 1978 (Walker, 2004). The $700 million plant had a capacity of 900 electrical megawatts (Walker, 2004). During normal operation, TMI-1 and 2 employed about 500 people (Walker, 2004). Only 3 months later, on the morning of March 28, 1979 at 4:00 a.m., the Three Mile Island Nuclear Power Station Unit 2 (TMI-2) reactor was operating at full power when a condensate pump tripped off and started a series of events that caused a partial meltdown in the core of the nuclear reactor (NSAC-80-1, 1980).

The heat in a nuclear power reactor comes from fission (splitting) of the nuclei of uranium atoms contained in fuel pellets in the uranium oxide core of the reactor (Walker, 2004). Fission occurs when a uranium nucleus captures a neutron and then splits; the sum of the fragmented fission products is less than the original mass and the "missing" mass is converted to energy. During fission, uranium atoms give off enormous amounts of energy that is converted to steam in a boiler (Walker, 2004). The steam turns a turbine and generator which manufactures electricity for distribution (NSAC-80-1, 1980). Unfortunately, the fission process produces considerable amounts of radioactive elements or fission products, the accidental release of these radioactive by-products are the primary danger of nuclear power (Walker, 2004).

Nearly all the commercial reactors in the United States are cooled by the circulation of water (coolant) through the core (Walker, 2004). In the event of a major loss of coolant from the core, the reactor could generate enough heat to devastate safety systems and structures and release hazardous levels of radiation into the atmosphere (Walker, 2004). Ensuring an adequate supply of water for cooling purposes is vital to avoid a meltdown (Walker, 2004). Nuclear facilities include safety systems to prevent an accident, and to limit the consequences if an accident occurred (Walker, 2004).

In a pressurized water reactor, the type installed at TMI-2, there are three cooling circuits (NSAC-80-1, 1980). The primary circuit circulates water through the reactor core (NSAC-80-1, 1980). The primary cooling water picks up the heat from the fission reaction and carries it out of the core to two steam generators (or boilers) (NSAC-80-1, 1980). These are 35-ft-tall tanks (heat exchangers) in which the primary water passes through a large number of small-diameter tubes, transferring heat to water flowing in the secondary circuit, which is outside these tubes (NSAC-80-1, 1980). Water in the secondary circuit boils to make steam which goes to the turbine generator unit (NSAC-80-1, 1980). The steam passes from the turbine to a condenser which is cooled in turn by water from the cooling towers (NSAC-80-1, 1980).

The water in the primary loop is kept from boiling by keeping it under high pressure, normally about 2,200 lb/in.^{2} (NSAC-80-1, 1980). A large vessel connected to the primary loop called the pressurizer is used to maintain this pressure (NSAC-80-1, 1980). The pressurizer is about half full of water, with a steam cushion in the top half (NSAC-80-1, 1980). As the water in the primary loop heats up or cools down, it expands or contracts by hundreds of cubic feet (NSAC-80-1, 1980). The steam cushion in the pressurizer takes up the slack, while maintaining pressure on the primary water (NSAC-80-1, 1980). The control system and the plant operators control the pressure exerted by the pressurizer by maintaining the temperature of the pressurizer with electric heaters and a cooling water spray (NSAC-80-1, 1980).

A relief valve is provided on the pressurizer to prevent overpressuring the system. This valve is a power (pressure)-operated relief valve (PORV) (NSAC-80-1, 1980). If this valve is opened to relieve the excess pressure, the steam or water flows to a drain tank (NSAC-80-1, 1980). If the drain tank becomes overfilled, a rupture disk is provided on that tank to relieve its pressure (NSAC-80-1, 1980). Also, the PORV has a backup, which is called a block valve; these redundant safety systems are intended to increase the reliability of the system, usually in the form of a fail-safe (NSAC-80-1, 1980).

The Setting

The moniker used to label the Nuclear Generating Station as *Three Mile Island* (refers to its length), which is located on a narrow island on the Susquehanna River near Middletown, Pennsylvania, just west of State Highway 441, about 10 miles south of Harrisburg, the capital of Pennsylvania (Walker, 2004). The Three Mile Island's 814 acres makes it the largest island in the river (Walker, 2004). Goldsboro, located about a mile due west across the river in York County, had fewer than 600 residents (Walker, 2004). Royalton, 2 miles north of the plants in Dauphin County, had a population of about 1,100 (Walker, 2004). One mile farther north lay Middletown, home to about 10,000 people, the closest sizable town to Three Mile Island (Walker, 2004). The population within a 5-mile radius of the site was about 38,000, within a 10-mile radius about 165,000, and within a 20-mile radius about 636,000 (Walker, 2004).

There are two nuclear reactor units on the island which are commonly referred to as TMI-1 and TMI-2 (Walker, 2004). The Three Mile Island facilities were a part of the assets of the General Public Utilities Corporation (GPU), a holding company based in Parsippany, New Jersey (Walker, 2004). GPU owned three operating subsidiaries that in turn owned the TMI plants; the principal owner and operator was the Metropolitan Edison Company, headquartered in Reading, Pennsylvania (Walker, 2004). TMI-1 was an 800-MW facility which had gone into commercial operation in September 1974 (Walker, 2004). TMI-2 plant had a capacity of 900 electrical megawatts and commenced commercial operation on December 30, 1978 (Walker, 2004).

The Disaster Unfolds

At 4:00 a.m. on Wednesday, March 28, 1979, a condensate pump (in the turbine water loop) tripped off-line and stopped operating at the TMI-2 facility; now the turbine's water could not remove heat from the steam generator (NSAC-80-1, 1980). Starving the system and lacking feed water supply, the two feed water pumps shut off automatically activating a safety circuit switch which automatically shut down the turbine generator (NSAC-80-1, 1980). When this happened, the control rods automatically dropped into the reactor stopping the fission process (Saygin, 2013). However, the radioactive fission products still produce massive amounts of heat so the temperature and pressure started to rise (Saygin, 2013). When the feed pump failed, three emergency auxiliary feed water pumps, which are provided as backups, automatically started, to keep the turbine water flowing (NSAC-80-1, 1980). Normally, these pumps would have provided the feed water needed by the steam generators; on this occasion, the block valves downstream from the auxiliary feed water pumps had been accidentally left closed after a required maintenance test operation (NSAC-80-1, 1980). Apparently the employees forgot to open the valve after the routine test, so the emergency water did not flow. As a result of the loss of feed water to the steam generators, more heat was being produced in the reactor than was being removed by the steam generators (NSAC-80-1, 1980). Consequently, the temperature of the primary water increased and the water expanded in volume; this caused a flow of primary water coolant into the pressurizer (NSAC-80-1, 1980). This flow compressed the steam cushion in the pressurizer and increased the pressure in the primary system (the nuclear portion of the plant) (NSAC-80-1, 1980).

Four seconds after the initial stoppage of the condensate pump, the water level and pressure in the pressurizer had increased so much that the PORV automatically opened (a valve located at the top of the pressurizer), to reduce the pressure (NSAC-80-1, 1980). Pressure continued to increase until, 9 sec after the first pump stoppage, the reactor shut down automatically in response to the high pressure (NSAC-80-1, 1980). Now the reactor was losing water and getting hotter. With the loss of water (and no air or steam in the pressurizer), the pressure in the primary circuit began to decrease, as normally anticipated, so the operators, in accordance with the procedures, turned on the high-pressure injection system (NSAC-80-1, 1980). The PORV should have closed when the pressure fell to proper levels, but it became stuck open (USNRC, 2013). Instruments in the control room, however, indicated to the plant staff that the PORV was closed (USNRC, 2013). As a result, the plant staff was unaware that cooling water was pouring out of the stuck-open PORV (USNRC, 2013). With the valve open, steam and water escaped the pressurizer; this water flowed into a drain tank (Saygin, 2013).

As coolant flowed from the primary system through the PORV, other instruments available to reactor operators provided inadequate information (USNRC, 2013). There was no instrument that showed how much water covered the core (Saygin, 2013). In case of an accident, a nuclear power plant has tanks of water with pumps that can quickly introduce water to cool the reactor (Saygin, 2013). One of these automatically started (Saygin, 2013). This was noted by the operators, but then they looked at the indicators for the pressurizer, these indicators were telling them that the pressurizer was full of water (which it was because of the steam in the reactor core area) (USNRC, 2013, Saygin, 2013). A full pressurizer means that the operators cannot control the pressure. As a result, plant staff assumed that as long as the pressurizer water level was high, the core was properly covered with water, therefore the operators, turned off the high-pressure injection system as was their normal procedure (NSAC-80-1, 1980; Saygin, 2013,).

As alarms rang and warning lights flashed, the operators did not realize that the plant was experiencing a loss-of-coolant accident, which will inevitably cause a reactor meltdown (USNRC, 2013). They took a series of actions that made conditions worse (USNRC, 2013). To prevent the pressurizer from filling up completely, the staff reduced how much emergency cooling water was being pumped in to the primary system (USNRC, 2013). These actions starved the reactor core of coolant, causing it to overheat (USNRC, 2013). As the escape of water and steam continued, the water level in the system fell (NSCA, 1980). The pressure in the pressurizer also fell so low that generation of steam started in the reactor core (NSCA, 1980). This had two major consequences; first it forced water into the pressurizer filling it completely, and second, steam rather than water surrounded some of the reactor fuel (Saygin, 2013). Steam does not conduct heat as well as water, so the fuel pellets heated up (Saygin, 2013).

About 100 minutes after the accident started, steam bubbles appeared in the coolant pumps, causing them to vibrate (Saygin, 2013). The water escaping through the stuck-open PORV reduced primary system pressure so much that the reactor coolant pumps had to be turned off to prevent dangerous vibrations caused by pumping a mixture of steam and water (NSAC-80-1, 1980). With no water flowing into the reactor and water and steam escaping the reactor, large portions of the reactor core became uncovered (Saygin, 2013). Without the proper water flow, the nuclear fuel overheated to the point at which the zirconium cladding (the long metal tubes that hold the nuclear fuel pellets) ruptured and the fuel pellets began to melt, resulting in a partial meltdown (NSAC-80-1, 1980, Saygin, 2013).

After the PORV remained in the stuck-open position, several important events took place.

1. First, the escaping water and steam filled the drain tank to which they are piped and, as flow continued, caused the rupture disk in that tank to burst (NSAC-80-1, 1980). This resulted in hot water and steam discharging into the reactor building (NSAC-80-1, 1980).
2. Second, as the primary loop pressure fell and the water in it started boiling, it was necessary to turn off the primary coolant circulating pumps since they would be severely damaged by pumping a mixture of steam and water (NSAC-80-1, 1980). Up to this time the pumps had been forcing water into the core, thus providing adequate cooling in spite of the loss of coolant through the stuck-open valve (NSAC-80-1, 1980). But when the pumps were shut off (the last one at about 1 hr and 40 min into the accident), core cooling was no longer adequate to prevent fuel damage by overheating (NSAC-80-1, 1980).
3. Third, the control room employees attempted various operations aimed at controlling the water level in the pressurizer (NSAC-80-1, 1980). None of these procedures eliminated the steam from the core; they had a reverse effect, adding more steam (NSAC-80-1, 1980). The fuel rods in the steam-filled regions boiled dry and then heated to extremely high temperatures (NSAC-80-1, 1980).

The combination of a rising water level in the pressurizer and a decreasing primary system pressure was new to the operators, and they did not understand what it signified until almost 2-1/2 hr later when they realized that the PORV had stuck open, so at 6:18 a.m., they closed a blocking valve introducing water into the reactor, thus ending the immediate emergency (NSAC-80-1, 1980). After the PORV block valve was closed, attempts to halt steam production in the core continued, but were hampered at various times by several problems:

- The coolant temperature was so high that the normal coolant circulating pumps could not be used due to cavitation (vibrations caused by pumping a mixture of steam and water) (NSAC-80-1, 1980).
- Because of the relatively small size of the PORV, it took several hours of injecting water with the high-pressure injection pumps and discharging it through the PORV to remove all the heat generated in the core (NSAC-80-1, 1980).
- The main condenser was out of service due to a problem with an oil-fired auxiliary boiler, and the State of Pennsylvania resisted discharging steam from the secondary loop to the atmosphere, as would have been the normal safe procedure (NSAC-80-1, 1980). Therefore, the steam generators could not remove heat from the primary loop (The state authorities apparently feared that the steam was harmfully radioactive, which was not the case.) (NSAC-80-1, 1980).
- The hydrogen interfered with heat transfer in the steam generators, so they were inefficient as heat sinks, though their steam could have discharged or condensed (NSAC-80-1, 1980).

In spite of these difficulties, the system was eventually cooled to the point that the reactor coolant circulating pumps could be started up again and the main condenser was restored to service (NSAC-80-1, 1980). About 16 hr into the accident, a stable condition was established, steam generation in the core had been stemmed, and the core temperature was declining (NSAC-80-1, 1980). However, between the time that the operators shut off the pumps and when the blocking valve to the PORV was closed, the core was uncovered enough to cause some uranium fuel to melt (Saygin, 2013). In fact, at the time of the accident, nobody thought that a major portion of the fuel melted (Saygin, 2013). When the reactor was opened months later, they were surprised to find that about 60% of the core had melted (Saygin, 2013).

While the reactor core was melting, the fuel rods in the steam-filled regions boiled dry and then heated to temperatures so high that their zircaloy cladding reacted chemically with the steam (NSAC-80-1, 1980). This chemical reaction produced hydrogen gas, which is combustible, and was eventually discharged via the stuck-open PORV and from the drain tank into the reactor building (NSAC-80-1, 1980). The damaged cladding failed and released radioactive fission products from the nuclear fuel itself (uranium dioxide) (NSAC-80-1, 1980). These radioactive fission products then mixed with the water, steam, and hydrogen, and were also discharged into the reactor building (NSAC-80-1, 1980). The hydrogen presented problems during and after this period (NSAC-80-1, 1980). The hydrogen which had escaped into the reactor containment building and mixed with the reactor building air atmosphere reached a concentration that could ignite (NSAC-80-1, 1980). Numerous sources of detonation (switches, etc.) were present and something ignited the hydrogen about 2:00 p.m.; the burn lasted for 6–8 sec, but did no damage to any systems in the building (NSAC-80-1, 1980).

After stabilization and 16 hr after the beginning of the accident, it was concluded that some of the hydrogen remaining in the reactor was in the form of a bubble in the top of the reactor pressure vessel (NSAC-80-1, 1980). Since there was no way of venting this bubble, it had to be removed slowly by passing water through the reactor and allowing the hydrogen to dissolve in the water; this process took several days (NSAC-80-1, 1980). During this period there was widely publicized concern that this hydrogen bubble might explode, causing a breach of the reactor vessel and maybe of the containment building (Pennsylvania Department of Environmental Protection, DEP.PA, 2015). Hydrogen must combine with oxygen to burn, but no oxygen was present in the reactor vessel, so the fear of an explosion was unfounded, but caused many of the public to evacuate the area around TMI (NSAC-80-1, 1980). It was later found that about half of the core melted during the early stages of the accident (USNRC, 2013). Although TMI-2 suffered a severe core meltdown, aftereffects outside the plant were minimal (USNRC, 2013). TMI-2's containment building remained intact and held almost all of the accident's radioactive material (USNRC, 2013).

Impact

Analysis after the accident has indicated that serious damage to the reactor fuel did not start until after about 1 hr and 40 min into the accident (NSAC-80-1, 1980). Had the stuck-open PORV been discovered earlier, no significant fuel damage would have occurred (NSAC-80-1, 1980). Actual rupture of a number of fuel rods apparently occurred about the same time the blocking valve to the PORV was closed (NSAC-80-1, 1980). A few weeks after the accident, the reactor heat had decayed sufficiently that the reactor coolant pumps could be shut off, and the heat could be carried away by natural circulation of the water (NSAC-80-1, 1980).

During the course of the accident some radioactive water and gases were emitted into the atmosphere of the auxiliary building and found their way into the plant through the auxiliary building ventilation system (NSAC-80-1, 1980). Radiation surveys were made around the plant during and after the accident at distances out to more than 40 miles (NSAC-80-1, 1980). Radiation doses to the population that would cause physical health effects were not detected (Kemeny Commission Report) (NSAC-80-1, 1980). However, mental trauma due to anxiety created by unnecessary and unwise publicity has been claimed by some persons (NSAC-80-1, 1980).

Federal and state authorities were initially concerned about the small releases of radioactive gases that were measured off-site by the late morning of March 28 and even more concerned about the potential threat that the reactor posed to the surrounding population (USNRC, 2013). They did not know that the core had melted, but they immediately took steps to try to gain control of the reactor and ensure adequate cooling to the core (USNRC, 2013). The NRCs regional office in King of Prussia, PA, was notified at 7:45 a.m. on March 28. By 8 a.m., NRC Headquarters in Washington, D.C., was alerted and the NRC Operations Center in Bethesda, MD, was activated (USNRC, 2013). The regional office promptly dispatched the first team of inspectors to the site and other agencies, such as the Department of Energy and the Environmental Protection Agency, also mobilized their response teams (USNRC, 2013). Helicopters hired by TMI's owner, General Public Utilities Nuclear, and the Department of Energy were sampling radioactivity in the atmosphere above the plant by midday (USNRC, 2013). A team from the Brookhaven National Laboratory was also sent to assist in radiation monitoring (USNRC, 2013). At 9:15 a.m., the White House was notified and at 11 a.m., all nonessential personnel were ordered off the plant's premises (USNRC, 2013).

By the evening of March 28, the core appeared to be adequately cooled and the reactor appeared to be stable (Corey, 1979). But new concerns arose by the morning of Friday, March 30 (USNRC, 2013). A significant release of radiation from the plant's auxiliary building, performed to relieve pressure on the primary system and avoid curtailing the flow of coolant to the core, caused a great deal of confusion and anxiety (USNRC, 2013). In an atmosphere of growing uncertainty about the condition of the plant, the governor of Pennsylvania, Richard L. Thornburgh, consulted with the NRC about evacuating the population near the plant (USNRC, 2013). Eventually, he and NRC Chairman Joseph Hendrie agreed that it would be prudent for those members of society most vulnerable to radiation to evacuate the area (USNRC, 2013). Thornburgh announced that he was advising pregnant women and pre-school-age children within a 5-mile radius of the plant to leave the area (USNRC, 2013).

The hydrogen bubble was a source of intense scrutiny and great anxiety, both among government authorities and the population, throughout the day on Saturday, March 31 (USNRC, 2013). The crisis ended when experts determined on Sunday, April 1, that the bubble could not burn or explode because of the absence of oxygen in the pressure vessel (USNRC, 2013). Further, by that time, the utility had succeeded in greatly reducing the size of the bubble (USNRC, 2013).

Some of the radioactivity was xenon and krypton (noble gases) and iodine from the partial meltdown (Dickerson, 2017). The gases could not be contained, so they soon leaked into the atmosphere, thus exposing the public to radiation from the radioactivity in the air (Dickerson, 2017). Although the release stacks on the auxiliary building contained radiation monitors, they were designed for much smaller releases (Dickerson, 2017). Therefore, the actual radioactivity that was released was never measured, but from later calculations, the scientific community estimated that about 17 million Curies escaped the reactor and transported to the auxiliary building (Dickerson, 2017). The auxiliary building served as something like a holding tank which allowed some of the radioactivity to decay before entering the atmosphere (Dickerson, 2017). As a result, a little more than half, 9 million Curies, made it to the environment (Dickerson, 2017).

Figure 24.2 Three Mile Island Nuclear Power Plant
Source: A. L. Spangler/Shutterstock.com

As a result of these noble gas releases, the public received some radiation dose (Dickerson, 2017). The NRC conducted detailed studies of the accident's radiological consequences, as did the Environmental Protection Agency, the Department of Health, Education and Welfare (now Health and Human Services), the Department of Energy, and the Commonwealth of Pennsylvania (USNRC, 2013). The actual dose received by any one person will never be known, but experts, according to testimony in the TMI Litigation, gave limits in the 25–50 mrem range (Dickerson, 2017). To put this into context, exposure from a chest X-ray is about 6 millirem and the area's natural radioactive background dose is about 100–125 millirem per year for the area (USNRC, 2013). In addition, researchers did not find any radioactive iodine from the accident in food and milk samples (Dickerson, 2017).

RELIEF AND FIRST RESPONSE EFFORTS

Fortunately, the scope of this accident was significantly less than the 1986 Chernobyl disaster. The authorities were concerned about the small releases of radioactive gases and the threat the reactor posed to the surrounding population. At the time the authorities were not aware that the reactor core had melted, but they took steps to try to gain control of the reactor and ensure adequate cooling to the core. The NRC's regional office in King of Prussia, PA, was notified at 7:45 a.m. on March 28. By 8 a.m., the NRC Headquarters in Washington, D.C., was alerted and the NRC Operations Center in Bethesda, MD, was activated. The regional office promptly dispatched the first team of inspectors to the site and other agencies, such as the Department of Energy and the Environmental Protection Agency, also mobilized their response teams. Helicopters hired by TMI's owner, General Public Utilities Nuclear, and the Department of Energy were sampling radioactivity in the atmosphere above the plant by midday. A team from the Brookhaven National Laboratory was also sent to assist in radiation monitoring. At 9:15 a.m., the White House was notified and at 11 a.m., all nonessential personnel were ordered off the plant's premises.

In the months following the accident, although questions were raised about the possible adverse effects of radiation on human, animal, and plant life in the TMI area, none could be directly correlated with the accident. Thousands of environmental samples of air, water, milk, vegetation, soil, and foodstuffs were collected by various government agencies monitoring the area. Very low levels of radionuclides could be attributed to releases from the accident.

The TMI-2 reactor was permanently shut down and all its fuel had been removed. The reactor coolant system was fully drained and the radioactive water decontaminated and evaporated. The accident's radioactive waste was shipped off-site to an appropriate disposal area, and the reactor fuel and core debris was shipped to the Department of Energy's Idaho National Laboratory.

AFTERMATH AND INVESTIGATION

A combination of equipment malfunctions, design-related problems, and worker errors led to TMI-2's partial meltdown and low emissions of radioactivity to the atmosphere (USNRC, 2013). In the months following the accident, although questions were raised about the possible adverse effects of radiation on human, animal, and plant life in the TMI area, none could be directly correlated to the accident (USNRC, 2013). Very low levels of radionuclides could be attributed to releases from the accident (Nuclear Regulatory Commission Special Inquiry Group, 1980). However, comprehensive investigations and assessments by several well-respected organizations, such as Columbia University and the University of Pittsburgh, have concluded that in spite of serious damage to the reactor, the actual release had negligible effects on the physical health of individuals or the environment (USNRC, 2013).

LESSONS LEARNED AND POLICY IMPACT

A combination of personnel error, design deficiencies, and component failures caused the Three Mile Island accident, which permanently changed both the nuclear industry and the NRC (Rogovin & Frampton, 1980). Public fear and distrust increased, NRC's regulations and oversight became broader and more robust, and management of the plants was scrutinized more carefully (USNRC, 2013). Careful analysis of the accident's events identified problems and led to permanent and sweeping changes in how NRC regulates its licensees, which in turn, has reduced the risk to public health and safety (USNRC, 2013). Itemized below are some of the major changes that have occurred since the accident:

- Upgrading and strengthening of plant design and equipment requirements (USNRC, 2013). This includes fire protection, piping systems, auxiliary feed water systems, containment building isolation, reliability of individual components (pressure relief valves and electrical circuit breakers), and the ability of plants to shut down automatically; (USNRC, 2013)
- Identifying the critical role of human performance in plant safety led to revamping operator training and staffing requirements, followed by improved instrumentation and controls for operating the plant, and establishment of fitness-for-duty programs for plant workers to guard against alcohol or drug abuse; (USNRC, 2013)
- Enhancing emergency preparedness, including requirements for plants to immediately notify NRC of significant events and an NRC Operations Center staffed 24 hr a day (USNRC, 2013). Drills and response plans are now tested by licensees several times a year, and state and local agencies participate in drills with the Federal Emergency Management Agency and NRC; (USNRC, 2013)
- Integrating NRC observations, findings, and conclusions about licensee performance and management effectiveness into a periodic, public report; (USNRC, 2013)
- Having senior NRC managers regularly analyze plant performance for those plants needing significant additional regulatory attention; (USNRC, 2013)
- Expanding NRC's resident inspector program, first authorized in 1977, to have at least two inspectors live nearby and work exclusively at each plant in the United States to provide daily surveillance of licensee adherence to NRC regulations; (USNRC, 2013)
- Expanding performance-oriented as well as safety-oriented inspections, and the use of risk assessment to identify vulnerabilities of any plant to severe accidents; (USNRC, 2013)
- Strengthening and reorganizing enforcement staff in a separate office within the NRC; (USNRC, 2013)

- Establishing the INPO, the industry's own "policing" group, and formation of what is now the Nuclear Energy Institute to provide a unified industry approach to generic nuclear regulatory issues, and interaction with NRC and other government agencies; (USNRC, 2013)
- Installing additional equipment by licensees to mitigate accident conditions, and monitor radiation levels and plant status; (USNRC, 2013)
- Enacting programs by licensees for early identification of important safety-related problems, and for collecting and assessing relevant data, so operating experience can be shared and quickly acted upon; and (USNRC, 2013)
- Expanding NRC's international activities to share enhanced knowledge of nuclear safety with other countries in a number of important technical areas (USNRC, 2013).

IMPLICATIONS FOR MODERN EMERGENCY MANAGEMENT

The main generic lessons learned and already being implemented by industry fall under the following broad categories (NSAC-80-1, 1980):

1. Increased scope of continuous technical support for reactor operation and safety surveillance (NSAC-80-1, 1980).
2. Added measures and criteria for selection and training of operators, including more comprehensive procedures for coping with plant upsets and accidents (NSAC-80-1, 1980).
3. Organization and procedures for insuring systematic and rigorous learning from operating experience (NSAC-80-1, 1980).
4. Equipment or design changes for improved ease of operation of existing plants (NSAC-80-1, 1980).
5. Evaluations of long-term improvements in reactor design (NSAC-80-1, 1980).
6. Establishment of emergency response organizations and defined decision-making processes (NSAC-80-1, 1980).
7. Establishment of practical goals for reactor safety (NSAC-80-1, 1980).

These generic lessons are being treated primarily by suitable actions by the utilities, with supplemental and supporting actions by reactor and equipment suppliers, new institutions such as NSAC and INPO, and enabling actions by the regulatory agencies (NSAC-80-1, 1980). The basic lessons learned, when applied intensively, can help prevent the full range of accidents involving core damage, environmental effects, and their social and fiscal consequences (NSAC-80-1, 1980).

A number of technological and procedural changes have been implemented by industry and the NRC to considerably reduce the risk of a meltdown since the 1979 incident. These include:

- Plant design and equipment upgrades, including fire protection, auxiliary feed water systems, containment building isolation, and automatic plant shut down capabilities; (Spencer & Loris, 2009)
- Enhanced emergency preparedness, including closer coordination between federal, state, and local agencies; (Spencer & Loris, 2009)
- Integration of NRC observations, findings, and conclusions about plant performance and management into public reports; (Spencer & Loris, 2009)
- Regular plant performance analysis by senior NRC managers who identify plants that require additional regulatory attention; (Spencer & Loris, 2009)
- Expansion of NRC's resident inspector program, whereby at least two inspectors live nearby and work exclusively at each plant; (Spencer & Loris, 2009)
- Expanded performance- and safety-oriented inspections;
- Additional accident safety equipment to mitigate and monitor conditions during accidents; and (Hathitrust Digital Library, 1979)
- Establishment of the Institute for Nuclear Power Operators, an industry-created nonprofit organization that evaluates plants, promotes training and information sharing, and helps individual plants overcome technical issues. (Spencer & Loris, 2009)

SUMMARY

On March 28, 1979, the TMI-2 reactor partially melted down, becoming the most serious accident in U.S. commercial nuclear power plant operating history, although its small radioactive releases had no detectable health effects on plant workers or the public (USNRC, 2013). Comprehensive changes to nuclear power plant operations immediately followed this incident concerning: emergency response, operator training, control engineering, and radiation protection, to name a few (Blandford & May, 2012). It also caused the NRC to intensify its regulatory oversight, all of which significantly improved U.S. reactor safety (USNRC, 2013).

The disaster at TMI-2 was caused by a sequence of equipment failures, human errors, and the inability of plant control room operators to acclimatize to the reactor's condition at specific times during the incident (NEI, 2014). A series of events caused the loss of cooling water to the reactor's heat-generating, radioactive, core leading to partial melting of the fuel rod casing (cladding) and the uranium fuel, releasing a small amount of radioactive material to the atmosphere (NEI, 2014).

With radioactive water escaping the containment building into one of the plant's auxiliary buildings, radiation alarms sounding, and mounting evidence of serious damage to the core, officials sealed the buildings and declared a state of emergency. The public was immediately informed of the emergency and communities around the plant were evacuated (Waugh, 2000).

According to Perrow (1984), such a disaster can be characterized as a type of "normal accident" that can transpire due to the complexities of the "tightly coupled" systems that maintain synergistic operational processes (Waugh, 2000). Today, the TMI-2 reactor is permanently shut down and all its fuel has been removed (USNRC, 2013). The reactor coolant system is fully drained and the radioactive water decontaminated and evaporated (USNRC, 2013). The accident's radioactive waste was shipped off-site to an appropriate disposal area, and the reactor fuel and core debris was shipped to the Department of Energy's Idaho National Laboratory (USNRC, 2013). In 2001, FirstEnergy acquired TMI-2 from GPU (USNRC, 2013). FirstEnergy has contracted the monitoring of TMI-2 to Exelon, the current owner and operator of TMI-1 (USNRC, 2013). The companies plan to keep the TMI-2 facility in long-term, monitored storage until the operating license for the TMI-1 plant expires, at which time both plants will be decommissioned (USNRC, 2013).

REFERENCES

Blandford, E. D., & May, M. M. (2012). *Lessons learned from "lessons learned": The evolution of nuclear power safety after accidents and near-accidents.* Cambridge, MA: American Academy of Arts and Sciences. ISBN: 0-87724-094-9

Corey, G. R. (1979, October). A brief review of the accident at Three Mile Island. *IAEA Bulletin, 21*(5), 54–59

Dickerson College's Three Mile Island web site. (2017). What went wrong? Retrieved from http://tmi.dickinson.edu/index.php/what-went-wrong

Hathitrust Digital Library. (1979, October). Report of The President's Commission on The Accident at Three Mile Island, The Need for Change: The Legacy of TMI. Retrieved from https://babel.hathitrust.org/cgi/pt?id=umn.31951d00830255k;view=1up;seq=47

Ireland, J. R., Scott, J. H., & Stratton, W. R. (1981, Summer/Fall). Three Mile Island and multiple failure accidents. *Los Alamos Science, 2*(2), 74–91.

Lochbaum, D. (2016). *Nuclear plant accidents: Three Mile Island.* July 26, 2016. Retrieved from http://allthingsnuclear.org/dlochbaum/nuclear-plant-accidents-three-mile-island

NEI—Nuclear Energy Institute. (2009, March). *The TMI 2 accident: Its impact, its lessons.* Washington, D.C. Retrieved from www.nei.org.

NEI—Nuclear Energy Institute. (2014, October). *Lessons from the 1979 accident at Three Mile Island.* Washington, D.C. Retrieved from www.nei.org; https://www.nei.org/Master-Document-Folder/Backgrounders/Fact-Sheets/The-TMI-2-Accident-Its-Impact-Its-Lessons

Nuclear Regulatory Commission Special Inquiry Group. (1980). *Three Mile Island: A report to the commissioners and to the public* (Vol. 2, Pt. 1). Washington, D.C.: Government Printing Office

NSAC-80-1. (1980). *Analysis of Three Mile Island—Unit 2 Accident.* Prepared by the Nuclear Safety Analysis Center Operated by the Electric Power Research Institute.

Pennsylvania Department of Environmental Protection, DEP.PA. (2015). *Three Mile Island (TMI) Lessons learned: Better protection for the public and the environment.* Commonwealth of Pennsylvania. Department of Environmental Protection. 2930-FS-DEP3129. Rev. 11/2015. Retrieved from www.dep.pa.gov

Perrow, C. (1984). *Normal accidents.* New York, NY: Basic Books.

Rogovin, M., & Frampton, G. T. (1980). *Three Mile Island—Volume 1: "A report to the commissioners and to the public."* Washington, D. C.: Nuclear Regulatory Commission Special Inquiry Group.

Saygin, H. (2013). Major nuclear accidents and their implications for the evolution of nuclear power. *The Turkish Model for Transition to Nuclear Power,* 53–81. Print.

Spencer, J., & Loris, N. (2009, March 27). Three Mile Island and Chernobyl: What went wrong and why today's reactors are safe. *The Heritage Foundation.* Retrieved from http://www.heritage.org/node/14531/print-display

U.S.NRC Backgrounder. (2013, February). *Three Mile Island Accident.* Rockville, MD: Nuclear Regulatory Commission (NRC) Public Document Room.

Walker, J. S. (2004). *Three Mile Island: A nuclear crisis in historical perspective.* Berkeley, CA: University of California Press.

Waugh Jr., William L. (2000). *Living with hazards dealing with disasters: An introduction to emergency management.* Armonk, NY: M.E. Sharpe.

The Hyatt Hotel Disaster (1981)

25

Figure 25.1 Hyatt Regency Skywalk Collapse

BRIEF OVERVIEW

On July 17, 1981, approximately 2,000 people were attending the Tea Dance, enjoying the band's music, the food and drinks, and the dance contest at the Hyatt Regency Hotel in Kansas City (Bernhardt, 2016). Unfortunately, the evening of celebration came to a crashing halt and would be remembered for the tragic deaths resulting from the most catastrophic failure of a structural connection in the United States (Bernhardt, 2016). The collapse of two of the three walkways spanning the atrium caused the death of 114 people and the injury of more than 180, and traumatized countless others (Bernhardt, 2016). The effects were felt throughout Kansas City and the United States and served as a wake-up call to the structural engineering community (Bernhardt, 2016).

Three suspended walkways spanned the atrium of the Hyatt Regency Hotel in Kansas City, MO, at the second, third, and fourth floor levels (Marshall et al., 1982). The fourth floor walkway was suspended from the atrium roof framing by a set of six hanger rods (Marshall et al., 1982). The second floor walkway was suspended from the fourth floor walkway which was directly above it (Marshall et al., 1982). The third floor walkway was offset from the other two and was independently suspended from the roof framing by another set of hanger rods (Marshall et al., 1982). In the collapse, the second and fourth floor walkways fell to the atrium floor, with the fourth floor walkway crashing down on top of the lower second floor walkway (Marshall et al., 1982).

Two factors contributed to the collapse:

1. inadequacy of the original design for the box beam–hanger rod connection, which was identical for all three walkways, and
2. a change in hanger rod arrangement during construction that essentially doubled the load on the box beam–hanger rod connections at the fourth floor walkway (Marshall et al., 1982).

As originally approved for construction, the contract drawings called for a set of continuous hanger rods which would attach to the roof framing and pass through the fourth floor box beams and on through the second floor box beams (Marshall et al., 1982). As actually constructed, two sets of hanger rods were used, one set extending from the fourth floor box beams to the roof framing and another set from the second floor box beams to the fourth floor box beams (Marshall et al., 1982).

Changing the hanger rod arrangement from a continuous rod to two sets of hanger rods drastically reduced the loading capacity of the walkways. With this change in hanger rod arrangement, the ultimate capacity of the walkways was so significantly reduced that they could barely resist their own weight and had virtually no capacity to resist additional loads imposed by people (Marshall et al., 1982).

INTRODUCTION

On July 17, 1981, at approximately 7:05 p.m., two suspended walkways within the atrium/lobby area of the Hyatt Regency Hotel in Kansas City, MO, collapsed, killing over 100 people and injuring almost 200 more (Marshall et al., 1982). Two of the injured subsequently died. In terms of loss of life and injuries, this was the most devastating structural collapse ever to take place in the United States (Marshall et al., 1982).

At the time of the collapse, the hotel had been in service for approximately 1 year (Marshall et al., 1982). The atrium/lobby was a large open area with three suspended walkways spanning the atrium at the second, third, and fourth floor levels (Marshall et al., 1982). The third floor walkway was independently suspended from the atrium roof trusses while the second floor walkway was suspended from the fourth floor walkway, which in turn was suspended from the roof framing (Marshall et al., 1982). In the collapse, the second and fourth floor walkways fell to the atrium first floor, with the fourth floor walkway collapsing on top of the second (Marshall et al., 1982). Most of those killed or injured were crushed either on the atrium first floor level or on the second floor walkway (Marshall et al., 1982).

THE SETTING

Planning for the Hyatt Regency project started in 1976 (Bernhardt, 2016). Walkways spanned across the atrium at the second, third, and fourth stories (Bernhardt, 2016). The atrium is centrally located between the 40-story tower section to the north and the function block directly to the south (Marshall et al., 1982). In addition to providing a direct traffic way between the tower section and the function block, the walkways were a dominant feature of the atrium and afforded a view of the atrium and terrace (Marshall et al., 1982). The distance that the walkways spanned was approximately 120 ft (Bernhardt, 2016). The 120-ft length was split into four equal spans of approximately 30 ft each (Bernhardt, 2016). The ends of the walkways were supported by the tower and conference center structures, but each interior span support consisted of a built-up box beam suspended from the roof structure by two steel hanger rods (Bernhardt, 2016). The walkway hangers were 1 1/4 in. (32 mm) diameter rods threaded top and bottom to receive a nut and washer (Marshall et al., 1982). The original design of a single continuous hanger rod arrangement was eliminated and replaced with two hanger rods with a 4 in (102 mm) offset at the fourth floor walkway and were likely the reason for the disaster (Marshall et al., 1982).

The Disaster Unfolds

On July 17, 1981, at approximately 7:05 pm at the Kansas City Hyatt Hotel's Tea Dance Party, with less than 10 people on the fourth floor walkway and less than 60 people on the second floor walkway, a disastrous rupture occurred (Bernhardt, 2016). Dancers crowded the lobby floor, with about 100 onlookers on a skywalk suspended over the floor, and another 50 or more on a skywalk above that. As many of the onlookers were swaying with the music, the support gave way, which then caused the 120-ft section of the upper skywalk to collapse onto the lower, and then both onto the dancers in the lobby (Waugh, 2000).

Immediately thereafter, a triage area and morgue were set up due to the presence and earnest assistance of about 20 radiologists who had been having dinner in the hotel. Unfortunately, the traumatic devastation resulted in an arduous process of trying to rescue those still trapped under the debris, as well as provide immediate medical assistance for victims experiencing significant injuries (Waugh, 2000). In addition, subsequent identification of victims became a challenge because their relatives and friends were unaware that they were at the hotel event. This resulted in on-scene authorities having to try and identify victims by matching information from vehicles parked around the hotel, and check-in records of the hotel clerk office (Waugh, 2000).

The rupture was caused by the force of the fourth floor walkway-to-roof hanger rod nut on the bottom surface of the box beam (Bernhardt, 2016). The impact from the bolt and nut on the upper flange of the box caused failure of the upper flange, with the bolt and nut slipping through the bolt hole (Bernhardt, 2016). The fourth floor walkway became disconnected from the roof support, and both walkways collapsed, leaving 113 people dead and 186 injured (Bernhardt, 2016; Luth, 2000; Marshall et al., 1982).

The National Bureau of Standards (NBS) researchers first arrived in Kansas City on July 21, 1981 (Marshall et al., 1982). During the course of the investigation, the NBS team of engineers and scientists inspected the Hyatt Regency atrium area and the warehouse where the walkway debris was stored (Marshall et al., 1982). Documents such as drawings, specifications, inspection reports, test reports, and construction logs, as well as photographs and videotapes became available to NBS from a number of sources (Marshall et al., 1982). These were useful in gaining information about the walkways as originally approved for construction and as modified during the construction process (Marshall et al., 1982).

As originally approved for construction by the Kansas City Codes Administration Office, the plans for the walkways called for a single set of hanger rods (attached to the roof framing) which would pass through the fourth floor box beams and on through the second floor box beams (Marshall et al., 1982). The box beams were made up of a pair of 8-in. steel channels with the flanges welded toe to toe and were to rest on hanger rod washers and nuts below each set of beams (Marshall et al., 1982). Under this arrangement each box beam would separately transfer its load directly into the hanger rods (Marshall et al., 1982).

However, during construction, shop drawings were prepared by the steel fabricator which called for the use of two sets of hanger rods rather than a single set (Marshall et al., 1982). The steel fabricator wanted to eliminate the need to thread a nut all the way up a continuous threaded hanger rod to the fourth floor during construction and also reduce the cost to make the long rod. One set of hanger rods extended from the fourth floor box beams to the roof framing and another set extended from the second floor box beams to the fourth floor box beams (Marshall et al., 1982). Under this arrangement all of the second floor walkway load was first transferred to the fourth floor box beams, where both that load and the fourth floor walkway load were transmitted through the box beam–hanger rod connections to the ceiling hanger rods (Marshall et al., 1982). As indicated by their stamps, these shop drawings were reviewed by the contractor, structural engineer, and architect (Marshall et al., 1982).

Impact

Efforts were made to establish as accurately as possible the loads on the walkways at the time of the collapse (Marshall et al., 1982). A television crew had videotaped parts of the walkway minutes before the collapse

(Marshall et al., 1982). By studying this videotape, NBS concluded that 63 people represented a credible estimate of combined second and fourth floor walkway occupancy at the time of the collapse (Marshall et al., 1982).

Based on field, laboratory, and analytical investigations, the NBS concluded that:

1. The collapse of the walkways occurred under loads that were substantially less than the design loads specified by the Kansas City Building Code (Marshall et al., 1982).
2. Observed distortions of structural components strongly suggest that failure of the walkway system initiated in the box beam–hanger rod connection at the east end of middle box beam in the fourth floor walkway (Marshall et al., 1982).
3. As constructed, the box beam–hanger rod connections, the fourth floor to ceiling hanger rods, and the third floor walkway hanger rods did not satisfy the design provisions of the Kansas City Building Code (Marshall et al., 1982).
4. The change in hanger rod arrangement from a continuous rod to interrupted rods essentially doubled the load to be transferred by the fourth floor box beam–hanger rod connections (Marshall et al., 1982).
5. The box beam–hanger rod connection would not have satisfied the Kansas City Building Code under the original hanger rod detail (continuous rod) (Marshall et al., 1982).
6. Under the original hanger rod arrangement (continuous rod) the box beam–hanger rod connections as shown on the contract drawings would have had the capacity to resist the loads estimated to have been acting at the time of the collapse (Marshall et al., 1982).
7. Neither the quality of workmanship nor the materials used in the walkway system played a significant role in initiating the collapse (Marshall et al., 1982).

RELIEF AND FIRST RESPONSE EFFORTS

First responders arrived on the scene quickly after the first calls for assistance were received. For the next 14 hr rescue operations would be ongoing, involving crews from 34 fire trucks, Emergency Medical Service units, and doctors and medical staff from five local hospitals. The collapse of the skywalks trapped many beneath an estimated 60 tons of wreckage to include steel, concrete, and glass. The weight of the debris was so significant that neither the equipment available at the Hyatt's nor possessed by the fire department proved capable to clear the wreckage.

A remarkable aspect of the emergency response was the number of volunteers who immediately converged on the scene. It was estimated that approximately 250 police, 250 fire service personnel, and over 100 paramedics had responded, along with volunteers from local hospitals, and community organizations. In addition, there was a very well-organized transportation process that included ambulances, helicopters, taxis, and private vehicles, which assisted in moving victims to hospitals (Waugh, 2000).

Additional volunteers from surrounding areas responded to the fire department's requests for help, including construction companies and building-supply stores, bringing hydraulic jacks, acetylene torches, compressors, and generators. Commercial cranes would be required to lift the sections of walkway off those trapped.

Dr. Joseph Waeckerle, former chief of Kansas City's emergency medical system, directed the rescue effort establishing a makeshift morgue in a ground floor exhibition area, using the hotel's driveway and front lawn as a triage area and helping to organize the wounded by greatest need for medical care.

Those people who could walk were instructed to leave the hotel to simplify the rescue effort; those mortally injured were told they were going to die and given morphine. Often, rescuers had to dismember bodies to reach survivors among the wreckage. One victim's right leg was trapped under an I-beam and had to be amputated by a surgeon, a task which was completed with a chainsaw.

One of the great challenges of the rescue operation was that the hotel's sprinkler system had been severed by falling debris, flooding the lobby and putting trapped survivors at great risk of drowning. As the pipes were connected to water tanks, not a public source, the flow could not be stopped. Mark Williams, the last person rescued alive from the rubble, spent more than nine and a half hours pinned underneath the lower skywalk with both of his legs pulled out of their sockets. Williams nearly drowned before Kansas City's fire chief realized that the hotel's front doors were trapping the water in the lobby. On his orders, a bulldozer was sent to break through the doors, which allowed the water to pour out of the lobby and thus eliminated the danger to those trapped. A fire hose was then placed over the broken pipe, redirecting the water outside the hotel. Additionally, the lobby was filled with concrete dust, and visibility was poor as the emergency workers had cut the power to prevent fires. Twenty-nine people were rescued due to the efforts of the first responders and volunteers.

AFTERMATH AND INVESTIGATION OF THE DISASTER

The following summarizes information and findings of the event (Marshall et al., 1982):

1. The collapse of the second and fourth floor walkways is believed to have occurred at approximately 7:05 p.m., Friday, July 17, 1981 (Marshall et al., 1982).
2. Analysis of the KMBC TV videotape indicates approximately 40 people on the second floor walkway shortly before the collapse (Marshall et al., 1982).
3. A reasonable upper bound for single-span occupancy is believed to be 16–20 people (Marshall et al., 1982).
4. It is concluded that a total of 63 people represents a credible upper-bound combined occupancy of the second and fourth floor walkways at the time of the collapse (Marshall et al., 1982).
5. All of the fourth floor to ceiling hanger rods pulled through both the lower and upper flanges of the fourth floor box beams (Marshall et al., 1982).

In the original arrangement, each hanger rod was to be continuous from the second floor walkway to the hanger rod bracket attached to the atrium roof framing (Marshall et al., 1982). The design load to be transferred to each hanger rod at the second floor walkway would have been one-half the sum of the dead load and the resultant live load for a single span (Marshall et al., 1982). An essentially identical load would have been transferred to each hanger rod at the fourth floor walkway (Marshall et al., 1982). Thus the design load acting on the upper portion of a continuous hanger rod would have been twice that acting on the lower portion, but the required ultimate capacity of the box beam–hanger rod connections would have been the same for both walkways (Marshall et al., 1982).

The hanger rod configuration actually used consisted of two hanger rods: the fourth floor to ceiling hanger rod segment as originally detailed and the second to fourth floor segment offset 4 in. (102 rom) inward along the axis of the box beam (Marshall et al., 1982). The washer and nut arrangement called for in the original connection detail was retained in the actual construction (Marshall et al., 1982). With this modification, the design load to be transferred by each second floor box beam–hanger rod connection was unchanged, as were the loads in the upper and lower hanger rod segments (Marshall et al., 1982). However, the load to be transferred from the fourth floor box beam to the upper hanger rod under this arrangement was essentially doubled, thus compounding an already critical condition (Marshall et al., 1982).

The following important points and missed opportunities can be found in Luth's chronology of events (Luth, 2000, Bernhardt, 2016).

1. The initial project engineer and senior designer, who were familiar with the background of the design, left the structural engineering firm early in the design process (Luth, 2000, Bernhardt, 2016). Their departure impeded the communication of the original design intent to those who completed the design (Luth, 2000, Bernhardt, 2016).
2. Similarly, the fabricator transferred incomplete shop drawings to an outside detailing firm for completion, further impairing the flow of information (Luth, 2000, Bernhardt, 2016). The outside detailer assumed the connection had been designed because it was shown on the shop drawings and not flagged for design check, as no loads were provided in the initially drafted sketch (Luth, 2000, Bernhardt, 2016).
3. The project manager conditionally approved a change request from the fabricator to use a double rod configuration in lieu of a single continuous rod (Luth, 2000, Bernhardt, 2016). This essentially doubled the load on the rod-to-box-beam connection at the fourth floor walkway (Luth, 2000, Bernhardt, 2016). The

structural engineering manager requested it be submitted as a formal change request, delaying the final review until a later date (Luth, 2000, Bernhardt, 2016). The fabricator did not formally submit the change from one continuous to two offset rods as requested by the structural engineer (Luth, 2000, Bernhardt, 2016).

4. The structural engineer's technician reviewed the shop drawings and questioned the yield stress of the steel hanger rod (Luth, 2000, Bernhardt, 2016). The project manager did not give the question his full attention but responded from memory (Luth, 2000, Bernhardt, 2016). If the question had drawn the structural engineer's focus, the deficiencies in the design might have been noticed (Luth, 2000, Bernhardt, 2016).

The damages awarded to victims and families of victims exceeded $100 million. A grand jury found no evidence of illegal action on the part of the design professionals (Bernhardt, 2016). The Missouri Board for Architects, Professional Engineers and Land Surveyors investigated and brought charges of gross negligence and misconduct against the structural engineer of record and the structural engineering project manager from the firm that had provided the structural design of the Hyatt (Missouri Board for Architects et al., n.d.). An administrative judge found them guilty of the charges (Bernhardt, 2016). Both engineers lost their licenses (Bernhardt, 2016; Pfatteicher, 2000; Roddis, 1993).

LESSONS LEARNED AND POLICY IMPACT

Structural failures are relatively common, but the American Institute of Architects (AIA) had previously warned of the possible danger attributed to "long span construction." The AIA had expressed concern about the complexity of construction projects, compromising infrastructure by using light materials and exotic designs, a lack of redundancy in using steel supports, pressure to keep construction costs low, and inadequate monitoring of construction to ensure compliance with safety standards, all of which could invariably lead to a structural failure (Waugh, 2000).

Gillum-Colaco, Consulting Structural Engineers (G.C.E.), the structural engineering design contractor, created only a partial design and left the most safety-critical design decisions to the fabrication/installation contractor, Havens Steel (Wander, 2008). Havens created the design without any documented engineering analysis and submitted it to G.C.E., which approved the design, without any documented engineering analysis either (Wander, 2008). NBS investigators were unable to find any significant recorded calculations of safety factors or yield strengths of the walkway connections (Wander, 2008). After weighing sections of the collapsed walkways, viewing footage of the crash, and re-creating sections of the walkways, NBS investigators determined that the load capacity on each connector was approximately only 60% of the Kansas City Building Code's required load capacity for that type of connection (Wander, 2008).

All parties involved had a responsibility to identify and recognize the walkway as a safety-critical suspended load, which warranted special consideration and care (Wander, 2008). During construction in 1979, the atrium roof had collapsed, prompting G.C.E. to ask Crown Center Redevelopment Corporation, the owners, for an on-site inspection of the entire site (Wander, 2008). Three different requests were denied due to the additional costs. According to G.C.E.'s safety inspector, as noted in the administrative hearings after the collapse, had he been instructed to inspect all the connections in the atrium (instead of just the roof), he most likely would have discovered the flaws in the walkway (Wander, 2008).

Clear delineation of accountability was absent with shared design responsibilities, numerous contractors and subcontractors, and overlapping design verification processes (Wander, 2008). Patty Berkebile Nelson Duncan Monroe Lefebvre (PBNDML), Architects Planners, Incorporated, the Architects, failed to exercise oversight of support contractors (Wander, 2008). And the Kansas City Division of Public Works Department failed to provide adequate oversight and evaluation of design documents when it approved of the original design, which NBS investigators said violated the building codes even before the design change (Wander, 2008). The Public Works Department denied that it had been notified of or had approved of the design change (Wander, 2008).

G.C.E. management failed to retain safety-critical design information when two key structural engineers, involved in preliminary design activities, left the company (Wander, 2008). The senior project designer and project engineer, both of whom had an exceptional knowledge of the design, left G.C.E. midway through the design process (Wander, 2008). The switch to the alternate design of the walkway support structure was never fully communicated to the new G.C.E. design engineers, and the downgrade in structural integrity went unnoticed (Wander, 2008).

A grand jury investigation into the collapse found no criminal actions linked to the accident (Wander, 2008). Nonetheless, after 2 years of civil suits involving all parties totaling more than $100 million, G.C.E. International Inc. had its license revoked (Wander, 2008). In addition, the two lead structural engineers working for G.C.E. were found guilty of gross negligence, misconduct, and unprofessional conduct in the practice of engineering (Wander, 2008). The board of the American Society of Civil Engineers finally placed the accountability for this disaster on the G.C.E. engineers and defined the necessity of determining individual roles in overlapping responsibilities (Wander, 2008).

Before the disaster, the city of Kansas City had a small subdivision of the Division of Public Works in place to review building plans (Wander, 2008). However, after the collapse, the city created a Codes Administration Department dedicated to reviewing building plans (Wander, 2008).

The walkway collapse underscores the need to incorporate Safety and Mission Assurance engineering support into design and design verification processes (Wander, 2008). Of particular importance is early incorporation of system safety activities, including:

1. identification of hazards; (Wander, 2008)
2. identification of control and/or mitigation measures; and (Wander, 2008)
3. specification of methods to verify controls have been implemented (Wander, 2008).

The Hyatt Hotel walkway collapse reminds engineers and program teams to develop an in-depth understanding of concepts such as yield, margin, and factors of safety when working with safety-critical structures

(Wander, 2008). This disaster also underscores the need for requirements ownership, requirements clarity, and requirements change control (Wander, 2008). Ambiguity associated with overlapping responsibilities, matrixed support relationships, and complex supply chains must be overcome by implementation of rigorous configuration management with formal requirement change boards that include independent engineering and assurance representation (Wander, 2008).

It is important to consider how big projects can fail when small details are overlooked (Wander, 2008). The need for disciplined design build processes that include appropriate independent reviewers to ensure that every single safety-critical detail is addressed with rigor and care becomes obvious during such a disaster (Wander, 2008). Many lessons can be drawn from this tragedy, including the following (Bernhardt, 2016; Luth 2000):

1. Connections should be designed by a qualified engineer at some point in the design and construction process (Bernhardt, 2016).
2. An internal quality review should be thorough—more than spot checking—and should include a formal check of details on the structural drawings (Bernhardt, 2016)
3. Questions posed during design and construction should not be disregarded, but should be given the utmost attention on a project with an accelerated schedule (Bernhardt, 2016).
4. When there are changes in personnel, steps should be taken that ensure a smooth transition and full transfer of knowledge about the design activities leading up to the personnel change (Bernhardt, 2016).
5. Even small, seemingly insignificant changes in concept should be handled through a process that compels the participants to focus on potential issues (Bernhardt, 2016).
6. Engineers in city building departments should not be depended upon for finding errors in the design (Bernhardt, 2016). Internal quality reviews should catch any errors before documents are issued for construction. (Bernhardt, 2016)

Additionally, many more valuable lessons can be extrapolated from the Hyatt collapse including issues about communication during the project, design quality control, design responsibility, shop drawing review, construction inspection, and structural observation (Bernhardt, 2016; Delatte, 2009; Morin & Fischer, 2005). Comprehensive quality control reviews during the design, effective shop drawing reviews, and vigilant structural observation during construction are three significant steps that structural engineers can use to ensure that the design conforms to accepted practice and that the fabricator and constructor understand and deliver a final product that meets the engineer's design intent (Bernhardt, 2016). Every project should have a clearly documented quality review process for design deliverables (House Committee on Science and Technology, 1984).

Major lessons learned from the Hyatt Hotel tragedy regarding quality control are (Bernhardt, 2016):

1. The originators of calculations and drawings need to provide a self-check of their work (Bernhardt, 2016).
2. An independent internal review should be required for calculations and deliverables (Bernhardt, 2016). This may be completed by the structural engineer of record or a senior engineer in the firm (Bernhardt, 2016). Some firms require drawings to be highlighted for compliance and redlined for disagreements (Bernhardt, 2016).
3. .The reviewer should have a commitment to quality for the firm's deliverables and should assume a sense of responsibility (Bernhardt, 2016). It should be required that the reviewer's comments be resolved by the originator of the work product (Bernhardt, 2016).
4. The review should not focus on only the details on the structural plans, but should envision the "big picture"—how the discipline-specific design fits into the whole project (Bernhardt, 2016). Coordination with other disciplines is necessary to provide fewer change orders during construction (Bernhardt, 2016).
5. The engineering firm should not depend on the client or a government agency to find errors (Bernhardt, 2016). The design and plans should be correct when delivered for a permit or issued for construction (Bernhardt, 2016).
6. Corrections made by a drafter need to be thoroughly reviewed to ensure that the corrections are accurate before submitting to the client (Bernhardt, 2016).

This review process is intensive, but essential, especially for more complex or innovative projects (Bernhardt, 2016). Only by adhering to this process can effectiveness and a high degree of accuracy in the final product be provided (Bernhardt, 2016). Self-check is important but equally important is that another set of experienced eyes review design calculations and details on a project (Bernhardt, 2016).

Shop drawing review is another step in the process of delivering a quality project (Bernhardt, 2016). The fabricator or vendor should submit detailed fabrication drawings and information for the materials that meet the designer's intent (Bernhardt, 2016). These submissions need to be approved by the engineer of record or a delegate before construction (Bernhardt, 2016). The reviewer needs to be someone who is familiar with the design and who is knowledgeable in the structural engineering requirements (Bernhardt, 2016). Shop drawing review should not be depended on as a quality review to find the designer's errors (Bernhardt, 2016). Rather, it is a step to ensure accurate communication of the design for fabrication before construction (Bernhardt, 2016).

Having a site presence during construction was requested by the structural engineer of record at the Hyatt three times (Bernhardt, 2016). Presence at the site by a qualified structural engineer may have finally prompted the attention to the critical detail that was required and the tragedy could have been averted (Bernhardt, 2016). As with shop drawings, observation of the construction by a structural engineer should not be used as a tool to find errors in the design, but it can be another layer of the safety net to ensure that the structural details are constructed as the engineer intended in the design (Baura, n.d.).

IMPLICATIONS FOR MODERN EMERGENCY MANAGEMENT

The Hyatt Hotel walkway disaster had a respectable emergency response time and within 8 min the first ambulance arrived on the scene. Triage was set up outside the hotel on the street with a physician. A smooth functioning recovery and assistance operation started to take place as leaders emerged from the catastrophe. Multiple patients per ambulance were transported to hospitals and local taxis were also used to carry patients to hospitals. Approximately 7 hr after the incident, injured bodies were still being extracted from the horrific scene, due to the difficult recovery and several surgical procedures were required in the following days.

Armbands with identifying titles are part of the Kansas City disaster plan; these were not available at the time of the collapse. Triage tagging using the METTAG (Medical Emergency Triage Tag) system was proposed for multiple casualty triaging in Kansas City. METTAG provides rescue personnel with the standard triaging identifying features required to allow them to perform a quick and accurate assessment of a mass casualty incident (MCI) victim's injuries and to easily record the data. In the event of an MCI requiring varying degrees of medical attention, it is vital that first responders have a quick, easy, and reliable method of assessing victims and assigning them with an appropriate triage or priority status, based on the severity of their injuries. Although called for in the Kansas City disaster plan, these tags were not available on the night of the Hyatt Regency walkway collapse. On-site volunteer physicians were not easily identifiable because of the lack of uniforms or identifying vests causing confusion. Several issues from an emergency response standpoint were apparent during the Hyatt Hotel walkway collapse:

- No METTAG system in place
- Lack of accountability
- Uncertain chains of command and supervision.
- Poor communication
- Inefficient use of available communications system in the hotel
- Lack of an orderly, systematic planning process.
- Poor coordination with volunteer rescue efforts and other first responders
- Lack of knowledge with common terminology during an incident.
- Incident Command Emergency Systems developed in the 1970s should have been planned, practiced, and implemented.

SUMMARY

Many factors could have affected the outcome of this tragedy (Bernhardt, 2016). Structural engineers should operate with a heightened level of discipline and conviction to ensure that projects are not deemed, or delivered, as complete without adherence to a proper quality control process (Bernhardt, 2016). Structural engineers should not avoid their responsibility to provide quality in their structural reviews of construction submissions and observation in the field during construction (Bernhardt, 2016).

Following the collapse of two suspended walkways within the atrium area of the Kansas City Hyatt Regency Hotel in Kansas City, MO, NBS carried out an extensive field investigation to reconstruct events preceding the collapse and to establish the condition of the walkways at the time of the collapse (Marshall et al., 1982). This included review of documents related to the design and construction of the walkways and measurements and inspections of the debris remaining after the collapse (Marshall et al., 1982). NBS also conducted comprehensive laboratory studies of materials and mockups representative of the walkway construction and laboratory studies of specimens removed from the debris (Marshall et al., 1982). These studies were supported by structural analyses of the walkways under the loading condition believed to have existed at the time of the collapse (Marshall et al., 1982). Conclusions drawn from this investigation are summarized in the following statements (Marshall et al., 1982):

1. Collapse of the walkways occurred under the action of loads that were substantially less than the design loads specified by the Kansas City Building Code (Marshall et al., 1982).
2. The ultimate capacity of box beam–hanger rod connections can be predicted on the basis of laboratory test results (Marshall et al., 1982).
3. Under the action of the loads estimated to have been present on the walkways at the time of the collapse, all fourth floor box beam–hanger rod connections were candidates for initiation of walkway collapse (Marshall et al., 1982).
4. Observed distortions of structural components strongly suggest that failure of the walkway system initiated in the box beam–hanger rod connection at the east end of the middle box beam in fourth floor walkway (Marshall et al., 1982).
5. As constructed, the box beam–hanger rod connections, the fourth floor to ceiling hanger rods, and the third floor walkway hanger rods did not satisfy the design provisions of the Kansas City Building Code (Marshall et al., 1982).
6. The change in hanger rod arrangement from a continuous rod to interrupted rods essentially doubled the load to be transferred by the fourth floor box beam–hanger rod connections (Marshall et al., 1982).
7. The box beam–hanger rod connection would not have satisfied the Kansas City Building Code under the original hanger rod detail (continuous rod) (Marshall et al., 1982).
8. Under the original hanger rod arrangement (continuous rod), the box beam–hanger rod connections as shown on the contract drawings would have had the capacity to resist the loads estimated to have been acting at the time of the collapse (Marshall et al., 1982).
9. Neither the quality of workmanship nor the materials used in the walkway system played a significant role in initiating the collapse (Marshall et al., 1982).

Bottom line, the inadequacy of city building inspection, coupled with how the design egregiously deviated from the original engineering drawings, along with the fact that the engineer who signed off on the supports was not licensed to do so by the State of Missouri, were significant factors that contributed to the structural failure of the hotel skywalks (Waugh, 2000). As a result, the skywalks were not built with the expectation that large numbers of people might dance or sway on them. In fact, there remains incontrovertible evidence that the skywalks were actually built for walking rather than for maintaining the structural resiliency necessary to support "dynamic loads" and the "harmonious motion" created by large groups of swaying people (Waugh, 2000).

REFERENCES

Bernhardt, R. P. (2016, August). Structural failures: Hyatt Regency Skywalk collapse remembered. *Structure Magazine,* pp. 25–28.

Baura, G. D. (n.d.). 1981: Kansas City Hyatt Regency Skywalk Collapse. Engineering ethics: An industrial perspective. Retrieved from http://www.asce.org/code-of-ethics

Delatte, N. J. (2009). Hyatt Regency Walkway. In Delatte, N. J (Ed.), *Beyond failure—Forensic case studies for civil engineers,* (pp. 7–25). Reston, VA: American Society of Civil Engineers.

Guyer, J. P. (2010). *Ethical issues from the Kansas City Hyatt Hotel Collapse.* Stony Point, NY: Continuing Education Department.

House Committee on Science and Technology. (1984). *House Reports No. 98-621, Structural Failures in Public Facilities.* House Report, Washington: U.S. Government Printing Office.

Luth, G. P. (2000, May). Chronology and Context of the Hyatt Regency Collapse. *Journal of Performance of Constructed Facilities*, Volume 14, 51–61.

Marshall, R. D., Pfrang, E. O., Leyendecker, E. V., Woodward, K. A, Reed, R. P., Kasen, M. B., & Shives, T. R. (1982). Investigation of the Kansas City Hyatt Regency Walkways Collapse. NBS Building Science Series 143. U.S. Department of Commerce. National Bureau of Standards, Ernest Ambler, Director. Issued May 1982, National Bureau of Standards, Washington, D.C., 20234.

Missouri Board for Architects, Professional Engineers and Land Surveyors v. Daniel M. Duncan, Jack D. Gillum, and G.C.E. International, Inc. Administrative Hearing Commission. 1985. Case No. AR-84-0239.

Roddis, W. M. K. (1993). Structural failures and engineering ethics. *Journal of Structural Engineering, 119*(5), 1539–1555.

Wander, S. (2008). System failure case studies: Two rods don't make it right. May 2008 Volume 2 Issue 2 National Aeronautics and Space Administration. ARES Corporation. Retrieved from http://pbma.nasa.gov

Waugh Jr., William L., (2000). *Living With Hazards Dealing With Disasters: An Introduction to Emergency Management.* M.E. Sharpe, Inc., Armonk, New York.

CPSIA information can be obtained
at www.ICGtesting.com
Printed in the USA
LVHW062204100220
646522LV00014B/284